CATALYSIS IN CHEMISTRY AND BIOCHEMISTRY
THEORY AND EXPERIMENT

THE JERUSALEM SYMPOSIA ON QUANTUM CHEMISTRY AND BIOCHEMISTRY

Published by the Israel Academy of Sciences and Humanities, distributed by Academic Press (N. Y.)

1st JERUSALEM SYMPOSIUM:	*The Physicochemical Aspects of Carcinogenesis* (October 1968)
2nd JERUSALEM SYMPOSIUM:	*Quantum Aspects of Heterocyclic Compounds in Chemistry and Biochemistry* (April 1969)
3rd JERUSALEM SYMPOSIUM:	*Aromaticity, Pseudo-Aromaticity, Antiaromaticity* (April 1970)
4th JERUSALEM SYMPOSIUM:	*The Purines: Theory and Experiment* (April 1971)
5th JERUSALEM SYMPOSIUM:	*The Conformation of Biological Molecules and Polymers* (April 1972)

Published by the Israel Academy of Sciences and Humanities, distributed by D. Reidel Publishing Company (Dordrecht, Boston and London)

6th JERUSALEM SYMPOSIUM:	*Chemical and Biochemical Reactivity* (April 1973)

Published and distributed by D. Reidel Publishing Company (Dordrecht, Boston and London)

7th JERUSALEM SYMPOSIUM:	*Molecular and Quantum Pharmacology* (March/April 1974)
8th JERUSALEM SYMPOSIUM:	*Environmental Effects on Molecular Structure and Properties* (April 1975)
9th JERUSALEM SYMPOSIUM:	*Metal-Ligand Interactions in Organic Chemistry and Biochemistry* (April 1976)
10th JERUSALEM SYMPOSIUM:	*Excited States in Organic Chemistry and Biochemistry* (March 1977)

VOLUME 12

CATALYSIS IN CHEMISTRY AND BIOCHEMISTRY THEORY AND EXPERIMENT

PROCEEDINGS OF THE TWELFTH JERUSALEM SYMPOSIUM ON QUANTUM CHEMISTRY AND BIOCHEMISTRY HELD IN JERUSALEM, ISRAEL, APRIL 2–4, 1979

Edited by

BERNARD PULLMAN

Université Pierre et Marie Curie (PARIS VI)
Institut de Biologie Physico-Chimique
(Fondation Edmond de Rothschild), Paris, France

D. REIDEL PUBLISHING COMPANY

DORDRECHT : HOLLAND / BOSTON : U.S.A.
LONDON : ENGLAND

Library of Congress Cataloging in Publication Data

Jerusalem Symposium on Quantum Chemistry and Biochemistry, 12th, 1979.
Catalysis in chemistry and biochemistry.

(The Jerusalem symposia on quantum chemistry and biochemistry ; v. 12)
Includes indexes.
1. Catalysis–Congresses. I. Pullman, Bernard, 1919– II. Title.
III. Series: Jerusalem symposia on quantum chemistry and biochemistry ; v. 12.
QD505.H47 1979 541'.395 79-19397
ISBN-13: 978-94-009-9515-4 e-ISBN-13: 978-94-009-9513-0
DOI:10.1007/978-94-009-9513-0

Published by D. Reidel Publishing Company,
P.O. Box 17, Dordrecht, Holland

Sold and distributed in the U.S.A., Canada, and Mexico
by D. Reidel Publishing Company, Inc.
Lincoln Building, 160 Old Derby Street, Hingham
Mass. 02043, U.S.A.

TABLE OF CONTENTS

PREFACE

The 12th Jerusalem Symposium was devoted to the exciting subject of Catalysis in Chemistry and Biochemistry. As usual in these Symposia Theory and Experiment have been considered simultaneously in discussions, involving renowned experts from many countries.

We wish to thank all those who made this meeting possible and contributed to its success : the Baron de Rothschild whose continuous generosity is at the source of these gatherings, the Israel Academy of Sciences and Humanities which offers us the cordial hospitality of its impressive and beautiful building and gardens and the Hebrew University of Jerusalem, faithful cosponsor of the Symposia.

THE MACROMOLECULAR ELECTROSTATIC EFFECT IN BIOCHEMICAL REACTIVITY OF THE NUCLEIC ACIDS.

BERNARD PULLMAN
Institut de Biologie Physico-Chimique, Laboratoire de Biochimie Théorique, associé au C.N.R.S., 13, rue P. et M. Curie, Paris 75005, France.

I. INTRODUCTION

The role of the macromolecule and of its overall structure upon the reactivity of its reaction site(s) represents one of the fundamental problems in the theory of catalysis. The most impressive and most frequently considered examples occur in the field of proteins where the problem generally consists of accounting for the enhancement of the reaction rate at the active site or at the coenzyme with respect to model systems, devoid of the macromolecular superstructure. Different effects have been considered as possibly responsible for the influence of the macromolecule on such localized-site reactions. I do not have the time to summarize or discuss them here. I would like, however, to underline the accent put recently in a number of studies, on the possibly primordial role in this respect of the electrostatic effect of the protein on its specific reactivities. Examples of such considerations may be found in the work of Hayes and Kollman on carboxypeptidase A [1][2], in the work of Warshel on lysozyme [3][4], in the work of Van Duijnen on the role of the α-helix dipole on the reactional properties of proteins, in particular papain [5][6] etc. (For general comments see also [7]). The different authors differ in the details of their approach, but have in common to limit their consideration to proteins.

It is, however, evident that similar effects may be considered also for other types of biopolymers in particular for nucleic acids in which it appears legitimate to ask what is the influence of the overall macromolecular structure of, say DNA, upon the properties of the different reaction sites of, say, its purine and pyrimidine bases.

It is this problem that we wish to treat here. I would like to state immediately that we shall limit ourselves to consider only the electrostatic aspect of this influence and shall neglect, for the while being, other possible aspects. This restriction introduces a severe limitation if we wish to compare the results of computations with experimental findings (vide infra). However, in as much as it may be estimated that the electrostatic effect is an important one, the study is obviously of interest if only as a step towards a more complete ex-

B. Pullman (Ed.): Catalysis in Chemistry and Biochemistry. Theory and Experiment. 1–10.

ploration.

II. THE ELECTROSTATIC MOLECULAR POTENTIAL

For the sake of exploring the electrostatic effect of the macromolecular structure of DNA upon the properties of its purine and pyrimidine bases we shall use the notion of the electrostatic molecular potential, about which Madame Pullman has lectured already here at a previous Jerusalem Symposium [8] . May I remind that under this denomination we consider the electrostatic potential created in the neighbouring space by the nuclear charges and the electronic distribution of the system. For a given wave function with the corresponding first order density function $\rho(i)$ the average value of such a potential V(P) at a given point P in space is given by [9] :

$$V(P) = \sum_{\alpha} \frac{Z_\alpha}{r_{\alpha P}} - \frac{\rho(i)}{r_{Pi}} d\tau_i$$

where Z_α is the nuclear charge of nucleus α. This quantity has the advantage of being directly obtainable from the wave function and thus does not suffer from the drawbacks inherent in the classical population analysis. Moreover its very definition makes it an interesting index for the study of chemical reactivity at least in the early stage of approach of an ionic reagent. Thus the interaction energy between the molecular electronic distribution (considered unperturbed) and an external point charge q placed at point P is qV(P), which is rigourously the first order perturbation energy of the molecule in the field of charge q. Taking q as a unit positive charge the interaction potential can be used directly for the study of proton affinities. (See e.g. [10-12] for such a study of the nucleic acid bases). Its utilization for the investigation of the interaction of the ligands with more complex cations such as metal or alkyl cations is subject to caution because of the increased importance in such cases of the complementary components of the binding energy (besides the purely coulombic one) namely the delocalization (polarization + charge transfer) and the exchange components. The problem has been investigated explicitly in our laboratory in relation with the interaction of the nucleic acid bases with Na^+ [13], Mg^{++} [14] and also with alkyl cations [15] .

With this (important) restriction in mind we would like to investigate, as the first stage of the exploration of the effect of the macromolecular superstructure, the modifications in the molecular potentials of the individual bases brought about by their pairing through hydrogen bonding in the Watson-Crick complementary scheme of DNA : adenine-thymine and guanine-cytosine. The usual way of representing these potentials is, as is well known, through the utilization of the electrostatic isopotential maps. Here we shall center our attention essentially on the regions surrounding the minima of the potentials in the most attractive regions, which are situated in the isolated bases in the molecular plane in the neighbourhood of N_1, N_3 and N_7 of adenine, N_7, O_6, N_3 of guanine, N_1, O_2 of cytosine and O_4, O_2 of thymine (in the order indicated for each base).

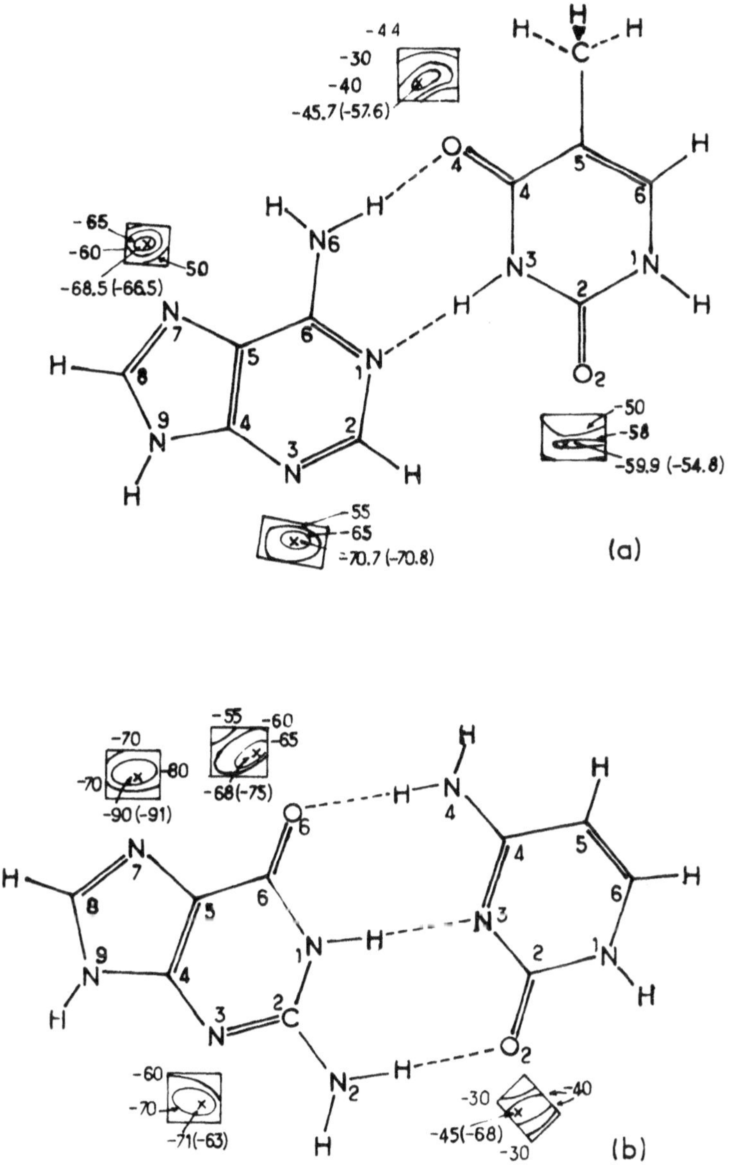

Figure 1.- The molecular potential in the most attractive in-plane regions of the complementary base pairs of DNA: (a) adenine-thymine; (b) guanine-cytosine. The potential minimum in the pair is indicated by a cross. The values in parentheses are the corresponding minima in the isolated bases. (The positions are the same except where indicated by a dark dot.

When the two bases of the pairs, adenine-thymine or guanine-cytosine, are placed in the appropriate orientation, a supplementary potential due to the partner appears in each zone and modifies the global distribution. The resulting potentials obtained for the two base pairs are given in Fig. 1 [16]. Only the regions surrounding the minima situated outside the regions of the hydrogen bond are given because these are the zones where the most negative potential wells remain. In the regions situated along the hydrogen bonds themselves (N_3 of cytosine, N_1 of adenine; O_6 of guanine, O_4 of thymine) the negative potential of the isolated base is essentially cancelled (or nearly so) by the strong repulsive potential of the partner [17]. The results indicate that : 1) the positions of the potential minima are very little modified with respect to those in the isolated molecules, 2) the order of the most attractive positions is

$$N_7(G) > N_3(G) \simeq O_6(G) > O_2(C) \quad \text{in the G-C pair}$$

$$N_3(A) > N_7(A) > O_2(T) > O_4(T) \quad \text{in the A-T pair}$$

with N_7 of guanine remaining the most attractive of all the positions.

An interesting feature to underline is that on the two pyrimidines, the oxygen atom involved in hydrogen bonding undergoes a decrease of its attractive character (stronger for cytosine). Moreover, in thymine, the oxygen which does not take part in the pairing undergoes an increase of its attractive character with respect to the situation in the isolated base, to the extent that this oxygen is, in the A-T pair, the most attractive of the two oxygens of thymine (the situation was inverse in the isolated base). We shall discuss the significance of this situation later.

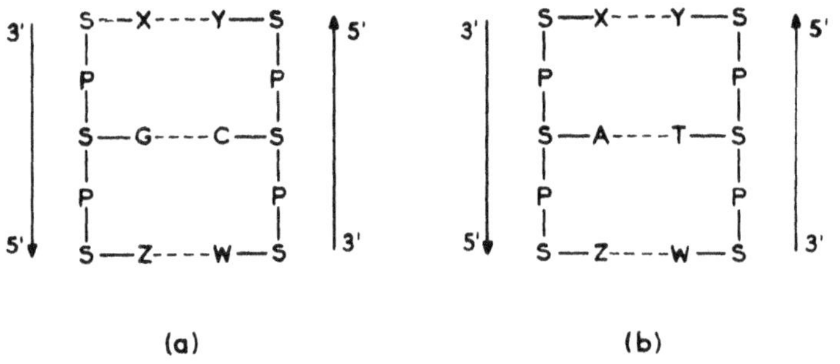

Figure 2.- Model compounds considered in the study of the electrostatic molecular potential in the vicinity of the base-pairs within B-DNA: a) in the vicinity of the G-C pair, B) in the vicinity of the A-T pair. S and P hold for deoxyribose and phosphate respectively; X, Y, Z and W could be either adenine, thymine, guanine or cytosine.

We may now pass to the essential step of this exploration by extending our computations to the evaluation of the potential around the complementary base-pairs, G-C and A-T, within the B-DNA double helix [18][19]. For this sake we have performed computations on the mini-helices, represented in Figures 2a and 2b which enable to study the influence of the sugar-phosphate backbone and of the adjacent base-pairs placed on both sides of the central G-C or A-T one, on the potential in the vicinity of this central pair. All the combinations of A-T and G-C have been considered for the X-Y and Z-W pairs of Figure 2. To facilitate the discussion let us look first at the case of the central G-C pair. Let us also consider first the effect of the two sugar-phosphate backbones of the helix upon the potential around this central G-C base pair. The effect of the adjacent base pairs will be considered later. The minima of the electrostatic potentials found are presented in Table I. For the sake of comparison we also present the minima found for the separate G and C bases and for an isolated G-C pair. The most striking effect associated with the incorporation of the base-pair into the helix is undoubtedly the very strong increase in the absolute values of the potential minima, translating a parallel increase of the affinity of the bases towards electrophilic agents. The situation obviously corresponds to the penetration of the strong potential generated by the phosphates [20,21] into the vicinity of the bases and its superposition upon the potential inherent to the bases themselves. A second effect is the modification produced in the ordering of the potential minima around the bases in the helix as compared to the isolated units. Thus e.g. it is seen that within the model the minima near N_3(G) and N_7(G) are close to each other, a result in contrast with the situation in isolated single and paired-bases, in which N_7(G) is largely favored compared to all the other sites.

TABLE I
Electrostatic potential minima (in kcal/mole) at different sites of the central G-C base pair in the mini B-DNA of Figure 2a taking into account the phosphate-sugar backbone. The stars point out the preferred sites. The numbering of the sites corresponds to the standard notations of nucleic acids, the letters within parenthesis refer to the nature of the base. The minima are in the plane of the base-pair.

Sites	Single bases	Paired bases	Paired bases within B-DNA
N3(G)	-64.5	-73.7	-241.9*
O6(G)	-76.6	-79.1	-204.4
N7(G)	-92.6*	-92.9*	-238.9
O2(C)	-70.6	-50.3	-217.8

We need to complete this study by evaluating the contributions of the adjacent base pairs surrounding the central G-C base pair on the electrostatic potential minima of that pair. Computations show that these effects are altogether quite small and do not modify much the absolute values of the last column of Table I. They may, however, influence the ordering of the minima. Thus, it is observed that either N_3(G) or N_7(G) is favored depending on the nature of the adjacent base-pairs. The same conclusion applies for the comparisons between O_2(C) and O_6(G). (For details see 18).

We may now consider the situation when the central base pair is A-T. Table II presents the effect of the two sugar-phosphate backbones of the helix upon the potential of the central A-T pair. In this approximation the effect of the adjacent base pairs is neglected. The most striking effect associated with the incorporation of the base pair into the helix is undoubtedly, as it was the case for the G-C pair, the very strong increase in the absolute values of the potential minima, reflecting the parallel increase of the affinity of the bases towards electrophilic agents. Because of the disappearance in the base pair of the potential minimum close to the N_1 atom of adenine in the free base it is not astonishing to find the deepest minimum to be in the vicinity of N_3 of adenine. Note also the deeper potential in the vicinity of O_2 of thymine as compared to its O_4. As it was the case with the G-C pair the effect of the adjacent base-pairs is minor and the addition of its contribution does not change the general aspects of the results described above (see 19).

TABLE II
Electrostatic potential minimum (in kcal/mole) at different sites of the central A-T pair in the mini B-DNA of Fig. 2b taking into account the sugar-phosphate backbone. The stars point out the positions of deepest minima. The numbering of the sites corresponds to standard notations of nucleic acids, the letters within parenthesis refer to the nature of the base. The minima are in the plane of the bases and base pairs.

Sites	Single bases	Paired bases	Paired bases within B-DNA
N7(A)	-67.5	-70.5	-218.7
N3(A)	-71.8*	-72.7*	-238.5*
O4(T)	-58.7	-47.7	-179.5
O2(T)	-55.9	-61.9	-228.3

The global minimum associated in the A-T pair with adenine is somewhat less deep than that associated with guanine in the G-C base pair and this correlates satisfactorily with the usual greatest reactivity of guanine towards electrophilic agents in nucleic acids [22]. Another interesting observation in relation to some recent experimental findings and statements by their authors concerns the oxygen atoms of the bases which are involved in hydrogen bonding to the complementary bases. Thus Singer et al. [23] express their surprise at the "unexpected finding" that certain alkylating agents which show a general preference for oxygen atoms, continue to attack the hydrogen bonded atoms O_6 of guanine, O_2 of cytosine and O_4 of thymine in the base pairs and that in fact the "Watson-Crick base-pairing does not significantly hinder alkylation of the oxygens". A glance at Tables I and II shows that this is, in fact, to be expected and that the potentials associated with these atoms are competitive with those of ring nitrogens. (The decision whether a given alkylating agent will attack oxygens or nitrogens of the bases is generally a function of the supplementary factors governing the association, examined elsewhere [15][24]). Similarly, one may notice in Table II, the relatively deep minimum associated in the A-T base pair with the non hydrogen-bonded O_2 of thymine. Recently Seeman et al. [25] expressed also their surprise to observe the binding of a Na^+ ion between two O_2 atoms from neighbouring uracil residues (on two strands) observed in the crystal structure of ApU. The existence of an appreciable affinity between Na^+ and the purine and pyrimidine bases has been demonstrated by quantum-mechanical computations [26] and is substantiated in a number of X-ray crystallographic results (for ref. see 26). In the present case it is certainly not surprising that the joined potentials of two O_2's are capable of attracting and fixing a Na^+ ion.

TABLE III
Electrostatic molecular potentials in the vicinity of the NH_2 groups

(Kcal/mole)

Isolated bases		Isolated base pairs	Base pairs in DNA (*)	
Guanine	0.6	-19.4	-171.5,	-170.9
Adenine	-15.6	-21.1	-158.2,	-157.9
Cytosine	-7.0	-13.8	-150.2,	-149.2

(*) The first number corresponds to the 3' side and the second to the 5' side.

Finally, a special mention is deserved by the potentials in the vicinity of the NH_2 groups of guanine, adenine and cytosine. These groups are generally considered as secondary sites for electrophilic attacks on these bases [27] and the energy minima associated with the isolated bases are in fact either very small as in adenine and in particular in cytosine, or may even correspond to a repulsion, as in the case of guanine (Table III). (These minima are located above and below the plane of the bases, in distinction to the minima in the vicinity of the ring N atoms and of the oxygens of the carbonyl groups which are located in the plane of the heteroaromatic ring). In the base pairs, the minima of the NH_2 groups become negative for all the bases, in the order (absolute value) : adenine > guanine > cytosine. It may be seen that this situation is still accentuated in the mini-helix, within which these minima become appreciable. Moreover, the depth of the minima is now in the order guanine > adenine > cytosine. The importance of this situation is linked with the recent interesting findings that the metabolites of carcinogenic polycyclic aromatic hydrocarbons, such as the 7,8-dihydrodiol-9,10-oxide of benzo[a]pyrene and the triol carbonium cation derived from it by the opening of the epoxide ring, considered by many as the probable ultimate carcinogenic forms of this hydrocarbon, interact in vitro and in vivo with the NH_2 groups of DNA in the decreasing order guanine > adenine > cytosine [28-31]. This order correlates with the affinity for electrophiles of these groups when incorporated within the model mini-DNA helices studied here and suggests that the electronic factors put into evidence here could be responsible for the observed ordering of the interaction (see, however, [32]).

It seems thus evident that the electrostatic effect of the whole macromolecular system represents undoubtedly an important factor in the determination of the reactivity of the different localized reaction sites of the purine and pyrimidine bases of the nucleic acids. Detailed comparison with experimental results may necessitate the taking into account of the complementary electronic (polarization, delocalization, exchange repulsion), steric and environmental (solvent, counterions) factors involved.

REFERENCES

[1] Hayes, D.M. and Kollman, P.A. : J. Amer. Chem. Soc. 98, 3335 (1976).
[2] Hayes, D.M. and Kollman, P.A. : J. Amer. Chem. Soc. 98, 7811 (1976).
[3] Warshel, A. and Levitt, M. : J. Mol. Biol. 103, 227 (1976).
[4] Warshel, A. : Proc. Natl. Acad. Sci., 75, 5250 (1978).
[5] Hol, W.G.J., Van Duijnen, P.T. and Beredsen, H.J.C. : Nature, 273, 443 (1978).
[6] Van Duijnen P.T., Thole, B.T. and Hol, W.G.J. : Biophys. Chem., 9, 273 (1979).
[7] Perutz, M.F. : Science, 201, 1187 (1978).
[8] Pullman, A. in : Chemical and Biochemical Reactivity, Proceedings of the 6th Jerusalem Symposium on Quantum Chemistry and Biochemistry (E.D. Bergmann and B. Pullman, Eds.) Reidel Publishing Co., Dordrecht, Holland, 1974, p.1.
[9] Scrocco,E. and Tomasi, J. : Topics in Current Chemistry, 42, 95 (1973).
[10] Pullman, A., in : Chemical and Biochemical Reactivity, 6th Jerusalem Symposium on Quantum Chemistry and Biochemistry (Bergmann, E.D. and Pullman, B. Eds.) Reidel Publishing Co., Dordrecht, Holland, 1974, p.1.
[11] Bonaccorsi, R., Pullman, A., Scrocco, E. and Tomasi, J. : Theoret. Chim. Acta, 24, 51 (1972).
[12] Bonaccorsi, R., Scrocco, E., Tomasi, J. and Pullman, A. : Theoret. Chim. Acta, 36, 339 (1975).
[13] Perahia, D., Pullman, A. and Pullman, B. : Theoret. Chim. Acta, 43 207 (1977).
[14] Perahia, D., Pullman, A. and Pullman, B. : Theoret. Chim. Acta, 42 23 (1976).
[15] Pullman, A. and Armbruster, A-M. : Theoret. Chim. Acta, 45, 429 (1977).
[16] Perahia, D. and Pullman, A. : Theoret. Chim. Acta, 48, 263 (1978).
[17] Pullman, A. and Berthod, H. : Theoret. Chim. Acta, 48, 269 (1978).
[18] Pullman, A., Zakrzewska Ch. and Perahia, D. : Int. J. Quant. Chem. Quant. Biol. Symp. 6, in press.
[19] Perahia, D. and Pullman, A. : Theoret. Chim. Acta, in press.
[20] Pullman, A. and Berthod, H. : Chem. Phys. Letters, 32, 233 (1975).
[21] Pullman, A. and Berthod, H. : Chem. Phys. Letters, 41, 205 (1976).
[22] See e.g. Pullman, A. in : Mecanismes d'Altération et de Réparation du DNA. Relations avec la Mutagénèse et la Cancérogénèse Chimique, Colloque International du CNRS, n°256, 1976, p. 103.
[23] Singer, B., Bodell, W.J., Cleaver, J.E., Thomas, G.H., Rajewski, M.F. and Thon, W. : Nature, 276, 85 (1978).
[24] Pullman, A. and Armbruster, A-M. : Theoret. Chim. Acta, 50, 359 (1979)
[25] Seeman, N.C., Rosenberg, J.P., Suddath, F.L., Kim, I.J.P. and Rich, A. : J. Mol. Biol., 104, 109 (1976).
[26] Perahia, D., Pullman, A. and Pullman, B. : Theoret. Chim. Acta, 43, 207 (1977).

[27] See e.g. Pullman, B. and Pullman, A. : Quantum Biochemistry, Academic Press, New York, 1963.
[28] Jerina, D.M., and Lehr, R.E. in : Microsomes and Drug Oxidation (Ullrich V. et al. Eds.) Pergamon Press, Oxford, 1977, p. 709.
[29] Weinstein, I.B., Jeffrey, A.M., Jennette, K.W., Blobstein, S.H., Harvey, R.G., Harris, C., Autrup, H., Kasai, H. and Nakanishi, K. : Science, 193, 592 (1976).
[30] Jeffrey, A.M., Jennette, K.W., Blobstein, S.H., Weinstein, I.B., Beland, F.A., Harvey, R.G., Kasai, H., Miura, I. and Nakanishi, K. : J. Amer. Chem. Soc., 98, 5714 (1976).
[31] Jennette, K.W., Jeffrey, A.M., Blobstein, S.H., Beland, F.A., Harvey, F.G. and Weinstein, I.B. : Biochemistry, 16, 932 (1977).
[32] Lavery, R. and Pullman, B. : Intern. J. Quant. Chem. Quant. Biol. Symp. n°6, in press.

DYNAMIC AND STATISTICAL ASPECTS OF ENZYME ACTION

Jill E. Gready
Physical Chemistry Laboratory,
South Parks Road,
Oxford, OX1 3QZ, UK.

The aim of this paper is to extend the current molecular concepts of enzyme structure and function based on a static view by incorporation of dynamic and statistical aspects. From the more complete description several new proposals about enzyme action, control and evolution are advanced : some general theoretical and experimental approaches for testing these suggestions are indicated. The arguments require reference to the statistical nature of the term 'mechanism' and to the principles of microscopic reversibility and detailed balance: these topics are examined in two short digressions.

1. INTRODUCTION

Until quite recently the standard microscopic (≡ molecular) description of enzymes has concentrated almost exclusively on aspects of static structure. As this model implies that only one 'structure' - or at most two or three 'structures' are involved in the binding and reaction process, the development of a macroscopic (i.e. bulk solution) picture has been severely hampered. Furthermore, analysis and interpreation of experimental data (kinetic, thermodynamic etc.) in terms of this limited microscopic view has led, in this author's opinion, to a number of misconceptions about enzyme function.

In this paper two extra levels of detail will be added to the description. Firstly, at the microscopic level, the dynamics of relatively small motions of the enzyme molecule will be considered[1]. Secondly, the macroscopic representation will be discussed in terms of a relatively large number of possible dynamical enzyme states whose relative contributions will depend on the *in situ* probability of occupation (population) of each state. While these dynamical and statistical considerations are of basic interest in characterizing completely the macroscopic behaviour of free enzyme or its various bound substrate complexes in solution, it is the combination of these two aspects in tracing the course of binding and, particularly, reaction

B. Pullman (Ed.): Catalysis in Chemistry and Biochemistry. Theory and Experiment. 11–23.

events which will be the major concern here.

The discussion will take the form of preliminary selective surveys of static, dynamical and statistical aspects of enzyme structure followed by brief reviews of the term 'mechanism' and the principles of microscopic reversibility and detailed balance. In section 5 a synthesis of the ideas from the earlier sections leads to a hypothesis about enzyme function which includes several new features. Some implications of this model for enzyme irreversibility, catalysis, evolution and kinetics are briefly discussed in section 6.

2. ENZYME STRUCTURE

2.1 Static

Most of the detailed information at the molecular level has come from X-ray diffraction studies of native enzymes or their complexes with substrates, products and in some cases,transition state/intermediate species. Here the information about relatively small local motions of atoms or groups has been averaged out, both by the experimental procedure (long data collection time) and by the analysis-the structure-refinement procedure. Also a non-uniform population with respect to gross conformational differences is precluded by the crystallization mechanics and by steric hindrance once crystallization has occurred. Even at the static level doubts are often expressed about possible discrepancies between the structure in solution and the poorly solvated crystal structure; recently it has become possible to resolve this problem in some cases using nmr techniques (Campbell et al., 1978). Also cases have been documented (McCammon et al., 1977) where X-ray structures obtained from different crystal forms show variations.

2.2 Dynamic

The discussion here is restricted to two recent developments - molecular dynamics simulation of protein movements and nmr results. Karplus and coworkers 1977a, 1979 have recently been studying the rapid (picosecond range) movements of the backbone and side-chain groups of basic bovine pancreatic trypsin inhibitor, using empirical potential functions and with the X-ray structure co-ordinates as the input geometry (Gelin and Karplus, 1975). Averaged results are compared with temperature factors available from the X-ray studies. Results for 8.8 ps (McCammon et al., 1977a) and 96 ps (Karplus and McCammon, 1979) simulations (after varying initial equilibration periods) have been published : one of the marked features of the latter study was the appearance of long-lived fluctuations in the positions of atoms of 5-15 kcal mol^{-1} with durations of 2-15 ps. One qualitative conclusion of the 8.8 ps study was that the fluctuations appeared diffusional in character to the extent that atomic displacements were dominated by collisions with neighbouring atoms: since the majority of such contacts were with non-bonded atoms this gave support to an earlier study (Gelin and Karplus, 1977b) implicating non-bonded contacts in the transmission of conformational changes from one part of the protein molecule to another.

However, transforming the physical description of the fluctuations from an atom model to a molecular one, these fluctuations might be thought of as the result of superposition of low frequency 'delocalised' vibrations. If it were possible to perform a normal mode analysis on a molecule as large and complex as a whole protein molecule, the resulting set of states would include a large number of very low frequency vibrations, corresponding physically to movements of sections of protein containing more than a few atoms, in addition to high frequency (i.e. $\sim 10^{13}\ s^{-1}$) vibrations associated with local bond stretching, bending etc. movements. On a cautionary note it is important to mention that hydrodynamic viscous forces leading to overdamping may prevent expression of some low-frequency periodic motions (cf. theoretical study of the hinge-bending mode in lysozyme (McCammon et al., 1976)) thus rendering their detection by spectroscopic means impracticable.

Notwithstanding alternative physical interpretations of the nature of the fluctuations, it is crucial for some of the arguments advanced in section 5 that the existence of such low frequency motions implies relaxation times for interconversion or redistribution of energy among these modes after a perturbation, characteristically orders of magnitude longer than those for rapid localised motions.

The second point - which is in fact related to the previous one - refers specifically to nmr evidence for the flipping of the aromatic side chains of tyrosine (TYR) and phenylalanine (PHE), although it is appropriate to note that there is also some evidence for rotation of bulky side-chains of other residues (tryptophan, leucine and valine) (Campbell et al., 1978). It is somewhat surprising that such bulky groups would have room to flip if they were located in the densely-packed interior of a (globular) protein - however, all studies to date indicate that some or all side-chains of TYR and PHE are able to flip or rotate about the $C_\beta - C_\gamma$ bond. This phenomenon may be rationalized in a number of ways. According to Campbell et al., 1978 flipping can occur when the protein structure around the side-chain relaxes : it is proposed that the relaxation arises when high-frequency, but limited structural, fluctuations of the interior medium are in phase. In terms of a molecular view the regional protein relaxation is the result of superposition of high (local) and/or lower (less local) frequency modes with a definite phase relationship. The other view already mentioned, is that the fluctuations have a diffusional character - at least on the picosecond time scale (McCammon et al., 1977a). The conceptual differences between these interpretations lie in the degrees of probabilistic or deterministic character accorded to the co-ordinated relaxation of a small region of protein : the proportions may vary both with the time scale and the extent of the protein movements so that no one theoretical description will be adequate for all ranges (see, for instance, models for the mechanism of complete folding and unfolding on the seconds to minutes time scale (Levitt and Warshel, 1975; Karplus and Weaver, 1976)). The actual flipping rates probably vary widely : by nmr techniques it is only possible to determine the rate relative to the nmr time-scale i.e. faster or slower. Detailed results of TYR ring orientation fluctuations (± 30° from the average : r.m.s. value of 12°) from the 96 ps molecular dynamics study are being published : a non-dynamical study of the energy barriers to rotation of aromatic side-chains has indicated the need for protein flexibility in order to account for the observed high flipping rates (Gelin and Karplus, 1975).

The questions which arise from this work are the extent of regular or irregular protein movements over long time scales and more extended regions of the protein volume and also the degree of variation in this respect from one protein to another - leaving aside the category of gross induced - or spontaneous conformational changes. How might these motions be detected? Side-chain rotation is a probe restricted to observing the behaviour of only small regions of protein although there is a suggestion of concerted flipping of groups separated by larger distances (Campbell et al., 1978).

2.3 Statistical

Having added a dynamical level to the static representation of an enzyme, the next question to consider is the relative contribution of the multitude of transient motional states in the various enzyme processes: it is these statistics which determine the bulk properties[2]. For the purpose of the following discussion (sections 5 and 6) the problem focussed on is : how long the macroscopic system takes to respond to a gross perturbation (eg. temperature jump, pressure jump, concentration jump, electric field perturbations etc. Hammes, 1968) or to a localised perturbation (eg. absorption of radiation by local centres in the protein; substrate binding)? If these perturbations are applied while substrate binding and reaction is occurring, what is the average extent to which the (physical) enzyme system relaxes on the time-scale of the binding and reaction events?

These questions require careful thought because there are thermodynamic constraints on the rate and manner in which the macroscopic system may relax - the rate of entropy production (Glansdorf and Prigogine, 1971) (in addition, of course, to energy conservation etc.). The relationship between what we may term microscopic relaxation rates - i.e. q.m. probabilities for molecular energy redistribution or time evolution of the molecular system - and bulk relaxation has recently been studied theoretically by Tabor et al., 1979 for dilute gas systems

3. MEANING OF 'MECHANISM'

A short analysis of the meaning of the term 'mechanism' is given here: most of the discussion of mechanism is deferred until later sections.

In general 'mechanism' is associated with description of the course of a chemical or biochemical reaction in terms of well-defined steps which order the bond breaking/making process. Each step is characterized by angles and distances of approach of reactive groups and even directions of electron movement. This description refers neither to the microscopic or macroscopic levels although superficially it would appear to be a molecular picture. Implicitly it is an average of the local high-frequency atomic motions and also some sort of statistical average of the dynamically/energetically accessible pathways by which the reaction may proceed - possibly only the predominant pathway. Obviously trouble with this intuitive idea will ensue if there are two similarly low energy pathways available but with very different spatial (or dynamical Kuntz, 1976) characteristics. Incorporation of pathway and statistical ideas also points out the simplified nature of the 'steps': steps represent (macroscopic) kinetic rate-limiting stages not time discontinuities on reactive pathways (as distinct from pathways which do not lead to reaction).

Mechanism as a statistical average of reaction pathways will be illustrated with particular reference to biochemical reactions. For enzyme-catalysed reactions mechanism may be a comparatively more accurate microscopic description than for the analogous reactions in solution. This results from the manifold of accessible reaction pathways being restricted in two ways. Firstly, spatial or orientation selection at the enzyme - substrate binding stage means that the range of angles and distances through which reactive groups may interact is greatly limited (and also that most of the entropic requirement for the activation energy is met Jencks, 1975). Secondly, the close interactions between substrate and protein groups at the active site further restricts the favoured pathways at a more intimate quantum mechanical level by exaggerating the differential between pathways on the potential energy scale and thus leading to preferential atomic and electronic movements.

These points are illustrated qualitatively in Fig. 1. The information conveyed in Fig. 1(c) is more detailed than the usual statement of enthalpic advantage leading to increased reaction rates for enzyme systems which co-operatively bind substrates to optimise localized destabilization of key reactive groups : here it is suggested that the activation energy for some of the microscopic pathways would be increased while the barrier of the 'predominant' pathway would undergo the postulated lowering - both relative to similar pathways on the lock-and-key system (Fig. 1(b)).

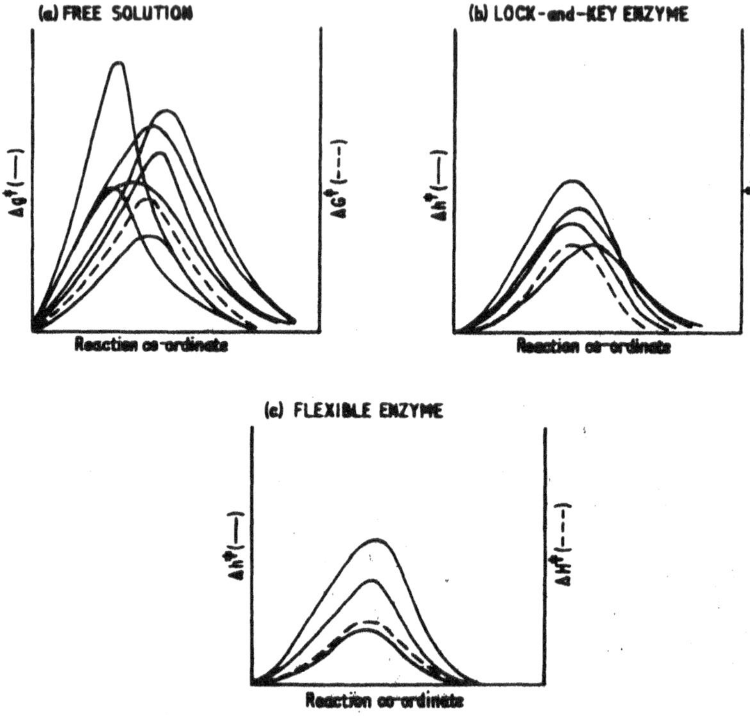

Fig. 1 : Transition-state (right-hand co-ordinate) and reduced microscopic pathways (left-hand co-ordinate) for hypothetical reaction under various conditions.

If mechanism is described in terms of a statistical distribution of pathways, the next consideration will be the structure of the statistical distribution. Although it may not be immediately apparent from Fig. 1, the lines of the microscopic pathways represent projections from a many-dimensional potential surface (eg. of an enzyme-substrate-product complex) onto a given physical co-ordinate - the reaction co-ordinate. Detailed discussion of the use of diagrams such as Fig. 1 is given elsewhere (Gready, 1979). The obvious inference from this is that for a given pathway to contribute to the total reaction in both directions the initial/final states (potential points), represented in the 'reduced' form in the left-hand and right-hand sides of Fig. 1, must occur.

The usual assumptions are that the proportion of species in a given state is determined by the energy of that state, and that the proportion of species reacting by a given pathway is determined by the activation energy (barrier height) of that pathway. The former assumption depends on the local equilibrium approximation : this requires that the populations of states of the reacting species be that which would obtain for the physical non-reacting system at equilibrium. That is : if any reactant state is depleted by reaction at an initial time t (N.B. either at or away from chemical equilibrium) then the rate of statistical relaxation is sufficiently rapid, compared with the rate of forward and backward reactions, that, at any later time $t + \Delta t$, the reactive processes may still sample the equilibrium distributions of initial and final states. The enzyme molecule is to be considered as a reactant. The second assumption also involves the local equilibrium approximation for the initial points of the pathways plus a 'local equilibrium' type approximation during progress of the pathway. The latter approximation may be phrased in general terms as implying that a microscopic system with a given initial energy will traverse reaction pathways consistent with this energy equally : several counter-phenomena are recognized in collision - theory studies - non-RRKM and non-ergodic behaviour (Hase, 1976) and dynamically-forbidden processes (Kuntz, 1976).

The implications of the view of mechanism as a statistical average will be considered in sections 5 and 6.

4. PRINCIPLES OF MICROSCOPIC REVERSIBILITY AND DETAILED BALANCE

The status of these two principles is widely misunderstood, especially by biochemists. Although the two terms are often confused microscopic reversibility will be taken to refer to the fundamental time-reversibility of microscopic event pathways either in a classical or qm description : this principle will not be at issue here although in Einstein's centenary year it is apt to note that doubts about its basis in the foundations of quantum mechanics are still being voiced (Penrose, 1979). Detailed balance refers specifically to the condition, at chemical equilibrium, that the number of reactive events occurring by any given forward pathway is matched by an equal number of back-reactions by the associated reverse pathway. Detailed balance is not a thermodynamic law but was introduced as an extra postulate in characterizing thermodynamic equilibrium in cyclic and multi-step reactions and used by Onsager in deriving the reciprocal relations for chemical reactions. Mostly detailed balance is invoked in more limited or derived forms such as:

rate of forward reaction for elementary step i
= rate of backward reaction for elementary step i, for all i
or $k_f/k_r = K$. ($K \equiv$ equilibrium constant).

Accounts of experimental and theoretical demonstrations of detailed balance have recently been reviewed (Boyd, 1977); the problems of experimentally verifying Onsager's reciprocal relations for chemical systems as distinct from physical systems have been noted in an earlier review (Miller, 1960). Theoretical proofs of detailed balance require a number of assumptions - apart from the principle of microscopic reversibility. The most rigorous statistical proof to date is that based on collision theory methods due to Light et al., 1969 which assumes binary collisions only, independent particle statistics and the local equilibrium assumption in two parts : a Maxwell distribution of velocities and a Boltzmann distribution for internal modes. Stochastic theory treatments have also been published which are slightly less restrictive than the statistical proofs in allowing disequilibrium among a small number of reactive states (Widom, 1965). Also, very recently a treatment has been advanced which requires that maximal entropy conditions be satisfied before introducing assumptions about the distribution of time-dependent degrees of freedom (Ben-Shaul and Tabor, 1979).

Before concluding this section it is worth stressing that the principle of detailed balance refers specifically to the situation of chemical equilibrium : in the biochemical literature the principle is often wrongly invoked to justify the notion that proposed mechanisms for forward and reverse reactions must be complementary under given reaction conditions - i.e. not necessarily at chemical equilibrium - for example, Fersht, 1977. That a reaction does proceed irreversibly towards chemical equilibrium - despite the fact that the individual microscopic reactions are assumed to be reversible - is interpreted as resulting from unequal numbers of forward and reverse collisions 'arising naturally' in the solution mixture. Taken together with the earlier remarks on the statistical nature of mechanism suggestions of the existence of necessary constraints on forward and reverse mechanisms away from equilibrium are unfounded. To reinforce these remarks the example of protein folding and unfolding is noted - very different mechanisms have been proposed for these two processes (Levitt and Warshel, 1975; Karplus and Weaver, 1976).

5. MODIFIED VIEW OF ENZYME ACTION

The ideas suggested in this section relate specifically to enzyme reactions although the possibility of some analogous effects in other reactions is not discounted. The proposals rest upon the exposition in sections 2 and 3 dealing with a complete enzyme description, and with mechanism : the comments in section 4 on the status of detailed balance are included to forestall possible objections based on cursory familiarity with that principle and to emphasize that it is only postulated to hold at equilibrium. The question of enzyme action at chemical equilibrium is discussed elsewhere (Gready, 1979).

The hypothesis may be stated as follows:
In some enzyme reaction systems complete conformational relaxation of sections of the enzyme involved in the catalytic process does not occur,

on average, on the time-scale of individual binding and reaction events - nor possibly during the interval between subsequent such events. This situation may significantly affect the enzyme-catalysed reaction by altering the statistical distribution of the actual reaction pathways as compared with that of a hypothetical local equilibrium distribution. The distribution of states from which the forward and backward reactions may proceed may also be affected.

In simplistic language this could be rephrased to imply the possibilities:
1. Transient retention of enzyme tension is involved in the reaction mechanism. 'Tension' is the dynamic equivalent to the static concept of 'strain' (eg. Jencks, 1975).
2. Longer-lived effects - memory effects - affects the binding/reaction of subsequent substrates.

Before proceeding to discuss some implications of these proposals a few comments are appropriate. So far this paper has neglected to provide the usual summary of the average rates of relaxation of molecular modes compared with the average rates of processes involved in chemical reactions : the crucial importance of the, as yet, poorly characterized conformational motions in enzymes in the current arguments indicates that caution should be exercised in using average numbers from small molecule chemistry. However, for completeness we may note the analyses of Eigen and Hammes, 1963 and Eigen, 1967 (also summarized by Fersht, 1977) who have considered upper limits on the rate constants of various processes. Thus association and dissociation constants for diffusion-controlled bimolecular complexes may be estimated as $10^{-9} - 10^{-11}$ s^{-1} M^{-1} and $10^{-9} - 10^{-12}$ s^{-1} M^{-1} respectively; unimolecular or intramolecular reaction is limited by the frequency of molecular vibrations ~ $10^{12} - 10^{13}$ s^{-1}; average ranges for electron and proton transfer rates are $10^{-10} - 10^{-15}$ s and $10^{-11} - 10^{-13}$ s respectively; and vibrational relaxation and internal conversion is of the order $10^{-9} - 10^{-12}$ s. We may add to this collection the average range for enzyme turnover numbers of $10 - 10^3$ s^{-1} although some systems have much higher turnovers - for example, carbonic anhydrase 10^6 s^{-1} and catalase 4×10^7 s^{-1} (Fersht, 1977).

Another piece of pertinent information is that there are many documented cases of substrate - induced conformational changes with rate constants in the range $10 - 10^4$ s^{-1} and some instances where discrepancies in rate constants indicate rate-determining protein isomerations (Fersht, 1977). Some interesting observations may be made from these numbers. Firstly turnover numbers represent only reactive collisions between substrates and enzyme : the total encounter rate will be close to the diffusion limit. Secondly when a substrate binds, its kinetic energy and also angular momentum must be taken up initially into a vibration or rotation in the complex - and some returned to the products when they dissociate. The efficiency of this process will depend on the dynamic state (and phase) of the active site it encounters : this effect constitutes part of the entropic activation energy. Secondly the above conformational changes refer to more substantial changes than those germane to the current proposals (see note 1).

In order to make a little clearer the sort of effects being hinted at in this paper two examples of vibrational trapping will be mentioned these will at least indicate the time-scales possible. A more comprehensive account of atypical relaxation rates and the conditions under which they occur will be given elsewhere (Gready, 1979). Rolfe et al., 1979 have noted the long history of theoretical studies on coupled non-linear oscillators : these may exhibit a variety of strange effects including quasi periodic energy exchange between initially exci modes of the system with no apparent evolution towards an asymptotic state - i.e. non-ergodic behaviour. Another peculiar example is a vibrational relaxation time of 1 second for an excited state of N_2 in a solid inert matrix at low temperatures (quoted in the review of Lin and Eyring, 1974).

6. SOME CONSEQUENCES

In this section some points arising from the proposal in the preceding section are briefly discussed: these questions are considered at greater length elsewhere (Gready, 1979).

6.1 Enzyme Irreversibility

The idea of 'one-way' enzymes has been around for a long time (Haldane, 1930). It has also been apparent from both lock-and-key, and the more recent co-operative, binding mechanisms that an enzyme active site cannot have optimum complementarity with respect to both substrates and products binding. On the other hand, substrates and products will inevitably show certain similarities in molecular structu so that the binding of one or other cannot be entirely prevented on a static picture. Application of the possibilities presented by the dynamical and statistical nature of the enzyme system would be one way of creating a more functionally irreversible enzyme if the reverse reaction were not biologically useful. This could be achieved straight forwardly by preferential acceleration of the forward reaction using the dynamical tension. Alternatively the reverse reaction might be inhibited by a perturbation in the distribution of enzyme states, cause by substrate binding to the active site or by the forward reaction, whi mitigates against product binding or reacting. In this case it will obviously be critical to study the enzyme under its physiological conditions of substrate/product concentrations.

6.2 Enzyme Evolution

In recent years evolutionary selection at the molecular level of an ever-increasing sublety has been proposed. The arguments in section 5 imply one further stage in complexity by suggesting that the time-dependent motions of the enzyme may be exploited for biological advantage. It is relevant to note that there are two major evolutionary forces to consider - maximized rate of a given reaction and control of the reaction rate in the context of other reactions, including the complementary backwards reaction. For some enzymes the former pressure predominates and in this case the factor which will ultimately limit the reaction rate is the rate of substrate and product diffusion (eg. triose phosphate isomerase Albery and Knowles, 1976).

The dynamic mechanism may promote the reaction rate and at the same time be involved in the control process. In the latter case its effect would be more short-lived than other control mechanisms such as allosteric feedback control. Also, if most of the dynamic effects are restricted to the region near the active site, enzyme development would probably be quicker on an evolutionary scale than more complex mechanisms. Its comparative simplicity would not make it suitable for all systems where control is required, however.

Since biological evolution tends to select a small number of secondary and tertiary structures to assist specific processes, we might focus attention on possible requirements for protein groups or larger structures involved in the effects proposed here. The tension effect would require a molecular 'shock absorber' which would not, however, immediately redistribute substrate impact energy. One of the longer-lived effects might operate via an isolated vibrational 'sink' which would absorb some of the energy from an exothermic reaction and make it available to promote a subsequent reaction : solvent accessibility and leakage to the rest of the protein will be paramount considerations in this case. Effectively we have a catalyst 'activated' by the reaction.

6.3 Enzyme Catalysis

The mechanisms proposed here add an extra level of complexity to the working model of an enzyme as a catalyst. At the same time they widen the gap from models of more conventional catalysts. The use of the term 'catalysis' to describe both sets of phenomena is becoming increasingly uncomfortable - particularly with respect to disparities in control.

6.4 Enzyme Kinetics

As kinetic methods study the bulk reaction under given experimental conditions the sort of concentration-dependent shifts in the statistics of the reaction system suggested here would be expected to affect measured rates in a complex manner. The proposed mechanisms also indicate that it is essential to study enzyme systems under conditions as close as possible to the physiological ones in order to detect the subtle time-dependent effects.

7. CONCLUSIONS

A number of interesting if speculative arguments have been advanced here. While suggestions of slow relaxation of proteins is not implausible, the idea of a positive role in enzyme function from substrate- or reaction-perturbed dynamical motions in some enzyme systems is an untested hypothesis.

Several types of study are indicated. More detailed experimental studies, particularly by nmr, of the dynamics of native proteins to detect low frequency motions together with studies in the presence of substrates to test the effects of binding and reaction on these motions. Long-time molecular dynamics simulations before and after introduction of substrates : demonstration of the existence of slow relaxation will not in itself, however, be proof of a positive role in catalysis or control. Vibrational trapping and shock absorber mechanisms will be more difficult to detect : an initial approach could be a search of simplified models of sections of proteins (preferably selected by comparative scrutiny of known structures of active sites) for features in common with known structures sustaining these effects from small molecule chemistry. Re-examination of kinetic data may be helpful in preliminary detection of enzymes exhibiting some of the postulated effects : kinetic analysis cannot, however, provide a definitive answer.

ACKNOWLEDGEMENTS

The author thanks Dr Graham Richards for helpful discussions and the Nuffield Foundation for financial support.

NOTES

1. Gross conformational changes will not be considered here. Also dynamical states of substrate or product molecules will not explicitly be mentioned : the extent to which these effects may contribute to the revised model in section 5 will be apparent from the discussion.

2. It will be convenient in the following discussion to switch backwards and forwards between number averages and time averages. The molecular dynamics studies are, so far, of too short duration to represent limiting time averages. Of course in very small systems

statistical treatments are entirely inappropriate and stochastic methods must be used (McQuarrie, 1967).

REFERENCES

Albery, W.J., and Knowles, J.R.: 1976, Biochem. 15, pp. 5627-5631.
Ben-Shaul, A., and Tabor, M.: 1979, Chem. Phys. Letters 61, pp.571-576.
Boyd, R.K.: 1977, Chem. Rev. 77, pp. 93-119.
Campbell, I.D., Dobson, C.M., and Williams, R.J.P.: 1978, Adv. Chem. Phys. 39, pp. 55-95.
Eigen, M., and Hammes, G.G.: 1963, Adv. Enzymol. 25, pp. 1-38.
Eigen, M.: 1967, Nobel Symposium 5, ed. S. Claesson, pp. 477-484.
Fersht, A.: 1977, Enzyme Structure and Mechanism. Freeman, Reading, UK.
Gelin, B.R., and Karplus, M.: 1975, Proc. Nat. Acad. Sci. USA 72, pp. 2002-2006.
Gelin, B.R., and Karplus, M.: 1977b, Proc. Nat. Acad. Sci. USA 74, pp. 801-805.
Glansdorff, P., and Prigogine, I.: 1971, Thermodynamic Theory of Structure, Stability and Fluctuations. Wiley-Interscience, London.
Gready, J.E.: 1979, The Molecular Basis of Enzyme Action - a Dynamic and Statistical View, to be published.
Haldane, J.B.S.: 1930, The Enzymes, Longmans, London.
Hammes, G.G.: 1968, Adv. Protein Chem. 23, pp. 1-57.
Hase, W.L.: 1976, in: Dynamics of Molecular Collisions, Vol. B, ed. Miller, W.H. Plenum Press, New York, pp. 121-169.
Jencks, W.P.: 1975, Adv. Enzymol. 43, pp. 219-410.
Karplus, M., and McCammon, J.A.: 1979, Nature 277, p. 578.
Karplus, M., and Weaver, D.L.: 1976, Nature 260, pp. 404-406.
Kuntz, P.J.: 1976, in : Dynamics of Molecular Collisions, Vol. B, ed. Miller, W.H. Plenum Press, New York, pp. 53-120.
Levitt, M. and Warshel, A.: 1975, Nature 253, pp. 694-698.
Light, J.C., Ross, J., and Shuler, K.E.: 1969, in: Kinetic Processes in Gases and Plasmas ed. Hochstim, A.R. Academic Press, New York, pp. 281-320.
Lin, S.H., and Eyring, H.: 1974, Ann. Rev. Phys. Chem. 25, p. 49.
McCammon, J.A., Gelin, B.R., Karplus, M., and Wolynes, P.G.: 1976, Nature 262, pp. 325-326.
McCammon, J.A., Gelin, B.R., and Karplus, M.: 1977a, Nature 267, pp. 585-590.
McQuarrie, D.A.: 1967, Stochastic Approach to Chemical Kinetics, Methuen, London.
Miller, D.G.: 1960, Chem. Rev. 60, pp. 15-37.
Penrose, R.: 1979, in : Einstein Centenary Volume eds. Hawking, S.W., and Israel, W. Cambridge University Press.
Rolfe, T.J., Rice, S.A., and Dancz, J.: 1979, J. Chem. Phys. 70, pp. 26-33.
Tabor, M., Levine, R.D., Ben-Shaul, A., and Steinfeld, J.I.: 1979, Mol. Phys. 37, pp. 141-158.
Widom, B.: 1965, Science 148, pp. 1555-1560.

EFFICIENCY AND MECHANISM IN INTRAMOLECULAR CATALYSIS

Anthony J. Kirby
University Chemical Laboratory, Cambridge CB2 1EW, UK

The potential efficiency of intramolecular catalysis, as measured by the effective concentration of the catalytic group, depends strongly on the mechanism. Nucleophilic catalysis can be very efficient indeed, but intramolecular general base, and probably general acid catalysis also, are intrinsically much less efficient. These generalisations are relevant to our understanding of the efficiency of catalysis by enzymes.

The way enzymes work has always held a special fascination for the organic chemist. They catalyse reactions he finds difficult or impossible, under the mildest conditions, and in the cheapest solvent available, with an efficiency which borders on the miraculous. Clearly enzymes have solved many of the problems that he has to face from day to day, so there must be valuable lessons to be learned from understanding enzyme chemistry.

This talk is about some of the work we have been doing in recent years on intramolecular catalysis, which bears on the problem of enzyme mechanism. I speak not as an enzymologist, but as an ordinary organic chemist interested in mechanism. That means that I approach the problem in a particular way, and I should start by defining this approach.

Enzyme chemistry starts when the enzyme meets the substrate, usually as a result of a random collision in solution.

$$\begin{matrix}\textbf{E}\text{nzyme} \\ +\ \textbf{S}\text{ubstrate}\end{matrix} \rightleftharpoons \textbf{E.S} \rightleftharpoons \textbf{E.P} \rightleftharpoons \begin{matrix}\textbf{E}\text{nzyme} \\ +\ \textbf{P}\text{roduct}\end{matrix}$$

BINDING MECHANISM — CATALYTIC MECHANISM

B. Pullman (Ed.): Catalysis in Chemistry and Biochemistry. Theory and Experiment. 25–36.

Enzymes which react free in solution in the cell have to select their substrates from the range of many hundreds of compounds present in the same solution, so the binding mech- anisms concerned - the way the molecules are brought togeth - must be highly efficient and selective. Especially when one considers that the product must not bind tightly, or turnover will be lost; though the structure of the product is identical to that of the substrate apart from the presen or absence of the bond involved in the reaction.

We are primarily interested in the catalytic mechanism the chemistry that goes on when the substrate is already bound to the enzyme, in contact with the catalytic groups o the active site. If the structure of the enzyme is known i sufficient detail, for example from X-ray crystallography, we know what these groups are, and can thus define the reac- tion concerned in terms of specific functional group inter- actions. At this point the problem becomes a problem of organic reaction mechanism, which can in principle be stud- ied by the methods of physical organic chemistry. It is no possible to apply most physical organic techniques with any confidence to enzyme-substrate reactions, so we have to rel on what we know about the relevant functional group inter- actions in systems which are simple enough to understand in detail.

Work on such simple systems has defined three mechanis of general importance (Kirby, 1979), which account for the reactions of all the common substrate groups with all the catalytic groups available on amino-acid side chains. These mechanisms are illustrated in the Scheme, in the context of ester hydrolysis. Nucleophilic Catalysis (NUC) is the famil iar double displacement mechanism in which an initial sub- stitution by the nucleophilic catalyst produces an inter- mediate reactive enough to be rapidly hydrolysed, thus supporting an overall reaction faster than direct attack by water. General base catalysis (GBC) is an intrinsically less effective mechanism, observed when nucleophilic catal- ysis is inhibited or prevented, in which the catalyst assist reaction by removing a proton from the attacking nucleophile as it attacks. And general acid catalysis (GAC) is the complementary mechanism where the acid catalyst donates a proton to the leaving group as it leaves.

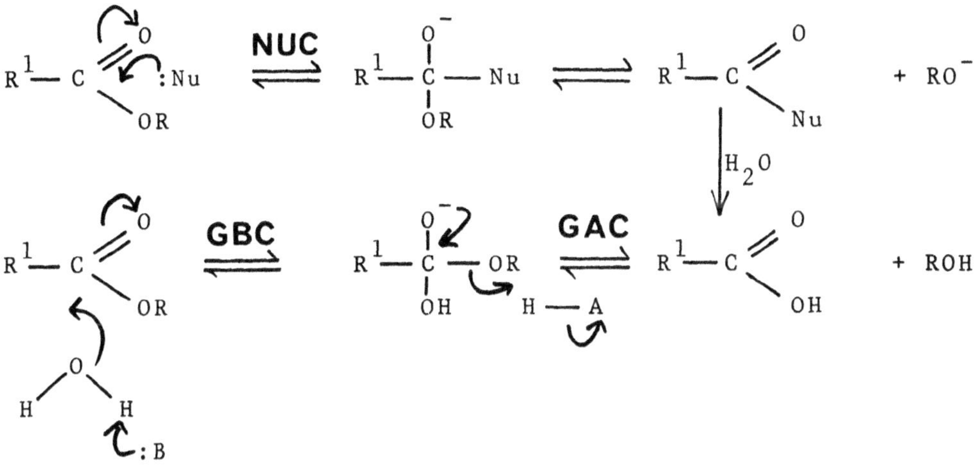

SCHEME

These mechanisms are normally observed for ordinary functional groups only in intramolecular reactions. These are faster than the corresponding bimolecular reactions because of the proximity of the groups concerned. Generally (Table 1) the closer the groups the faster they react; though they can get too close, as shown by the result with the malonate half-ester (Kirby and Lloyd, 1976). Here catalytic efficiency falls, and the mechanism changes from intramolecular nucleophilic (**1**) to general base catalysis (**2**), because the nucleophilic mechanism would involve an unfavourable four-membered cyclic anhydride intermediate.

Table 1. Relative Rates of Hydrolysis of Phenol Esters at pH 7.

	Relative Rate, 25°	Mechanism	Effective Concn. of COO^-
$PhOOCCH_3$	1	(GBC)	-
$PhOOC(CH_2)_3COO^-$	150	NUC	25M
$PhOOC(CH_2)COO^-$	23,000	NUC	4000M
$PhOOC\ CH_2\ COO^-$	50	GBC	10M
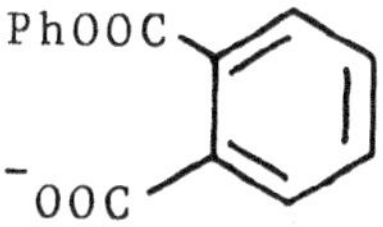	$1.1x10^6$	NUC	$2x10^5M$

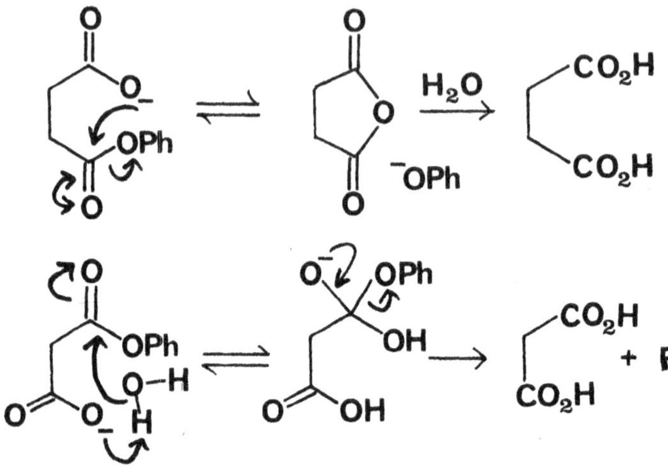

(1)

(2)

Enzyme-substrate reactions involve a subtle interplay of catalytic and binding mechanisms, which cannot readily be separated: the catalytic mechanism and its efficiency depends crucially on the way the reacting groups are brought together. The same is evidently true of intramolecular reactions, and one experimental approach to the problem of the efficiency of enzyme catalysis is to vary structure in intramolecular reactions, to try to understand how catalytic efficiency can depend on the way groups are brought together.

First we need to define efficiency. The relative rates of hydrolysis of Table 1 give some indication for a limited series of related compounds, but more generally useful is the effective concentration. This is obtained by dividing the observed first order rate constant for the intramolecular reaction by the second order rate constant for the corresponding intermolecular reaction. For example, the rate constant (k_1) for the hydrolysis of aspirin (**3**) is $1.1 \times 10^{-5} s^{-1}$ (Fersht and Kirby, 1967). Intermolecular catalysis of this reaction by acetate ion (**4**) can also be detected, and the second order rate constant measured ($k_2 = 8.4 \times 10^{-7} M^{-1} s^{-1}$). Both reactions involve general base catalysis, so the effective concentration of the carboxylate group of aspirin, acting as general base, is $k_1/k_2 = 13M$.

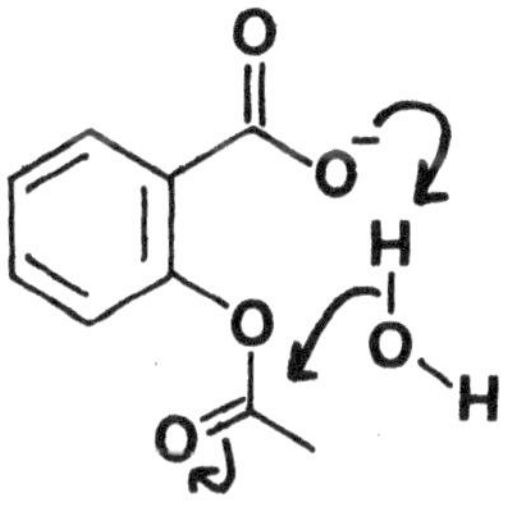

(3)

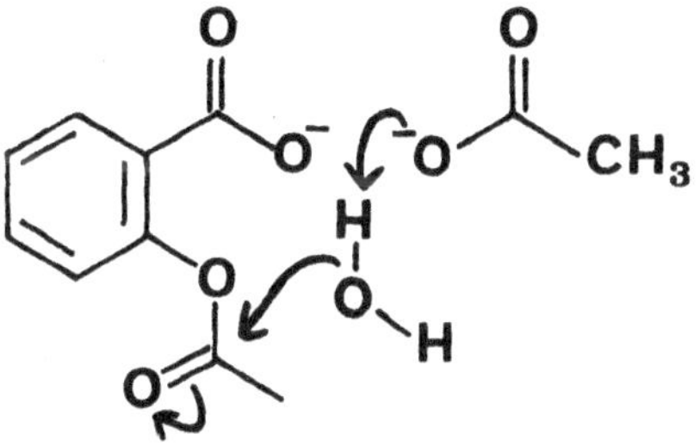

(4)

This is a typical value for intramolecular general base catalysis. We have looked at a series of compounds hydrolysing by this mechanism (Table 2), and find that the effective concentration of the carboxylate group in this system is notably insensitive to structural variation, and numerically small. (Kirby and Lloyd, 1976).

Table 2. Structure and Reactivity in Intramolecular General Base Catalysis

	Angle α	Effective Concn. of COO^-.
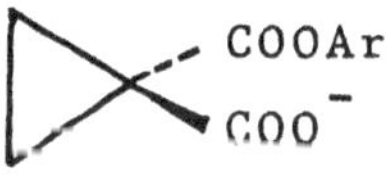	118.4	60M
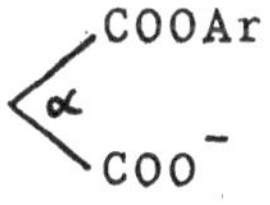	110.0	25M
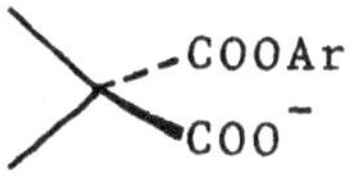	106.2	11M

Scattered data from a wide range of other systems are consistent with the conclusion that the effective concentration of a neighbouring general base is usually of the order of 1-10M, and never more than 100M.

Intramolecular nucleophilic catalysis, on the other hand, can be very much more efficient, as the data in Table l have already indicated. And systematic structural variation can lead to very high effective concentrations indeed. The data in Table 3 show how the reactivities of a series of maleamic acids vary by over nine powers of ten, as a result of changes in geometry comparable to those which changed the effective concentration of a neighbouring general base by less than an order of magnitude (Kirby and Lloyd, 1976).

Table 3. Structure and Reactivity in Intramolecular Nucleophilic Catalysis

	Angles α,β	Effective Concn. of COOH
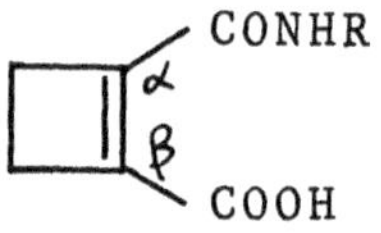	132.1,133.4°	> 10M
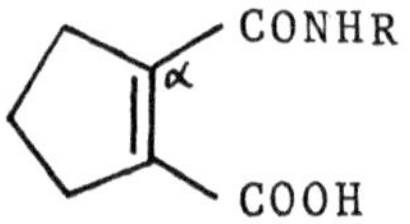	127.7,131.5°	> 100M
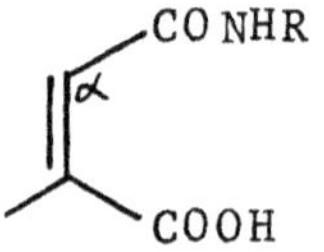	126.8,131.7°	> 10^7M
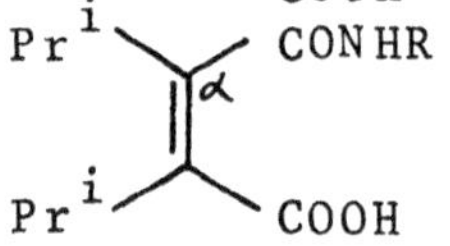	121.0,121.7°	> 10^{10}M

The most reactive compound in this series shows extraordinary reactivity: under conditions where ordinary amides have half-lives of hundreds of years, dialkylmaleamic acids have half-lives of less than one second (Kirby and Lancaster We are thus dealing with simple reactions going at rates comparable with those of some related enzyme catalysed reactions. It is highly significant, therefore, that the mechanisms of intramolecular catalysis change with increasing reactivity.

The data in Table 4 are for ester hydrolysis. The most reactive system in Table 1 was monophenyl phthalate, which is hydrolysed with intramolecular nucleophilic catalysis, though the mechanism of the corresponding bimolecular reaction is general base catalysis. Up to this point the simple alkyl esters have been quite stable, but monometh

Table 4. How Mechanism Depends on the Reactivity of the System

	Catalysis	Mechanism	Effective Concn(M)
$CH_3CO_2Me + CH_3CO_2H$	no catalysis		(1)
$CH_3CO_2Ph + CH_3CO_2H$	by $CH_3CO_2^-$	GBC	(1)
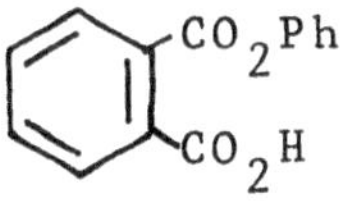	by CO_2^-	NUC	2×10^5
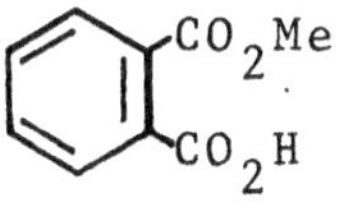	by CO_2H	NUC + GAC	2×10^5
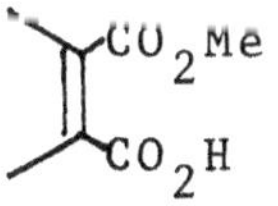	by CO_2H and CO_2^-	NUC + GAC NUC	$\sim 10^{10}$ $\sim 10^{10}$

phthalate is hydrolysed at a significant rate, now at low pH, in a reaction involving intramolecular catalysis by the COOH group. The most reactive ester acids known are the half-esters of dialkylmaleic acids. Methyl hydrogen dimethylphthalate, for example, has a half-life of just 30 seconds at 37° at low pH. And in this system we see for the first time a rapid hydrolysis of the alkyl ester anion, which turns out to be simple intramolecular nucleophilic catalysis (as in **1** above) (Aldersley, Kirby and Lancaster, 1974). Evidently as we make the system more reactive new mechanisms, or at least new rate determining steps of old mechanisms appear. So if one is ultimately interested in the mechanisms of enzyme reactions, which are faster than almost any we have seen so far, it is clearly important to study the most reactive systems available.

Nucleophilic catalysis, then, can give rise to very large rate accelerateions, while general base catalysis is far less efficient. The picture for general acid catalyis is less clear. In principle general acid catalysis is the microscopic reverse of general base catalysis, as is clear from the following example taken from the Scheme. Therefore

one expects general acid catalysis also to be a relatively inefficient mechanism. In many cases this is true, but in one important class of reactions intramolecular general acid catalysis appears to show very large rate enhancements. The data in Table 5 (Fife, 1975) are for acetal hydrolysis, and show accelerations comparable with those observed for nucleophilic catalysis.

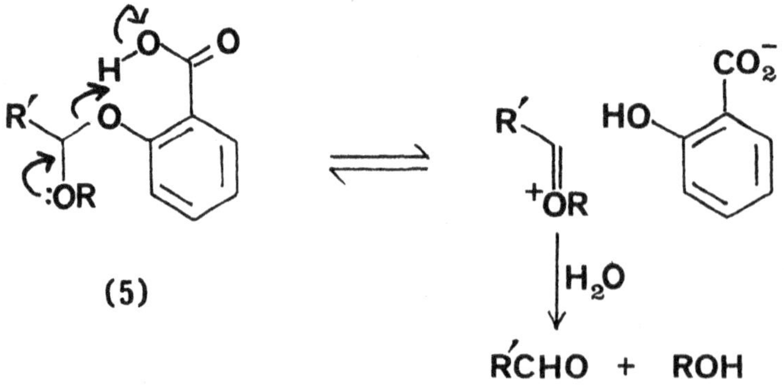

Table 5. Structure and Reactivity in Intramolecular General Acid Catalysis (5)

	Rate Enhancement	Effective Concn.
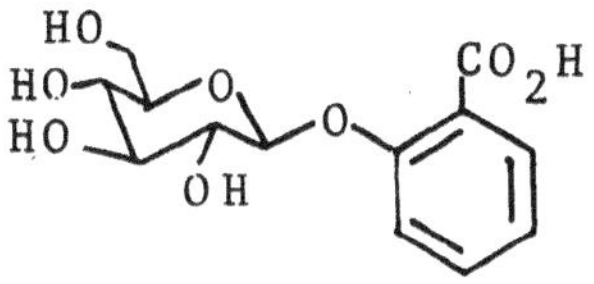	10^4	
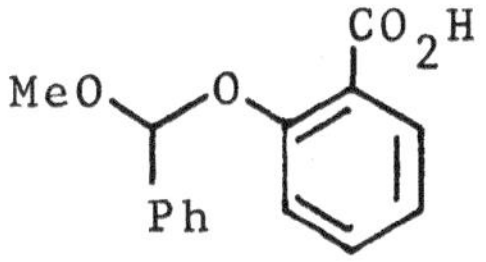	10^5	330M
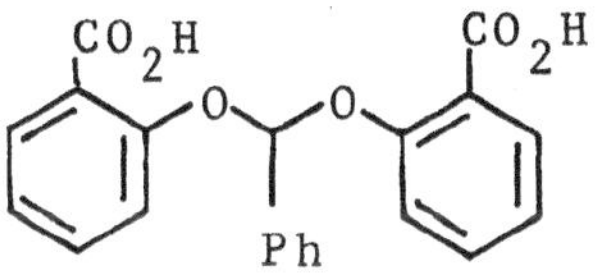	$3x10^9$	3-400M

Presented in this way these figures are misleading. First, the rate enhancements quoted refer to comparisons of hydrolysis rates, involving either the corresponding methyl esters or the para-substituted isomers. It is not possible to measure the effective concentration of the COOH group in these reactions, because intramolecular catalysis is so efficient that the intermolecular reaction is not observed. But rate constants are available for the general acid catalysed hydrolysis of one or two methyl esters, and the effective concentrations calculated using these data give effective concentrations of only 3-400M.

It is also significant that all the compounds showing high reactivity (and it should be stressed that a large number do not) are derivatives of salicylic acid. We have suggested before that these compounds with the leaving group conjugated with the neighbouring general acid (**6**) are a special case (Craze and Kirby, 1974). For example, Fife (1975) has shown that catalysis disappears in the corresponding saturated system (**7**). This could be because the leaving group is not good enough: the hydrolysis of dialkylacetals commonly shows little or no general acid catalysis.

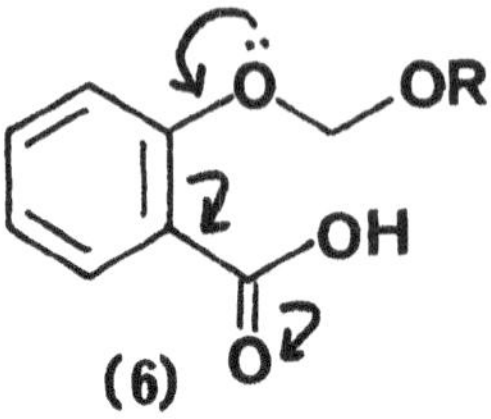

(6)

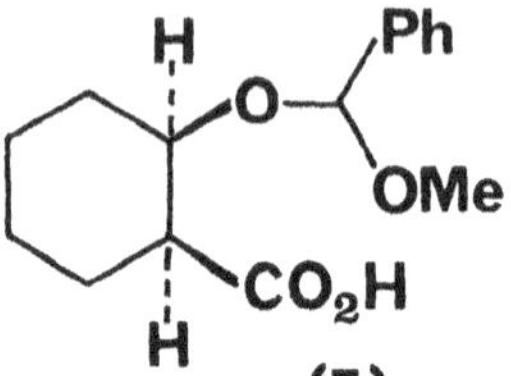

(7)

We have tested this point by a comparison of another pair of compounds, (**8**) and (**9**). The hydrolysis of (**8**) shows efficient intramolecular general acid catalysis, by the mechanism shown (Kirby and Osborne, unpublished).

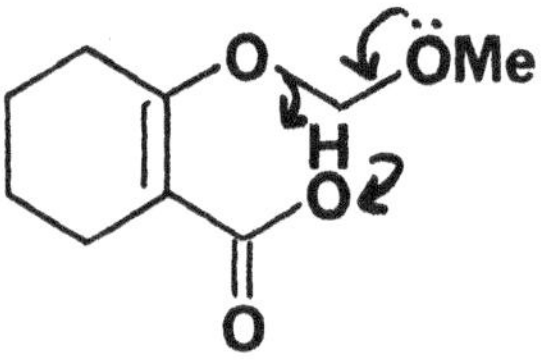

(8)

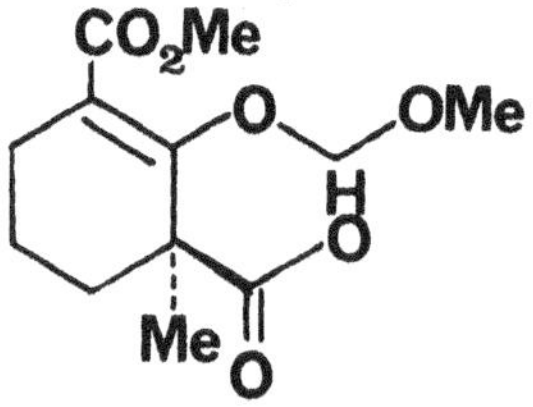

(9)

But the related enol acetal (**9**), which has an identical leaving group, and differs from (**8**) only in that the COOH is no longer conjugated with the leaving group (and in the small change in their geometrical relationship) shows only very weak intramolecular catalysis.

We are still looking at these problems. Clearly COOH groups acting as general acids in glucosidase enzymes like lysozyme are not conjugated with the groups they are helping to leave, and it is important to establish whether the high reactivity of acetals derived from salicylic acid is relevant to enzyme catalysis. Even if it is, our preliminary results suggest strongly that this mechanism also is rather insensitive to structural variation, and thus is not likely to give rise to very large effective concentrations. The preliminary data in Table 6 refer to the intramolecular general acid catalysed hydrolysis of three related acetals (Kirby and Osborne, unpublished), and show the expected small range of reactivity in this limited series of compounds. We do not so far have enough data to estimate the effective concentration of the carboxyl group in these systems.

Table 6. Structure and Reactivity in Intramolecular General Acid Catalysis

	Relative Rate of COOH-catalysed hydrolysis
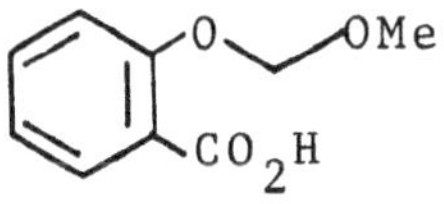	1.0
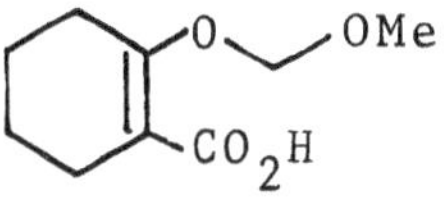	2.0
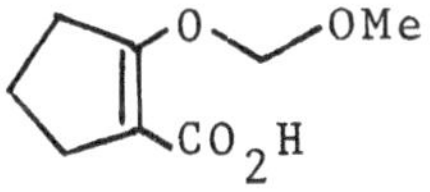	0.08

Conclusion. The evidence from intramolecular catalysis shows rather clearly that high efficiency depends primarily on the way heavy atom centres are brought together to make new covalent bonds. It is a reasonable surmise that the same is true in enzyme catalysis, though many other factors not considered here also make important contributions (Fersht, 1977). General acid-base catalysis plays an essential but supporting role, maintaining and indeed increasing high efficiency; but its additional contribution to efficiency is relatively modest.

References

Aldersley, M. F., Kirby, A. J., and Lancaster, P. W.:1974, J. Chem. Soc., Perkin Trans. 2, pp. 1504-1510.

Craze, G.-A., and Kirby, A. J.: 1974, J. Chem. Soc., Perkin Trans. 2, pp. 61-66.

Fersht, A. R., and Kirby, A. J.: 1967, J. Am. Chem. Soc., 89, 4857-4863.

Fersht, A. R.: 1977, 'Enzyme Structure and Mechanism', W. H. Freeman, San Francisco.

Fife, T. H.: 1975, Adv. Phys. Org. Chem., 11, pp.1-122.

Kirby, A. J.: 1979, Comprehensive Organic Chemistry, Ed. Barton, D. H. R., and Ollis, W. D., Pergamon Press, vol. 5 pp. 389-460.

Kirby, A. J., and Lancaster, P. W.: 1972, J. Chem. Soc., Perkin Trans. 2, pp. 1206-1214.

Kirby, A. J., and Lloyd, G. J.: 1976, J. Chem. Soc., Perkin Trans. 2, pp. 1753-1761.

CRYOENZYMOLOGY IN REVERSE MICELLES

BALNY Claude, HUI BON HOA Gaston and DOUZOU Pierre
U-128 INSERM, BP 5051, 34033 Montpellier cédex, France, and Institut de Biologie Physico-Chimique, 13 rue Pierre et Marie Curie, 75005 Paris, France.

ABSTRACT

Solutions of homogeneous and optically transparent micromicelles in nonpolar solvents of low viscosity, are used to investigate enzyme systems in pure aqueous media at subzero temperatures. The properties of such water microspheres are presented and their behavior studied as a function of their chemical composition, in a large temperature range. Enzymatic reactions are described. Rapid mixing of very low viscosity micromicellar solutions to record fast kinetic are discussed in comparison with results obtained in bulk water or in mixed hydro-organic solvents.

INTRODUCTION

There is a great potential interest in the use of subzero temperatures for the study of enzyme-catalyzed reactions and of their kinetic, thermodynamic and structural parameters. Up to now, enzymology at subzero temperatures (sometimes called cryoenzymology) has been mainly concerned with the use of solvent mixtures of low freezing point and therefore fluid in the explored range of temperatures (1)(2)(3). In such conditions, reactions normally occurring in a few seconds can be sufficiently decreased in rate to obtain their temporal resolution step by step, and analysis of most of their intermediates by all available spectroscopic techniques. Fast kinetic studies are then carried out on elementary steps (1).

Recently, new procedures have been successfully sought out to perform cryoenzymology : the so-called cooling-trapping procedure (4) combines use of cooled fluid media to mix reactant and of frozen media to trap unstable intermediates. Solid states formed by vitrified aqueous solutions of hydrophilic polymers (5) might provide aqueous microenvironments for enzymes investigated at low temperatures. Water in oil emulsions in which water is held stable and unfreezable for hours as low as -40°C under the form of droplets (μm in dimensions) have been proposed and used (6)(7)(8) in spite of severe limitations. Thus the low temperature procedure becomes diversified and a new alternative

B. Pullman (Ed.): Catalysis in Chemistry and Biochemistry. Theory and Experiment, 37–50.

using water microspheres (nm in dimensions) in homogeneous and optically transparent nonpolar solvents of low viscosity has been quite recently tried (9)(10). Preliminary results have shown a number of advantages in using such procedure, and the present paper describes new results able to provide further information about the perspectives and problems of these micromicellar solutions as media to investigate enzyme-reactions at subzero temperatures.

1. EXPERIMENTAL

1.1. Preparation and properties of micromicellar solutions of enzymes

A microemulsion is a clear, transparent and stable system consisting of essentially monodisperse oil in water (O/W) or water in oil (W/O) droplets with diameters generally in the range of 10-200 nm. Micro emulsions are transparent because of their small particle size, they are spherical aggregates of oil or water dispersed in the other liquid and stabilized by an interfacial film of one or more surfactants.

Micelle formation in nonaqueous solution (polar solvents) although recognized for some time, has been investigated systematically only in recent years both experimentally and theoretically. Amphiphilic surfactants, characterized by possessing in the same molecule a hydrophilic group which tends to be water soluble and hydrocarbon insoluble, and a lipophilic group which tends to be hydrocarbon soluble and water insoluble, tend to be miscible with both water and hydrocarbon. Suitable amphiphilic surfactants dissolved in a hydrocarbon are able to solubilize water which, in the absence of these compounds, is insoluble in the hydrocarbon. In such systems, water and hydrocarbon can be regarded as solubilized with amphiphile, this amphiphile acting as cosolvent.

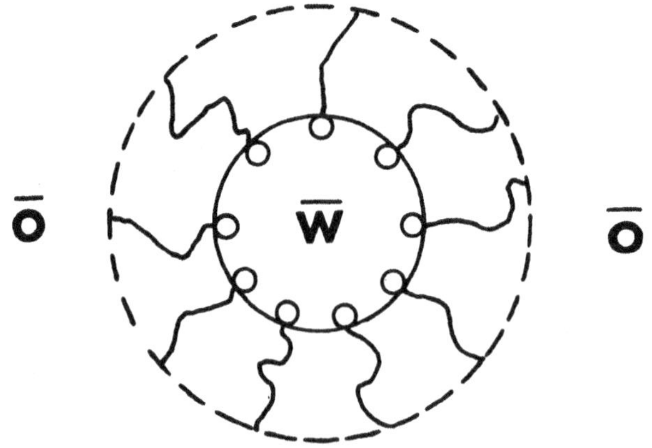

Figure 1. Possible reverse micelle in nonpolar solvent.
$\bar{W}$: water pool; $\bar{O}$: nonpolar phase.

The polar groups on the amphiphile aggregate and form an aqueous core in the presence of water. Such micelles are termed inverse, reverse or reverted micelles. A possible structure is shown in fig. 1.

It appears from a survey of literature that the essential properties of micelles in nonpolar solvents are understood, namely their stability and variations of size, their dissociation behavior and their solubilizing capacities. Reverse micelles can dissolve relatively large amounts of water (1-10 % vol/vol depending on emulsion formulae) as well as polar solutes and of course water soluble compounds. Consequently they can act and be used as media for a number of reactions, including enzyme-catalyzed reactions. Very few attempts to investigate such reverse micelles at subzero temperatures are known, in spite of the fact that hydrocarbon solutions present very low freezing points.

Amphiphilic compounds. Amphiphiles are characterized by possessing in the same molecules two groups which differ markedly in their solubility relationships. These are a hydrophilic group which tends to be water soluble and hydrocarbon insoluble, the typical hydrophilic groups being :

$$-OH, -CO_2Na, -SO_3K, -NMe_3Br, -O(CH_2-CH_2-O)_nH$$

Typical lipophilic groups, which tend to be hydrocarbon soluble and water insoluble are :

$$-C_nH_{2n+1}, -C_{17}H_{33}, C_8H_{17}O_2C\underset{|}{C}H-CH_2-CO_2-C_8H_{17}$$

According to the relative magnitudes of each of these functions, amphiphiles may range from predominantly hydrophilic, water soluble and hydrocarbon insoluble, to predominantly lipophilic, hydrocarbon soluble and water insoluble. Each amphiphile is then characterized by an hydrophilic-lipophilic balance (H-L-B), and when for $C_nH_{2n+1}COOK$, n = 1 to 8, they are predominantly hydrophilic whereas, when $n \geq 12$, they are predominantly lipophilic.

The best amphiphiles, showing micelle formation and solubilization of water and hydrocarbons, are those in which both the hydrophilic and lipophilic functions are strong and not too unequally matched (n = 8 to 20 with $C_nH_{2n+1}COONa$). A large number of commercially available amphiphiles meet such conditions. Their polar "heads" can be non ionic and ionic. The most popular amphiphile is an anionic one : sodium diisooctyl sulfosuccinate (Aerosol OT, termed AOT, Fluka compound) and has been used in the present work.

Suitable amphiphiles dissolved in a hydrocarbon are able to soluble water which, in the absence of these amphiphiles is insoluble in the hydrocarbon. Amphiphiles provide the easiest way of getting monolayers of surfactants surrounding a water core (often termed water pool) and it is correct to regard water and hydrocarbon as solubilized with amphiphiles.

Nonpolar solvents. High-grade alkane solvents (n-hexane, n-heptane, dodecane) as well as benzene and tetrachloroethylene, carbon tetrachloride are most often used to dissolve amphiphiles (3-10 % wt/vol) under stirring at room temperature or moderate heating. Silicone oils are used for the investigation of enzymes. All these solvents present low freezing points (<-50°C) and remain colorless or become translucent at subzero temperatures depending on the type and concentration of amphiphile surfactant. Information about the physical chemical properties of surfactants in nonpolar organic solvents is meager and it is assumed that there is an equilibrium between monomers and micelles, with the concentration of the monomer remaining constant above a critical micelle concentration. Indeed, in the absence of traces of water, aggregation is sometimes precluded and in many instances water promotes the formation of larger and more stable reverse micelles.

Micelles. In solutions of amphiphiles in hydrocarbon solvents, there is a tendency for groupings of molecules to arise in which like is associated with like. These local groupings fluctuate under the influence of the thermal motion and are in equilibrium with neighboring molecules. These essentially transitory groupings are termed micelles. Theories and models for aggregation in nonpolar solvents have been based largely on models derived from aqueous micelles. However, it has been found that the number of monomers involved in most surfactant aggregates in nonpolar media is relatively smaller than in aqueous solution, and that the monomer ↔ n mer type association is unlikely to represent the behavior of surfactant (12).

It is obvious that a large variety of phases can exist in equilibrium with each other. Each phase might involve a different micellar structure in equilibrium with each other. Micelles are therefore of a statistical character, and it is essential to guard against a general picture of micelles as persistent entities having well-defined geometrical shapes. Any local change in conditions, e.g., concentration or temperature, will immediately alter them. These very mobile structures can dissociate extremely rapidly and such property will be of essential importance to undertake the rapid mixing of solubilizates in micromicellar solutions.

In conclusion of this brief survey let's define reverse micelles as microemulsions made of spherical aggregates of water dispersed in oil and stabilized by an interfacial film of one (or more) amphiphilic surfactants such that the water droplets range in diameter from 10 to 20 nm. These microemulsions are clear, transparent and stable systems of low viscosity and freezing point, micelles are not permanent entities but very mobile structures dissociating and aggregating extremely rapidly when local changes in conditions are induced. Due to their average size and water content, such micelles can solubilize a number of solutes including macromolecules and therefore enzymes; they can then act as media for reactions and be investigated at subzero temperates. However they represent unique media and necessitate an analysis of the nature and properties of surfactant-entrapped water.

Water in micelles is expected to behave very differently from ordinary water because of extensive binding and orientation effects induced by the polar heads forming the water core. The effective polarity, acidity, and microscopic viscosity of the surfactant-entrapped water pools should be different from those in bulk water and these specific properties should provide a unique medium for interactions and reactions of water-soluble substrates. In particular, these properties permit to expect substantial changes in reaction kinetics.

1.2. Solubilization of proteins by the water pools

Reverse micelles are able to solubilize relatively large amounts of water in a varicty of nonpolar organic solvents and the water is accomodated in the polar centers of the aggregates where it forms spherical or near-spherical pools the sizes of which are controlled by the water/surfactant ratios. Polar solutes are expected to be localized in water pools and numerous observations have amply substantiated this expectation (12)(13)(14). In fact the extent of solute solubility depends on the polarity of the solute, the nature and concentration of the surfactant, the polarity of the bulk nonpolar organic solvent, and the amount of entrapped water and the size of the solubilizate.

Systematic spectroscopic assays were carried out in this Laboratory with cytochrome *c* (Sigma, type VI), horseradish peroxidase (Sigma, type VI), catalase (Calbiochem), bacterial cytochrome P450 (supplied by I.C. Gunsalus). Aqueous solutions of these proteins were prepared at +4°C prior to their addition under sonication of several seconds to dry apolar solutions of AOT. Two different apolar solvents were used, i.e. n-heptane and silicone oils of various molecular weight and viscosity, their freezing point being -90°C and -50°C respectively. Micromicellar solutions with viscosities < 1 cP were prepared with silicone oils of low molecular weight as well as with alkane.

An Aminco-Chance DW2 spectrophotometer equipped to investigate subzero temperatures was used to carry out spectroscopic assays on solubilized enzymes as well as to monitor enzyme-catalyzed reactions (15).

Most of these solutions containing proteins are homogeneous, optically transparent down to 250 nm and of low freezing point (< -50°C) but show turbidity well before this freezing point; the lowest temperature reached without appearance of turbidity was -40°C with solutions of cytochrome *c*. Selected proteins appear to be soluble in reversed micelles depending largely on their molecular weight, water content and in fact [water]/[surfactant] ratio, but their absorption spectra are markedly influenced by this water to surfactant ratio. The absorption spectrum of cytochrome *c* present abnormal extinction coefficient values even for a large water to surfactant ratio (upper than 25).

Enzymes in water pools are surrounded by a large number of surfactant molecules and should be, therefore, effectively shielded from the bulk apolar solvent and then substracted from its denaturing effect.

In fact, there is no general rule and protection of proteins cannot be taken as granted : some of them can be maintained almost indefinitely in their native state (cytochrome *c*), some others are stable only if temperature is lowered (bacterial cytochrome P_{450} at -10°C), whereas others are partially or totally (and progressively) altered (myoglobin loosing its heme).

Using the parameters for the absorption spectra (ε and λ_{max}) of several haemoproteins *vs* different surfactant and water concentrations permit to determine best medium conditions to accomodate them. Absorption spectrum of horseradish peroxidase in micellar solution obtained for an $[H_2O]/[AOT]$ ratio of 19 indicates that the diffusion of this solution is then relatively small in spite of the size of protein and micelles. Solutions are thus homogeneous, optically transparent down to 250 nm and of very low macroviscosity (< 1 cP). Alterations of the parameters for the absorption spectra have been found at lower water content (< 5 % vol/vol) and might result from restricted polarities of water (16). Since the higher the water content, the higher the temperature at which micromicellar solutions become turbid, a compromise must be sought out to accomodate the enzymes and to carry out their investigation at subzero temperatures. Each type of enzyme needs to be examined individually to find out medium conditions insuring both its normal spectral response and its experimentation in a suitable range of subzero temperatures.

It is of course much more easier to solubilize properly enzymes in catalytic concentrations (10^{-9} M) than in concentrations allowing spectroscopic studies (10^{-5}, 10^{-6} M). In the last case, the water content must be high (> 5 % vol/vol) to accomodate enzyme molecules, changes in spectroscopic parameters (ε, λ_{max}) and abnormal diffusion of solutions can be observed, indicating the difficulty to obtain "true" solutions comparable to those obtained in bulk water. Other data obtained with proteins of different molecular weight (cytochrome *c*, mv : 12,500; horseradish peroxydase, mw : 45,000 and catalase, mw : 240,000) indicate that there is no clear correlation between this molecular weight and diffusion of solutions.

Thus the uniqueness of the aqueous media in micromicellar reactions results in spectral changes which must be carefully examined and modified by suitable changes in medium conditions. The profound influence of the type and concentration of surfactant used, of water/surfactant ratios, molecular weight of proteins should be emphasized. Effects of these parameters are complex and, as yet, unpredictable.

2. RESULTS AND DISCUSSION

2.1. Enzyme activity *vs* water to surfactant ratio

It has been generally found that reaction rates in the presence of reverse micelles are quite different from those observed in bulk water.

There is first a strong and complex relationship between enzyme activity and the concentration of water entrapped in micelles. Such relationship has been studied in some details by Misiorowski, Poon and Wells with snake venenom phospholipase A_2 which can be extremely reactive in ethereal solutions of phosphatidylcholine used both as surfactants and potential substrates (17)(18). It has been found that the enzyme becomes active only when the hydration core is large enough to accomodate the enzyme, the onset of activity as well as half-maximal activity value being recorded at selected concentration in water. On the basis of data obtained it has been assumed that at least four distinct micellar species can exist in ethereal solutions of phosphatidylcholine : 1) an anhydrom species, 2) a Ca^{2+} cation associated species, Ca^{2+} playing a catalytic role in phospholipase activity, 3) a moderately hydrated species, and 4) a highly hydrated species. The enzyme can only express its activity when present in micellar species 3. Further investigation has shown that the enzyme is only active when dissolved within a micelle which contains bound water whose properties closely resemble those of pure water.

Tryptic enzyme activity towards benzoyl-L-arginine ethyl ester (Bz Arg OEt, highest quality grade from Fluka) has been studied as a function of the water to surfactant ratio in micromicellar solutions. The reaction was initiated by mixing in the sample cuvette 50 µl of enzyme micellar solution and 950 µl of a substrate micellar solution. In standard experiments, the AOT was 1.5 % (wt/vol) in silicone oil (viscosity 0.65 cP) and the water phase 1 % (vol/vol). The enzyme stock solution (0.1 mM) was made in 1 mM HCl and stored at +4°C; the substrate was dissolved in the buffer. For the substrate measured in micellar solution $\Delta\varepsilon_{255-290\,nm}=1700\ M^{-1}cm^{-1}$. The kinetic parameters k_{cat} and $K_{m(app)}$ were determined from Lineweaver-Burk plots using a substrate concentration range from 10 to 100 mM in the aqueous phase.

The pH-activity profile of the tryptic hydrolysis of the Bz Arg OEt in micellar solutions accomodating the enzyme shows a marked shift compared to that obtained in bulk water. This shift very probably arises from the influence on the enzyme of the assembly of negatively charged polar heads of a polyanionic environment, resulting from AOT (10).

Relatively high buffer concentrations, ≃ 100 mM, are required to provide a suitable pH for enzyme activity. Activity measurements are passable over a broad temperature range. The activation energy (11.5 ± 0.5 Kcal/mol) is essentially identical to the value obtained in bulk water over the narrower temperature range of 30 to 4°C. However substantial effects are observed on substrate affinity as compared to those found in bulk water and in the cosolvent, 1:1, vol/vol, ethylene glycol-water. The respective $K_{m(app)}$ values are 3.5 10^{-3} M for the micelles [concentrations in the water phase; 1% (vol/vol) water], 4 10^{-6} M in bulk water and 8.4 10^{-5} M in cosolvent (19); the k_{cat} (V_{max}) values are 8.7 sec^{-1} in bulk water, 11.0 sec^{-1} in mixed solvent (ethylene glycol-water 1:1 vol/vol), and 9 sec^{-1} in micellar solution. Such changes should be expected when one considers the number of parameters that can influence enzymes and substrates in reverse micellar systems. Still very little is known of the details of mechanisms contributing to rate enhancement or retardation. One may cite as probable mechanisms the partitioning of enzyme and/or substrate between surfactant-trapped water molecules and those occupying the core of the micelle, differential interactions (electrostatic, hydrolytic, nucleophilic, electrophilic) altering the free energy of activation of the overall process as well as K_m and V_{max} value

The magnitude of rate effects could be drastically altered upon changing the enzyme system and, for a given enzyme, upon changing the nonpolar solvent, the nature and concentration of the surfactant and the concentration of added water. In the case of AOT used as surfactant, it can be seen in figure 2 that the onset of trypsin activity, its half maximal and maximal activities depend on the ratio of the water to the surfactant, the optimal activity being obtained with a concentration of 10 molecules of water for one molecule of AOT.

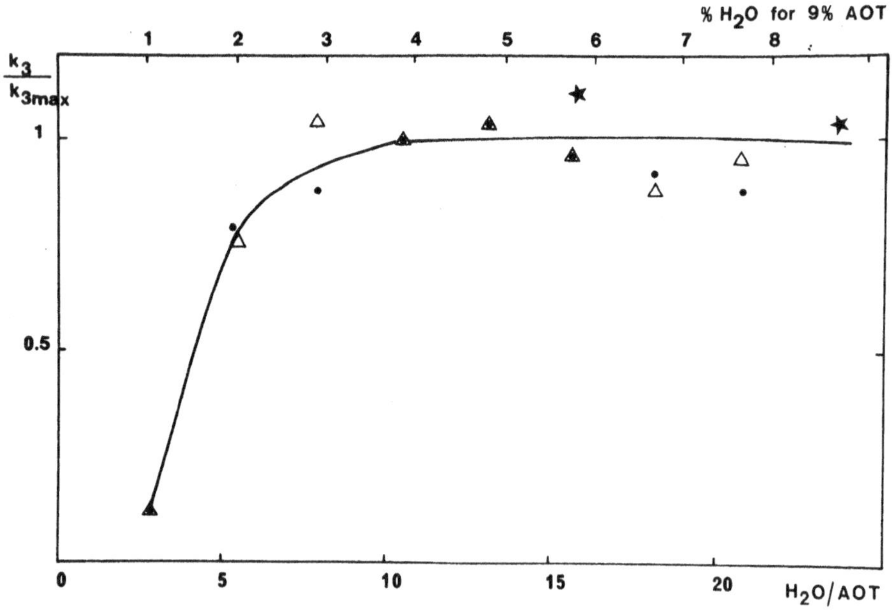

Figure 2. Variation of k_{cat} (expressed in the k_3/k_{3max} form) for tryptic hydrolysis of Bz Arg OET at +20°C in micelles as a function of H_2O/AOT, these two concentrations being expressed in $M.l^{-1}$. (●) in silicone oil containing 9 % wt/vol AOT, 200 mM Tris HCl aqueous buffer pH 9.0; (Δ) in silicone oil containing 9 % wt/vol AOT, aqueous buffer concentration 100 mM; (★) in silicone oil containing 3 % wt/vol AOT, aqueous buffer concentration 100 mM. The upper scale indicates the percentage of water phase (vol/vol) for 9 % wt/vol AOT in silicone oil (viscosity 0.65 cP).

2.2. Behavior and use of micromicellar solutions at subzero temperatures

At room temperature, most micromicellar solutions are homogeneous and optically transparent. They are in fact "colorless" down to 350-250 nm according to the type of surfactant used. These solutions must retain these optical properties in a given range of subzero temperatures to be used as reaction medium for cryoenzymologic studies. The temperature at which transparency is changing can be easily determined and solubility diagrams recorded. There are three different regions appearing as temperature is lowered, colorless, blue translucent and turbid (20). The extent of these regions is remarkably dependent on the particular surfactants, solvents and solutes.

The effects of added solutes on the solubilization of water in non aqueous solutions involving Aerosol OT (AOT) has been treated in only a few papers (20)(21)(22) and changes in the minimum temperature of solubilization as well as region ranges have been measured. It has been found that the precipitate was composed of surfactant and water. Most of such kind of investigations have been carried out in a range of temperatures aboves 0°C, while it is known that many formulas of micromicellar solutions permit to keep homogeneity and transparency below 0°C. We explored the behavior of some micellar solutions in the range of subzero temperatures with the main following results : Micellar solutions containing AOT were examined as a function of temperature to determine the range over which they remain homogeneous. Among the three regions previously described (colorless, blue translucent and turbid), the first two are homogeneous and stable, they represent the regions of water solubility; the third, appearance of turbidity, is dependent on the water/surfactant ratios.

$[H_2O]/[AOT]$ in M/M	H_2O %, v/v	AOT %, w/v	Precipitation temperature ± 1°C
5.5	2	9	-42
8.2	3	9	-35
8.2	1	3	-34
10.9	4	9	-29
13.6	5	9	-22
16.3	6	9	-17
16.3	2	3	-16.5
16.3	1	1.5	-17.5
24.6	9	9	+4

Table I. Dependence of AOT precipitation as a function of H_2O/AOT ratios for various media compositions. Aqueous phase, 100 mM Tris-HCl buffer, pH 9.0; nonpolar organic solvent, silicone oil (viscosity 0.65 cP). Measurements were carried out by light scattering at 350 nm.

The extent of these regions, particularly of the turbid one, are remarkably dependent on the concentration parameters; they depend also on the particular surfactant and solvent used. We found a remarkable difference in the solubility diagrams of anionic and neutral surfactants the latter show colorless regions only at a higher temperature range (> -15°C under the best conditions) whereas AOT is suitable to temperatures as low as -45°C. The presence of enzyme and of electrolytes (neutral salts, buffers) does not markedly affect the solubility of water in the non aqueous solutions of AOT nor appreciably modify the minimum temperature of the transparent region.

It has been found after separation of precipitates that practically all the enzyme and water molecules entrapped in micelles remained in the supernatant. Experiments were carried out with cytochrome *c* in micro-micellar solutions precipitating at +4°C and -17°C, respectively, cooled down to -35°C and then incubated during 10 minutes. Absorption spectra of the supernatant were recorded and corresponding absorption coefficients are shown in Table II as well as in figure 3.

Medium composition	H_2O/AOT in M/M	Absorbance before cooling	Absorbance of the supernatant collected
100 % aqueous buffer		0.59 (86)	
9 % (v/v) aqueous buffer, 9 % (w/v) AOT	24.6	0.69 (106)	0.80 (123)
6 % (v/v) aqueous buffer, 9 % (w/v) AOT	16.3	0.70 (107)	0.76 (117)

Table II. Absorbance at 403 nm of cytochrome *c* in 100 mM cacodylate buffer, pH 7.0 and in micellar solutions, at +20°C. Silicone oil viscosity 0.65 cP. Cytochrome *c* concentration : 6.5 µM. In parenthesis are quoted the corresponding absorption coefficients in mM^{-1} cm^{-1}.

An unknown amount of surfactant and water certainly precipitated but sufficient concentrations of both remain in solutions to accomodate the protein. Thus the supernatant can be used after removal of the precipitate, and even further cooled. A stepwise procedure can thus be used to attain lower temperatures and to retain transparent media and enzyme activity. Such possibility is of course of great practical significance to investigate reverse micellar solutions of enzymes at subzero temperatures.

Present results were obtained with the AOT as detergent but as any other amphiphiles might be suitably tried to determine the best possible

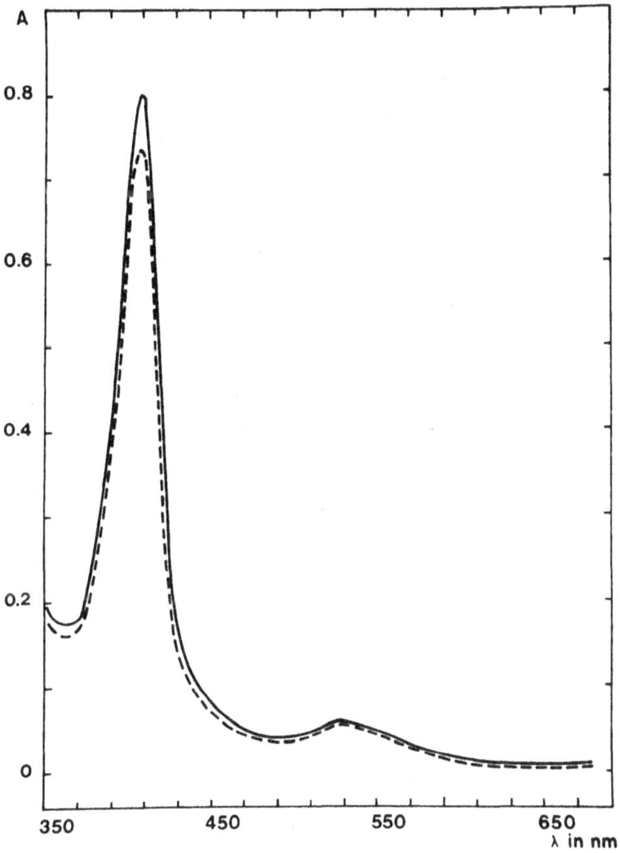

Figure 3. Absorption spectra of cytochrome *c* in micellar solution. (---) in 9 % (wt/vol) AOT, 9 % (vol/vol) aqueous buffer solution, silicone oil 0.65 cP, before cooling; (——) supernatant solution collected at -35°C. Both spectra were recorded at +20°C. Cytochrome *c* concentration in the 100 mM sodium cacodylate buffer pH 7.0 : 73 µM.

conditions of investigation. Our feeling is that the present procedure, which is at its early stage, should be diversified by use of different amphiphiles and emulsion formula and that any new investigation of a selected enzyme should involve systematic trials in these micromicellar solutions.

2.3. Rapid mixing of micromicellar solutions

Since a number of enzyme-catalyzed reactions can be investigated into micromicellar solutions of low viscosity and that, in general, the only substantial effect of micelles on the catalytic is an apparent increase in Km as a consequence of partitioning of substrates between the water pools and the bulk nonpolar solvent, mixing of such solutions might permit to carry out kinetic studies. It is known that with aqueous organic solvents a technical barrier is caused by the difficulty of mixing two solutions of high viscosity (> 50 cP), such values being attained at subzero temperatures. On the other hand reaction rates of diffusion controlled reactions can be altered with such viscosity values.

Since micelles are not persistent entities but very mobile structures dissociating extremely rapidly to adopt new geometrical shapes through changes in conditions, e.g., in concentration or temperature, mixing of micromicellar solutions and therefore of reactants could be obtained in very short time-ranges and then fast kinetic studies could be applied.

Observations carried out with acido-basic indicators and recorded by stopped-flow have shown that the mixing of micromicellar solutions is completed in a time range lower than 10 msec, which is the dead time of the apparatus described elsewhere (23).

Recent kinetic studies on the peroxidatic reaction involving the enzyme horseradish peroxidase (HRP) clearly confirm the above results : recordings of the two first steps of the reaction, namely,

$$E(HRP) + H_2O_2 + AH_2 \leftrightarrow \text{Compound I} \rightarrow \text{Compound II}$$

where AH_2 is an electron donor (ascorbic acid), compound I is the enzyme substrate compound $E{-}H_2O_2$, compound II is the compound in which the heme (Fe^{3+}) has been reduced (Fe^{2+}).

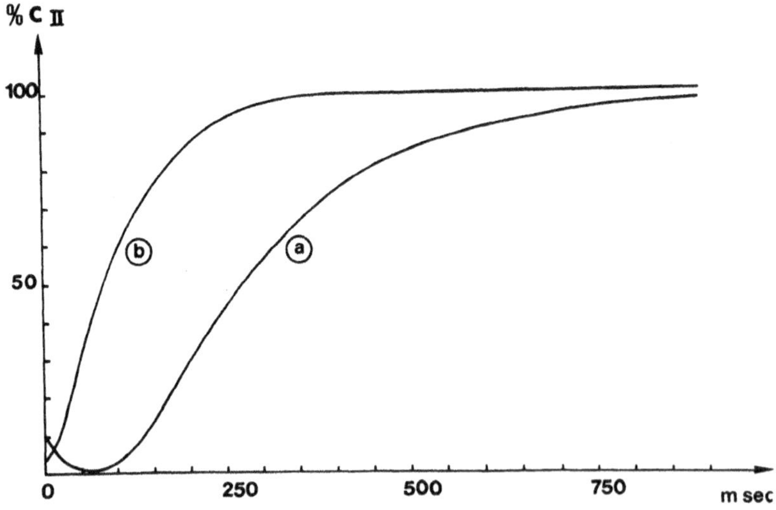

Figure 4. Kinetic traces of the formation of compound II (horseradish peroxidase) at 10°C : (a) in aqueous solution : 100 mM sodium cacodylate buffer, pH 7.0; (b) in reverse micelles (AOT : 10 % wt/vol; heptane; 100 mM sodium cacodylate buffer, pH 7.0 in water : 6 % vol/vol).

In figure 4, kinetic traces obtained in bulk water and in the mixed solvent ethylene glycol-water (50:50 vol/vol) at the same temperature (4°C) clearly show that the rates of formation of compound II are different in each selected medium. Table III reports the rate constants obtained in these different conditions.

t = +10°C		λ = 421 nm
	(a) H_2O	(b) reversed micelle
	caco 100 mM pH 7	AOT 10 % (wt/vol) Heptane 6% (v/v) caco 100 mM pH 7
HRP	$3\ 10^{-6}$ M	$2.5\ 10^{-5}$ M
H_2O_2	10^{-3} M	10^{-3} M
Asc.	10^{-3} M	$5\ 10^{-3}$ M
k_{cII}	$5.26\ 10^{3}\ M^{-1}sec^{-1}$	$2\ 10^{3}\ M^{-1}sec^{-1}$

Table III. Rate constant values for the formation of compound II (horseradish peroxidase) determined by stopped-flow experiments. (a) in bulk water, (b) in reverse micelles.

Thus it is possible to obtain the rapid mixing of micromicellar solutions and then to carry out fast kinetic recordings, but the uniqueness of reverse micelles determines marked changes in such kinetics, as mixed solvents were doing.

CONCLUSION

In conclusion of this brief survey on micromicellar solutions as possible media to investigate enzyme systems in aqueous microenvironments at subzero temperatures, we must say that this procedure represents just but an alternative characterized both by interesting perspectives and severe limitations due to a number of unexpected and unwanted side-effects.

REFERENCES

(1) Douzou, P.: 1977, in "Cryobiochemistry, an introduction", Academic Press, London.
(2) Douzou, P.: 1977, Adv. in Enzymol. 45, pp. 157-272.
(3) Fink, A.L.: 1976, J. Theor. Biol. 61, pp. 419-445.
(4) Chance, B., Graham, N., and Legallais, V.: 1975, Anal. Biochem. 67, pp. 552-579.
(5) Franks, F., Asquith, M.H., Hammond, C., Skaer, H., and Echlin, P.: 1977, J. Microsc. 110, pp. 223-238.

(6) Douzou, P., Debey, P., and Franks, F.:1977, Nature 268, p. 466.
(7) Douzou, P., Debey, P., and Franks, F.:1978, Biochem. Biophys. Acta 523, pp. 1-8.
(8) Douzou, P., Balny, C., and Franks, F.:1978, Biochimie 60, pp.751-758.
(9) Balny, C., Keh, E., and Douzou, P.: 1978, Biochem. Soc. Trans. London 6, pp. 1277-1278.
(10) Douzou, P., Keh, E., and Balny, C.: 1979, Proc. Natl. Acad. Sci., U.S.A. 76, pp. 681-684.
(11) Pilpel, N.: 1963, Chem. Rev. 63, pp. 221-234.
(12) El Seoud, O.A., Fendler, E.J., and Fendler, J.H.: 1974, J. Chem. Soc. Faraday Trans. 70, pp. 450-458.
(13) El Seoud, O.A., Fendler, E.J., and Fendler, J.H.: 1974, J. Chem. Soc. Faraday Trans. 70, pp. 459-462.
(14) Fendler, E.J., Chang, S.A., and Fendler, J.H.: 1975, J. Chem. Soc. Perkin Trans. 2, pp. 482-489.
(15) Maurel, P., Travers, F., and Douzou, P.: 1974, Anal. Biochem. 57, pp. 555-563.
(16) Menger, F.M., Donohue, J.A., and Williams, R.F.: 1973, J. Am. Chem. Soc. 95, pp. 286-288.
(17) Misiorowski, R.L., and Wells, M.A.: 1974, Biochemistry 13, pp. 4921-4927.
(18) Poon, P.H., and Wells, M.A.: 1974, Biochemistry 13, pp. 4928-4936.
(19) Maurel, P., Hui Bon Hoa, G., and Douzou, P.: 1975, J. Biol. Chem. 250, pp. 1376-1382.
(20) Kon-No, K., and Kitaharia, A.: 1970, J. Colloid. Interface Sci 33, pp. 124-132.
(21) Wentz, M., Smith, W.H., and Martin, A.R.: 1969, J. Colloid. Interface Sci. 29, pp. 36-41.
(22) Aebi, C.M., and Wiebush, J.R.: 1959, J. Colloid. Interface Sci. 14, pp. 161-165.
(23) Hui Bon Hoa, G., and Douzou, P.: 1973, Anal. Biochem. 51, pp. 127-136.

THEORETICAL STUDIES ON MODELS OF THE ACTIVE SITE OF CARBONIC ANHYDRASE

D. Demoulin and A. Pullman
Institut de Biologie Physico-Chimique
Laboratoire de Biochimie Théorique, associé au C.N.R.S.
13, rue Pierre et Marie Curie - 75005 Paris.

ABSTRACT

Theoretical *ab initio* pseudopotential SCF computations performed on a model of the Glu 106 - Thr 199 - water H-bonded chain, free and bound to the zinc cation at the active site of carbonic anhydrase show that whereas the Glu - Thr - water form of the unbound chain is stable, the same chain, when bound to zinc, is more stable in the Glu - Thr - OH^- form, and the proton-relay process to that form occurs with no barrier.

Further computations show that the OH^- + CO_2 reaction proceeds readily on zinc at the active site with no activation energy and a small exothermicity.

The implications of these findings are discussed.

I. INTRODUCTION

The enzyme carbonic anhydrase catalyzes with great efficiency the interconversion between CO_2 and bicarbonate (1) which can be formulated by either equilibrium :

(1) $$CO_2 + H_2O \rightleftharpoons CO_3H_2 \rightleftharpoons CO_3H^- + H^+$$

(2) $$CO_2 + OH^- \rightleftharpoons CO_3H^-$$

Although the enzyme has been studied to a considerable extent (2)(3), the precise mechanism of its action is still under debate, even as concerns the formulation (1) or (2). X-ray structural studies (4) have shown that the active site (Figure 1) contains a zinc ion, liganded in a distorted tetrahedral geometry by three hystidyl residues and a solvent (water) molecule, which appears to be hydrogen-bonded to Thr 199, itself hydrogen-bonded to Glu 106, a partially buried group.

Enzyme activity is controlled by a group of pK about 7, close to the zinc ion, the high pK form of the enzyme acting in the hydration reaction. The nature of the ionizing group corresponding to this

B. Pullman (Ed.): Catalysis in Chemistry and Biochemistry. Theory and Experiment. 51–66.

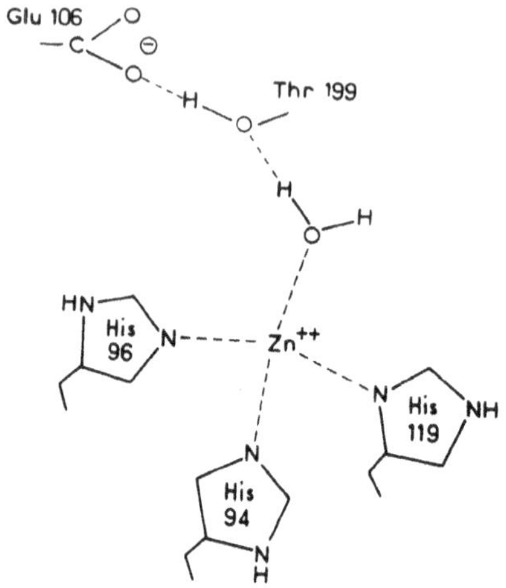

Figure 1. The active site of carbonic anhydrase

pK is under much debate. One strongly favored hypothesis is that water bound to zinc ionizes and that the resulting zinc-bound hydroxide acts as a nucleophile towards CO_2, either directly or through a water molecule (5,6,7,8). Another recent proposal (8) is that the buried carboxyl of Glu 106 ionizes at pK = 7; the Glu 106 - Thr 199 hydrogen-bonded chain would then accept a proton from the zinc-bound water at some stage of the reaction with CO_2. Other proposals include ionization of a zinc-bound imidazole (9) and activation, by the resulting imidazolate, of either a water molecule (10) or of CO_2 itself (11). The role of another imidazole not directly bound to zinc has also been strongly advocated (12).

The exact location and nature of the binding of the substrate CO_2 are not known. A recent X-ray crystallographic study (8) of a complex of human carbonic anhydrase (13) with imidazole has shown that the imidazole (which is a competitive inhibitor in the hydration reaction) binds to the enzyme as a fifth distant ligand to the zinc ion (with Zn - N = 2.7 Å). On this basis the authors proposed a similar binding for the substrate CO_2, a proposal compatible with earlier infrared (13) and NMR (14) observations.

In a set of previous *ab initio* molecular orbital studies (15-18) we have discussed some of the theoretical problems raised by these proposals. We studied first (15-16) the binding characteristics of Zn^{++} with the molecules of CO_2, water and imidazole (ImH) and with the anions hydroxide and imidazolate. We then investigated (17-18) tetraliganded zinc complexes such as $Zn^{++}(NH_3)_3(OH_2)$and $Zn^{++}(NH_3)_2$ (ImH)

TABLE I
Computed deprotonation energies (kcal/mole) of ligand L (16-18)

L	H_2O	Imidazole
in vacuum	380	364
in Zn^{++}...L	82	168
in $Zn^{++}(NH_3)_3(ImH)(H_2O)$	171	207

(OH_2) and their deprotonated counterparts (with OH^- and Im^-) as models of the tetraliganded zinc at the active site. The results of these studies can be summarized as follows :

a) imidazole alone deprotonates more easily than water alone but binding to zinc strongly lowers both ionization energies, in such a way that zinc-bound water ionizes more easily than zinc-bound imidazole. This result remains true in tetraliganded zinc complexes (see Table I for a summary).

b) the zinc-bound hydroxide ion in such complexes retains the major part of its negative charge.

c) electrostatic calculations (17-18) showed that CO_2 can bind as distant ligand to the tetraliganded zinc complexes, and that when the zinc-bound water is ionized to a zinc-bound hydroxide ion, the carbon

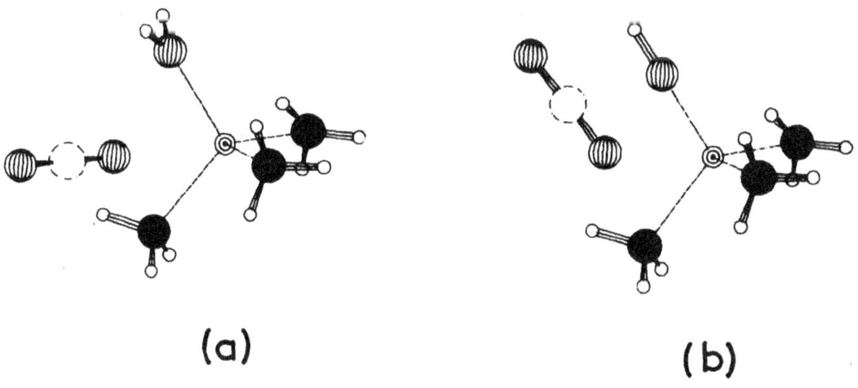

Figure 2. Optimal positions of CO_2 at the active site modelled by (a): $Zn^{++}(NH_3)_3OH_2$; (b) $Zn^{++}(NH_3)_3OH^-$ (reference 17).

atom of CO_2 can approach closer to the oxygen atom of the zinc-bound hydroxide than it does to the oxygen of a zinc-bound water (see Figure 2), a further indication that the hydroxide retains an appreciable nucleophilic character.

d) binding to zinc, even as a distant ligand, polarizes the CO_2 molecule in such a way as to open a channel of depleted electron density towards the carbon atom, thereby favoring the approach of a nucleophile (18).

We present here further theoretical investigations aiming at a better understanding of a possible mechanism of the reaction at the active site. In the first part of this work we consider the possible role of the existence of a hydrogen-bonded chain Glu - Thr - water close to the active site, by studying the equilibrium structure of a model of the active site containing the tetraliganded zinc plus the side chains Glu 106 and Thr 199. In the second part we consider the energetics of a reaction between OH^- and CO_2 taking place on zinc at the active site by computing a possible path for the reaction. We further compare some aspects of carbonic acid and bicarbonate binding in a tetraliganded zinc complex and discuss their implications.

II. COMPUTATIONAL DETAILS

All the calculations were carried out with the *ab initio* SCF molecular orbital method (19), using a pseudopotential to replace the core electrons of the atoms (20-21). The method and the gaussian basis sets used have been presented in details in our previous papers (16,17). The valence orbital basis sets are of double zeta quality, so as to insure enough flexibility in the description of the intermolecular complexes. A set of diffuse *p* functions has been added when necessary to describe anions as discussed in details in reference (16).

TABLE II
Examples of reaction energies (kcal/mole) computed without (A) and with (B) diffuse *p* functions on the oxygen(s) of the anion.

Reaction	(A)	(B)	Exp[l] (23)
$OH_2 \rightarrow OH^- + H^+$	445.5	379.5	390.0
$HCOOH \rightarrow HCOO^- + H^+$	368.8	331.6	345.2
$H_2CO_3 \rightarrow HCO_3^- + H^+$	360.6	319.6	
$Zn^{++}(NH_3)_3(OH_2) + HCOO^- \rightarrow$ $Zn^{++}(NH_3)_3(OH^-) + HCOOH$	-182.0	-173.5	

Table II gives a numerical illustration of the situation : it is seen that without diffuse functions on the anionic species the dissociation energies are too high, while when diffuse functions are added the values are closer to the experimental ones, although slightly underestimated. As concerns reaction energies for proton transfer between anion and neutral molecule, they are less dependent on the presence of diffuse functions than the dissociation energies.

III. STUDY OF A MODEL OF THE ACTIVE SITE INCLUDING Thr 199 and Glu 106

On the basis of the relative arrangement of the Glu 106, Thr 199 and solvent molecule bound to zinc at the active site, Kannan et al. (8) proposed the existence of the hydrogen-bonded chain depicted in Figure 1 and suggested its possible involvment in the catalytic process, writing two possible mechanism, one where the H-bonded chain is essentially holding the zinc-bound OH^-, without being itself involved in proton transfer, the other in which at some stage of the interaction with CO_2 the zinc-bound water has transferred a proton to Glu 106 through the hydrogen-bonded chain in a manner reminiscent of the "charge-relay system" in the serine proteases (22).

We have thus deemed of interest to consider from a theoretical point of view the possible role of the hydrogen-bonded chain. For this purpose we shall study the energetics of a hydrogen bonded chain water - threonine - glutamate, on the one hand alone, on the other hand when its water end is bound to a zinc cation otherwise triliganded as seen at the active site.

For computational convenience we model the side chain of Glu 106 ($-CH_2-CH_2-COO^-$) by the formate ion, and that of Thr 199 ($-CHOHCH_3$) by methanol. A justification of this simplification for the problem considered is the fact that the gas-phase proton affinities of $HCOO^-$ and $CH_3CH_2COO^-$ differ only by 2.1 kcal/mole, and those of CH_3O^- and $CH_3CHO^-CH_3$ by 3.3 kcal/mole (23).

To model the tetraliganded zinc at the active site we have adopted the complex $Zn^{++}(NH_3)_3(OH_2)$, with the distance Zn-N = 2.1 Å and Zn-O = 2.0 Å found earlier, since this model was found (17) to provide an adequate approximation of the larger model $Zn^{++}(NH_3)_2(ImH)(OH_2)$ for the purposes considered.

In a first step, working without diffuse p functions, we have obtained optimized structures for hydrogen-bonded chains of increasing complexity : OH_2-CH_3OH, $OH_2-CH_3OH-HCOO^-$, $Zn^{++}(NH_3)_2OH_2-CH_3OH-HCOO^-$ (I, III, VI in Table IV), and for their counterparts in which the protons in the hydrogen bonds have been moved completely to the right (II, V, IX in Table IV). The first set of structures have reasonable intermolecular bond length and bond energies, considering the basis set used. As ionic species appear, the hydrogen-bond distances decrease and the stabilization energies increase strongly, as expected for such systems (see for instance (24)).

TABLE III
Geometries (Å) and energies (kcal/mole) for the hydrogen bonded chains (A : no diffuse p functions, B : diffuse p functions on all the oxygen atoms).

System		R(O...O)[a]	BE[d]	ΔE	
				(A)	(B)
I	HOH....$\overset{CH_3}{O}$H	2.8	-9.8	0.0	
II	HO^-....H$\overset{CH_3^+}{O}$H	2.2		+50.1	
III	HOH....$\overset{CH_3}{O}$H.... $^-$OCOH	2.7, 2.6	-40.8	0.0	0.0
IV	HO..H..$\overset{CH_3}{O}$..H..OCOH	2.55, 2.55			+21.3
V	HO^-....H$\overset{CH_3}{O}$....HOCOH	2.4, 2.5		+40.9	+32.5
VI	$Zn^{++}(NH_3)_3\overset{H}{O}$H....$\overset{CH_3}{O}$H.... $^-$OCOH	2.7, 2.6	-178.7	0.0	0.0
VII	$Zn^{++}(NH_3)_3$ $\overset{H}{O}$.H...$\overset{CH_3}{O}$.H... $^-$OCOH	2.625,2.575[b]			-7.1
VIII	$Zn^{++}(NH_3)_3$ $\overset{H}{O}$..H..$\overset{CH_3}{O}$..H..OCOH	2.55, 2.55[c]			-12.4
IX	$Zn^{++}(NH_3)_3$ $\overset{H}{O}^-$....H$\overset{CH_3}{O}$....HOCOH	2.4, 2.5		-35.8	-29.8

(a) When two numbers are given, the first one corresponds to the first O-O distance from the left.
(b) O.H = 1.117, 1.127 as indicated
(c) H..O = H..O = 1.275.
(d) With respect to full dissociation into the isolated species.

If one compares now the relative stabilities of the hydrogen-bonded species with the protons unmoved and with the protons completely moved to the right, an interesting result appears : for the chains not bound to zinc (III,V) the structure involving water on the left end and formate on the right end is appreciably more stable than the structure VI (OH_2). In order to strengthen our confidence in the validity of this result, we have repeated the calculations, including a set of diffuse p functions on each of the four oxygen atoms of the hydrogen-bonded chains using the optimal geometries found earlier. It is seen that the same conclusion is maintained : in the free chain, the structure with OH^- and HCOOH (V) is destabilized by 32.5 kcal/mole with respect to the structure involving OH_2 and $HCOO^-$ (III), whereas upon binding to zinc at the active site, the structure involving OH^- and HCOOH (IX) becomes more stable by 29.8 kcal/mole than that involving OH_2 and $HCOO^-$ (VI).

In order to learn more about the facility of the proton transfer, we have also computed intermediate structures with the protons transferred half the distance (IV and VIII) and with the protons transferred one quarter of the way in the case of the zinc complex (VII). Although a complete optimization of the geometry is missing, the numbers show clearly that there is no barrier for the simultaneous transfer of the protons and that the passage from VI to IX requires no activation energy. Hence these computations indicate that in a gas-phase situation where the tetraliganded zinc would be hydrogen-bonded to the Thr-Glu chain in the way suggested by the X-ray crystallographers (8), a zinc-bound hydroxide ion would be produced spontaneously in situ, the hydrogen-bonded chain with its terminal COO^- providing the necessary acceptor and the threonine serving as proton-relay.

Concerning the validity of these results and their relevance to a real system (as opposed to a gas phase situation to which the computations apply) let us point out the following : a) the energy differences observed between the initial and final structures appear large enough to remain at least in the same direction even upon serious refinements of the computations : (see for instance the influence of the introduction of the diffuse p-functions on the oxygens in Table III); b) The same can be said of the most important result, namely the reversal of the order of stability upon binding to zinc; c) as concerns the absence of a barrier for the proton relay one might argue that improvement of basis sets may lead to find a barrier where a computation with a small basis showed none (25) : however in the present case, we start already with a double-zeta basis + diffuse p functions, sufficient to account correctly for the proton affinities and hence for a possible barrier; d) last but not least, we think that a good support for the validity of the conclusions concerning the zinc-bound hydrogen-bonded chain is provided by the fact that the results obtained for the free water - threonine - glutamate chain are in agreement with common sense and with the relative proton affinities of $HCOO^-$ and of OH^-. We thus tend to conclude that the reversal in the stability order in the presence of the otherwise triliganded zinc is also representative of the real system.

It appears to be the result of i) the labilization of the water proton by binding to zinc (15,17,18) and ii) the availability of a relay system whereby the proton can be transferred to an appropriate proton acceptor. Note that both components of the system are necessary to make the phenomenon possible : the zinc complex and the H-bonded chain with the adequate proton acceptor.

At this point we reach the conclusion that a zinc-bound hydroxide can be produced at the active site if the geometrical arrangement considered is realized. Let us for the moment accept that this situation occurs and consider the further fate of the zinc-bound hydroxyde in a possible reaction with CO_2.

IV. THE REACTION $OH^- + CO_2$ AT THE ACTIVE SITE

It is known experimentally that the reaction

$$OH^- + CO_2 \rightarrow CO_3H^-$$

in water solution is exothermic and requires an activation energy of 13.2 kcal/mole (26). Computations of the reaction in vacuum (27) indicate a high exothermicity and no activation energy. What is however relevant to the catalytic rôle of the enzyme is the comparison between the reaction in solution and the reaction at the active site. In other words a computation of the activation energy for the reaction at the active site should show a lowering or a suppression of the activation energy as compared to its value in solution. A computation of the real system is of course impossible and we are quite conscious of the limitations of model calculations. However, we believe that it is possible with reasonable models to obtain at least meaningful trends. Thus we have decided to consider the reaction :

$$Zn^{++}(NH_3)_3(OH^-) + CO_2 \rightarrow Zn^{++}(NH_3)_3(HCO_3^-)$$

where the addition occurs between a zinc-bound OH^- and CO_2 to produce a zinc-bound HCO_3^-, the main purposes of this computation being to ascertain a) the possibility of OH^- bound to zinc to act as a nucleophile, b) the possibility that the activation energy for such a reaction is low or absent.

The computation cost being already quite considerable for such a system we have not included the Thr-Glu chain here, assuming the OH^- to be produced. Furthermore we have limited our "reaction path" to a small number of carefully chosen points as follows : for the starting point we choose the geometrical arrangement at the active site obtained in our previous electrostatic search of the most favorable position of a molecule of CO_2 approaching a tetraliganded zinc complex $Zn^{++}(NH_3)_3$ OH^-. The geometry of the complex itself has been given in reference (18), and we indicate here only the bond length and angles relevant to a mutual disposition of Zn^{++}, OH^- and CO_2 (Table IV). The final point is the optimized structure for $Zn^{++}(NH_3)_3$ (HCO_3^-) with the product HCO_3^- bound to the zinc in the bidentate fashion schemati-

TABLE IV
Reaction $Zn^{++}(NH_3)_3(OH^-) + CO_2 \rightarrow Zn^{++}(NH_3)_3(HCO_3^-)$ (a)

	O_1C	∞	3.0	2.0	1.7	1.43
Å	$Zn\ O_1$	1.8	1.8	1.85	1.9	2.0
	$Zn\ O_2$	∞	3.38	2.72	2.37	2.05
	CO_3	1.16	1.16	1.20	1.22	1.25
	O_3CO_2	180	180	147	139	129
degrees	O_1CO_3		87	108	110	116
	$Zn\ O_1C$		102.4	98	94	86.5
ΔE (kcal/mole) (b)		0	-10.3	-9.6	-11.3	-13.3

(a) O_1 is the OH^- oxygen, O_2 the oxygen of CO_2 closest to zinc at the starting point, O_3 the other oxygen of CO_2.

(b) Computations performed with basis set A.

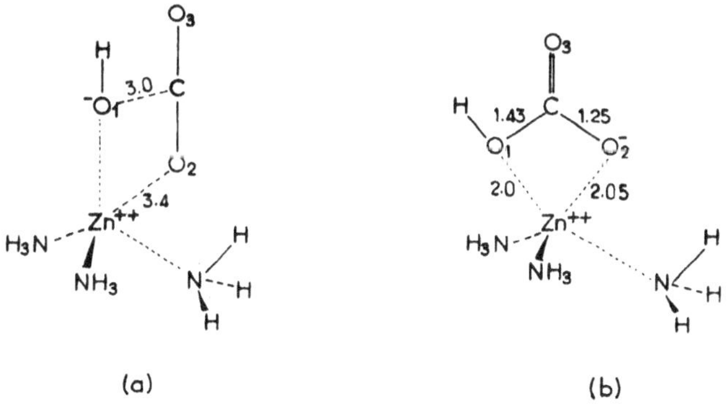

Figure 3. Initial (a) and final (b) geometries for $Zn^{++}(NH_3)_3(OH^-)$ $CO_2 \rightarrow Zn^{++}(NH_3)_3(HCO_3^-)$.

zed in Figure 3. (Test calculations to explore the possibility of reaching a product bound to the zinc through one oxygen only, so as to relieve the possible strain associated with a bidendate structure, gave configurations of comparable stabilities). The intermediate points along the reaction path have been chosen so that the geometrical parameters for the (OH^- + CO_2) fragment coïncide with those of reference (27) at every $O_1...C$ distance. The corresponding geometrical parameters describing the relative position of $Zn^{++}(NH_3)_3$ and the (OH^- + CO_2) fragment were interpolated from their values in the initial geometries. The values are indicated in Table IV together with the corresponding computed SCF energies.

The energy results indicate that the reaction

$$Zn^{++}(NH_3)_3(OH^-) + CO_2 \longrightarrow Zn^{++}(NH_3)_3(HCO_3^-)$$

is exothermic by 13.3 kcal/mole with respect to the reactants at infinity. (With the same basis set, the computed exothermicity of the gas phase OH^- + CO_2 reaction is 99 kcal/mole).

In the starting geometry (where CO_2 can be considered to serve as a fifth distant ligand to zinc (17-18)), the stabilization energy is already -10.3 kcal/mole. As the C...O bond forms and the CO_2 molecule deforms, we find a slight decreased stabilization of 0.7 kcal/mole at an O1-C distance of 2Å, then the stabilization increases towards the product. The energy values indicate that there is practically no activation energy and that reaction can go smoothly to the products. Thus to the two questions raised at the beginning of this section, the answer is that : a) A hydroxide ion bound to zinc at the active site does keep enough nucleophilic character, not only to let the CO_2 approach the site as shown by the former electrostatic study (18), but further to start and pursue the reaction. b) The reaction at the active site can proceed with very small or negligible activation energy.

A third result of importance is the small exothermicity of the catalyzed reaction which makes reversibility possible.

V. DISCUSSION

As concerns the relevance of these findings to the mechanism itself of the enzyme functionning let us add some further features : in order to account for the exceptionnally high turnover rate of the enzyme, the product of the reaction must be readily displaced by a solvent molecule and the whole process repeated. We find that the zinc-bound HCO_3^- has a binding energy to $Zn^{++}(NH_3)_3$ of 229.4 kcal/mole, a rather large value but not surprising for a Zn^{++}-anion interaction (15-16). Since the corresponding binding energy of H_2O to the zinc complex is -56.8 kcal/mole, the generally assumed displacement of HCO_3^-

by water is unlikely unless a special mechanism in the enzyme intervenes to displace the product as soon as formed (see for instance (28)). Another possibility might be that the HCO_3^- anion is transformed into a zinc-bound H_2CO_3 by capture of a proton which could be released from the enzyme itself : upon reaction of OH^- with CO_2 the hydrogen-bond to the Glu-Thr chain is likely to be disrupted, allowing a return to the $-COO^-...HO-$ form with release of a proton, either to the buffer or directly to the zinc-bound bicarbonate ion, transforming it into the zinc-bound acid. We have searched for the optimal structure of a zinc-bound CO_3H_2 and found the monodendate structure of Figure 4 with Zn-O = 1.9 Å. Interestingly, in this structure CO_3H_2 is bound to the zinc complex by -55.3 kcal/mole, a value very close to the corresponding binding energy of water (-56.8 kcal/mole), and thus permitting an easy exchange of ligands at the active site.

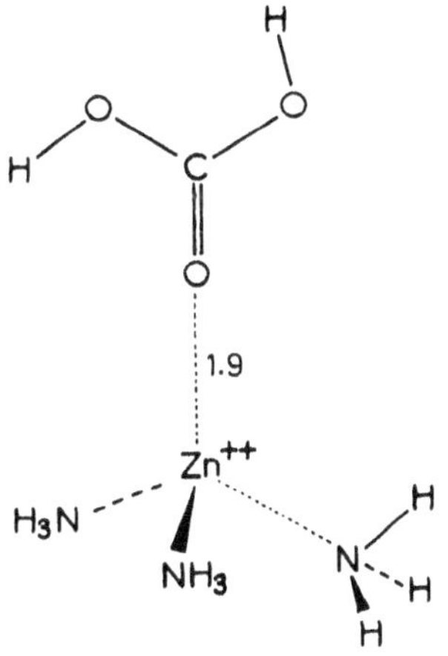

Figure 4. Optimal geometry of H_2CO_3 bound to $Zn^{++}(NH_3)_3$.

In this research we have been led to follow the point of view of the hydroxide mechanism rather than that of a direct attack of CO_2 by the zinc-bound water itself. We think that the possibility of production of OH^- at the active site and the absence of a barrier for the reaction found in our computations speak in favor of this mechanism. It must be noted that even in the absence of the hydrogen-bonded chain, it might be conceived that a zinc-bound hydroxide is produced on zinc, possibly with the help of the buffer (see (29) and references therein) serving as a proton sink for the zinc-labilized proton of water. Then the reaction described in section IV would proceed similarly.

It would of course remain to be seen whether a direct attack of the zinc-bound non ionized water on CO_2 could instead proceed at the active site to form directly H_2CO_3. The reaction in solution is known to require a large activation energy of 18 kcal/mole (30), but it cannot be excluded that the barrier would be lowered for the reaction taking place on the zinc complex : since binding to zinc labilizes the water proton appreciably (16), the proton-transfer step which is likely the major source of the barrier, should be facilitated. We have computed that the energy of the active site-CO_2 complex where CO_2 is placed in the position found by the electrostatic computation (Figure 2a) is -9.8 kcal/mole (with respect to the reactants at infinity), whereas the (active site $-CO_3H_2$) complex in the final geometry of Figure 4, has a stability of -12.9 kcal/mole. This shows an exothermicity comparable to that found for the zinc-hydroxide reaction. The question of the activation energy would however require a special study. We do not think however that a model computation similar to that carried out for OH^- at the active site could yield a value of the activation energy as small as that obtained in the same conditions for OH^- because the barrier computed in vacuum (31) yields a rather large value of 54 kcal/mole.

In summary, the present theoretical studies have shown i) that the Glu-Thre-solvent hydrogen-bonded chain observed as a prominent feature (32) of the structure of both carbonic anhydrase C and B at the active site can provide a proton-relay system to produce in situ a zinc-bound hydroxide ion ii) that such a zinc-bound OH^- would react readily with CO_2 at the active site with little or no activation energy. Insofar as the model studies can be taken as representative of the situation in the enzyme cavity these results seem compatible with a hydroxide mechanism for the mode of action of carbonic anhydrase.

Acknowledgements. This work has benefited from a C.N.R.S. ATP contract n° 2646.

REFERENCES

1 Meldrum, N.U. and Roughton, F.J.W. : 1932, J. Physiol. London, 75, pp. 15-21.
2 Lindskog, S., Henderson, L.E., Kannan, K.K., Liljas, A., Nyman, P.O. and Strandberg, B., in the Enzymes, P.D. Boyer Ed., Acad. Press, N.Y. (1971) vol. 5, pp. 587-665.
3 Coleman, J.E., in Inorganic Biochemistry, G.L. Eichhorn Ed., Amsterdam (1973), vol. 4, pp. 488-548.
4 Nostrand, B., Vaara, I. and Kannan, K.K., in the Isozymes, C.L. Market Ed., Acad. Press, N.Y. (1975) vol. I, pp. 577-599.
5 Prince, R.H. and Woolley, P.R. : 1972, Angew. Chem. Intern. Edit. 11, pp. 408-417.
6 Woolley, P.R. : 1975, Nature, 258, pp. 677-682.
7 Coleman, J.E. : 1976, J. Biol. Chem. 242, pp. 5212-5219.
8 Kannan, K.K., Petef, M., Fridborg, K., Cid-Dresdner, H. and Lövgren, S. : 1977, FEBS Letters, 7, pp. 115-119.
9 Appleton, D.W. and Sarkar, B. : 1974, Proc. Natl. Acad. Sci. US, 71, pp. 1686-1690.
10 Pesando, J.M. : 1975, Biochemistry, 14, pp. 681-688.
11 Mac B. Harrowfield, J., Norris, V. and Sargeson, A.M. : 1976, J. Amer. Chem. Soc. 98, pp. 7282-7289.
12 Pocker, Y. and Meany, J.E. : 1965, Biochemistry, pp. 2535-2541.
13 Riepe, M.E. and Wang, J.H. : 1968, J. Biol. Chem. 243, pp. 2779-2787.
14 Campbell, I.D., Lindskog, S. and White, A.J. : 1974, J. Mol. Biol. 90, pp. 469-489.
15 Demoulin,D., Pullman, A. and Sarkar, B. : 1977, J. Amer. Chem. Soc., 99, pp. 8498-8500.
16 Demoulin, D. and Pullman, A. : 1978, Theor. Chim. Acta, 49, pp. 161-181.
17 Pullman, A. and Demoulin, D. : Proceedings of the Kiev Conference in Quantum Chemistry, Quantum Biology and Quantum Pharmacology, Int. J. Quant. Chem., in press.
18 Pullman, A. and Demoulin, D. : 1978, Proceedings of the Nagoya Conference on water and cations in biological systems, K. Yagi Ed., in press.
19 Roothaan, C.C.J. : 1951, Revs. of Modern Physics, 23, pp. 69-89.
20 Melius, C.F., Goddard III, W.A. and Kahn, L.R. : 1972, J. Chem. Phys., 56, pp. 3342-3359.
21 Topiol, S., Moskowitz, J.W., Melius, C.F., Newton, M.V. and Jaffri, J. : Courant Institute of Mathematical Sciences, ERDA research and Development Report (COO-3077-105).
22 Blow, D.M., Birktoft, J.J. and Hartley, B.S. : 1969, Nature, 221, pp. 337-340.
23 Kebarle, P. : 1977, Ann. Rev. Phys. Chem. 28, pp. 445-476.
24 Port, G.N.J. and Pullman, A. : 1975, Int. J. Quant. Chem., Quant. Biol. Symposium 1, 21.
25 Broer, R., Van Duijnen P. Th. and Niewpoort, W.C.: 1976, Chem. Phys. Letters, 42, pp. 525-529.

26 Pinsent, B.R.W., Pearson, L. and Roughton, F.J.W. : 1956, Trans. Farad. Soc., 52, pp. 1512-1520.
27 Jönsson, B., Karlström, G. and Wennerström, H. : 1978, J. Amer. Chem. Soc., 100, pp. 1658-1661.
28 Yeagle, P.L., Lochmüller, C.H., and Henkens, R.W. : 1975, Proc. Natl. Acad. Sci. USA, 72, pp. 454-458.
29 Jonsson, B.H., Steiner, H. and Lindskog, S. : 1976, FEBS Letters, 64, pp. 310-314.
30 Magid, E. and Turbeck, B.O. : 1968, Bioch. Biophys. Acta, 165, pp. 515-524.
31 Jönsson, B., Karlström, G., Wennerström, H., Forsen, S., Roos, B. and Almlöf, J. : 1977, J. Amer. Chem. Soc., 99, pp. 4628-1632.
32 Pocker, Y. and Sarkanen, S. : 1978, Advances in Enzymology, 47, pp. 149-274.

DISCUSSION

G. COHEN : The array of residues in the active site of carbonic anhydrase, indicated schematically on your slide, glu 106- thr 199 - H_2O.. Zn(his)$_3$, suggests to me a possible analogy to the charge-relay system in serine proteases, and covalent involvement of the threonine hydroxyl. In the serine enzymes, reaction - in that case hydrolysis - proceeds in two steps, acylation of the serine hydroxyl and hydrolysis of the acyl enzyme. A similar sequence may be formulated for the formation of carbonic acid.

Instead of a scheme in which the threonine hydroxyl is used only for proton transfer, one may be formulated in which it is also acylated by carbon dioxide, i.e. the oxygen actq as a nucleophile. In the first step threonine hydroxyl may be converted to the half carbonic acid, as protons are transferred from zinc bound water to CO_2 and from threonine hydroxyl to glutamate anion, eq.1

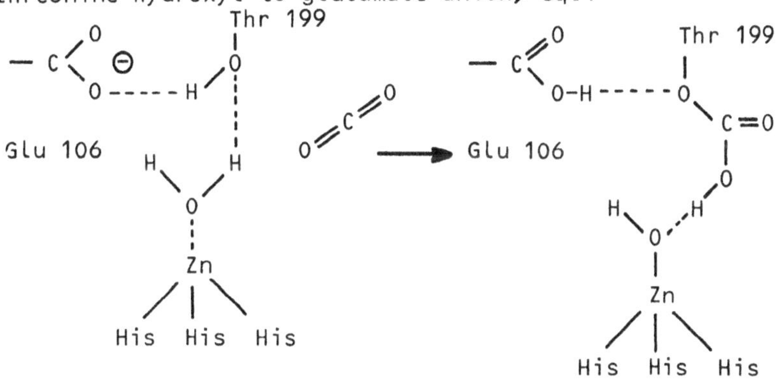

(1)

In a second step a water molecule may hydrolyze the half carbonic acid and the protons are transferred back to their initial positions, eq.2

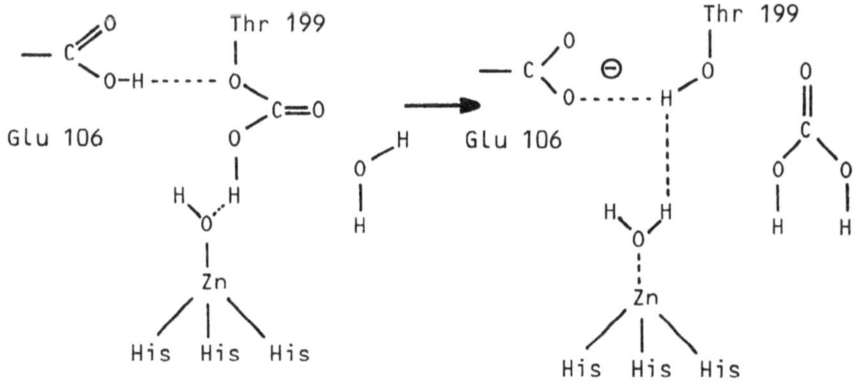

(2)

In an alternate formulation the water molecule may be formulated as hydrogen-bonded to : Zn-OH, eq.3.

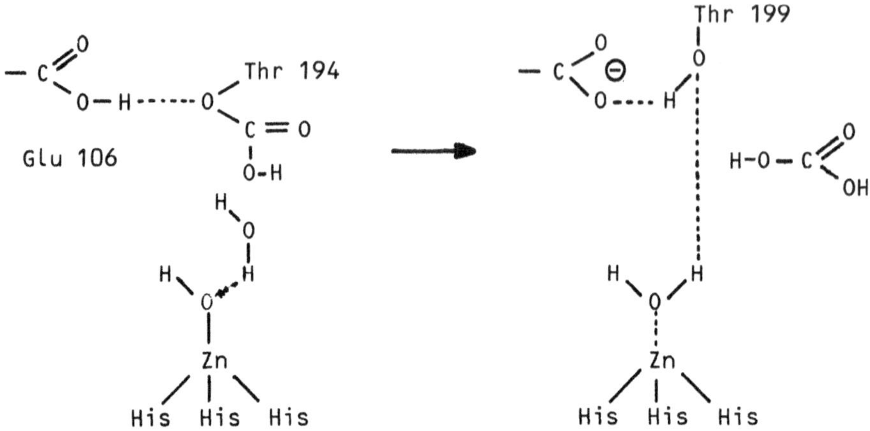

(3)

Such a process (eq.2 or 3) may have the advantage of not requiring rupture a Zn-O bond and of allowing ready release of the carbonic acid.

Mme A. PULLMAN : This is an interesting suggestion, to which of course, I cannot have a ready-made answer. From the theoretical point of view it is a framework quite different from the one we have considered.

THE MECHANISM OF CARBONIC ANHYDRASE -- A SOLVATION PROBLEM?

Bo Jönsson
Physical Chemistry 2, Chemical Centre,
POB 740, Lund, Sweden

ABSTRACT

The catalytic activity of the enzyme carbonic anhydrase has been studied by statistical and quantum mechanical methods. The coordination number for the zinc ion in the active site has been found to be five, three histidine residues and two water molecules, with another water molecule very close to the first solvation sphere. The zinc-hydroxide mechanism for the enzymatic reaction is supported by the results obtained from a quantum mechanical study of the reaction $OH^- + CO_2 \rightleftarrows HCO_3^-$, indicating that the specific solvation in the active cleft of carbonic anhydrase may be the clue to the catalytic activity.

INTRODUCTION

The enzyme carbonic anhydrase (CA) has been the subject of an enormous amount of research since it was first discovered in 1932. The molecular weight of the enzyme is 30 000, and it contains one Zn^{2+} ion, which is coordinated to three histidine residues[1-2]. The zinc ion, which is essential for the enzymatic activity is situated in a rather deep cavity. One of the few metal ions, possibly the only one, which can replace zinc without any significant loss of enzymatic activity[3] is Co^{2+}. Apart from the reversible hydration of CO_2, CA also catalyzes aldehyde hydration and ester hydrolysis reactions[3]. The enzymatic reactions are affected by the presence of sulfonamides and imidazole derivatives as well as a number of anionic inhibitors[3-4]. Two human forms of CA, B and C, exist, and their crystal structures are known to a resolution of ~2 Å .

Despite the considerable work which has been done, a detailed mechanistic description of the CO_2 hydration and HCO_3^- dehydration reactions is still lacking. In discussing the enzymatic reaction a number of other questions have been raised: for example, the coordina-

B. Pullman (Ed.): Catalysis in Chemistry and Biochemistry. Theory and Experiment. 67–76.

tion number for the zinc ion in the active cleft, and how to estimate the limits set by diffusion for an enzymatic reaction.

In order to gain further insight into the enzymatic hydration of CO_2, we have considered the CO_2+H_2O system from a theoretical point of view using statistical and quantum mechanical methods. We have on one hand used a general approach to the problems related to the H_2O+CO_2 system[5-10], but also tried to answer the more specific questions which arise in the study of the enzymatic reactions[11-12]. What is reported here are results obtained from both of these approaches. In the following three sections the limits set by diffusion for a reversible enzymatic reaction, the zinc-water coordination number in human carbonic anhydrase B and the reaction $OH^- + CO_2 \rightleftarrows HCO_3^-$ respectively, will be discussed.

DIFFUSION LIMITS FOR A REVERSIBLE ENZYMATIC REACTION

It has been discussed to some extent in the literature whether the products of the enzymatic hydration of CO_2 are $HCO_3^- + H^+$ or H_2CO_3[13-15]. It has been argued that both products are in disagreement with restrictions set by diffusion. The basis for these statements is the result obtained by solving the time independent diffusion equation,

$$\Delta(rc) = 0, \tag{1}$$

for a system, in which the reactants are supposed to diffuse to a spherical symmetrical enzyme of radius R, where they react. The boundary conditions commonly used for the differential equation are,

$$c(\infty) = c_o = \text{the bulk concentration} \tag{2a}$$

$$c(R) = 0 \tag{2b}$$

that is, the concentration on the enzyme surface is set to zero. The maximum value for the second order rate constant then becomes

$$k_2^{lim} = 4\pi N_a DR, \tag{3}$$

where D is the diffusion coefficient and N_a is the Avogadro number. This expression was originally derived by Smulochowski[16] for the rate of crystal growth in solutions. That this is too crude an approximation for a reversible enzyme reaction is seen by considering the reaction scheme

$$A + E \rightleftarrows AE \rightleftarrows BE \rightleftarrows B + E. \tag{I}$$

Here A and B are reactants and products respectively and E is an enzyme If the binding of the substrates to the enzyme is negligible, that is, if the Michaelis-Menten constants are large relative to the substrate concentrations, the rate per enzyme molecule can be written as

$$\frac{dc_A}{dt} = -\frac{dc_B}{dt} = -\vec{k}c_A + \overleftarrow{k}c_B \; . \qquad (4)$$

Instead of using the boundary condition (2b) it is a better approximation to assume that the two species A and B are in equilibrium at the enzyme surface, with an equilibrium constant K[11]. Solving the diffusional equations for A and B with this boundary condition and the restriction that the flow of the two species should be equal leads to an effective rate constant for the forward reaction of

$$k_{eff} = k_2^{lim} K/(K + D_A/D_B) \qquad (5)$$

From the equation (5) it is seen that when $K << D_A/D_B$, D_A and D_B beeing the respective diffusion coefficients, the effective rate constant will be much smaller than the previous estimate based on Smulochowski´s boundary conditions.

With these types of boundary conditions, it is possible to show that the experimentally determined hydration and dehydration rates for carbonic anhydrase are in conflict with the limits set by diffusion! This apparent contradiction can be resolved by introducing buffer molecules, which are usually present both *in vivo* and *in vitro*, into the reaction scheme. In this way the reaction rates become dependent on the buffer concentration, which indeed has also been shown experimentally[17]. It is also concluded that the products for the hydration process are most likely to be $HCO_3^- + H^+$.

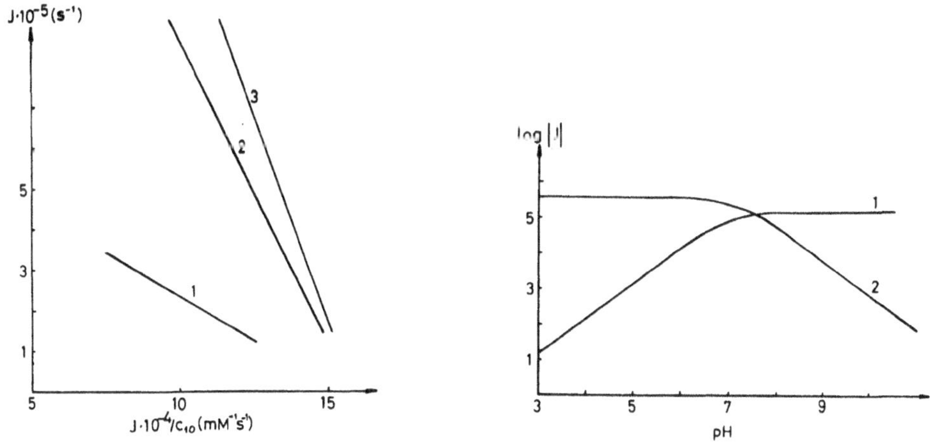

Figure 1. a) Dependence of the rate on buffer concentration. 1=1 mM, 2=10 mM and 3=100 mM buffer with pK_a=6.2. J=reaction rate/enzyme and c_{10}=CO_2 bulk concentration. b) The reaction rate/enzyme as a function of pH. 1=hydration and 2=dehydration.

THE SOLVATION OF THE ZINC ION IN CARBONIC ANHYDRASE

In the X-ray diffraction study by Kannan *et al.*[2] it was concluded that the active cleft contained a few water molecules; however, it was not possible to determine the coordinates for these molecules. The number of water molecules coordinated directly to the zinc ion has not been unambigously determined as yet. The coordination number for Zn^{2+} in aqueous solution is usually either 4 or 6 depending on the ligands[18-19].

One possible way to approach this problem is to perform Monte Carlo (MC) simulations on the enzyme plus a sufficient number of water molecules to solvate the active cleft. In order to do this, it is necessary to know the relevant intermolecular potentials and the enzyme coordinates. In this work, the intermolecular potentials needed have been obtained from *ab initio* MO-LCAO-SCF calculations available in the literature[10,20-21]; the enzyme coordinates used are those determined in ref. 2.

In order to reduce the computer time, the MC-simulation was performed using a smaller part of CA. That is, the active cleft and the amino acids bordering it, including the zinc ion, were "cut out" (see figure 2). Altogether 479 atoms were considered in the MC-procedure.

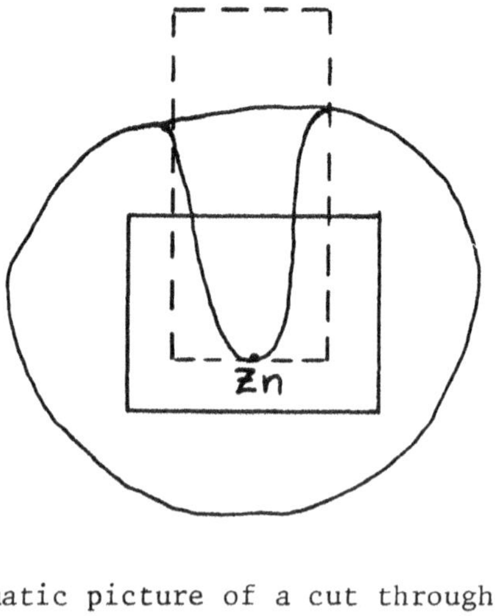

Figure 2. Schematic picture of a cut through the active cleft of carbonic anhydrase. All amino acids within the indicated cylinder (full lines) were included in the Monte Carlo simulation. The smal cylinder (dashed lines) defines the boundaries used in the MC-simulation. The dimensions of the large cylinder was r=7 and h=16 and of the small cylinder r=4.5 and h=22 Å.

These 479 atoms were kept at fixed positions and only the water molecules, 20 in number, were allowed to move during the simulation, which was performed at 300 K. The box was introduced in order to avoid solvation of the "outside" of the enzyme piece.

The result from the simulation, which consisted of 350 000 moves for the equilibration and another 80 000 moves under which the different statistical averages were calculated, are most easily summarized in figure 3. Integration of the peaks shows that the first three correspond

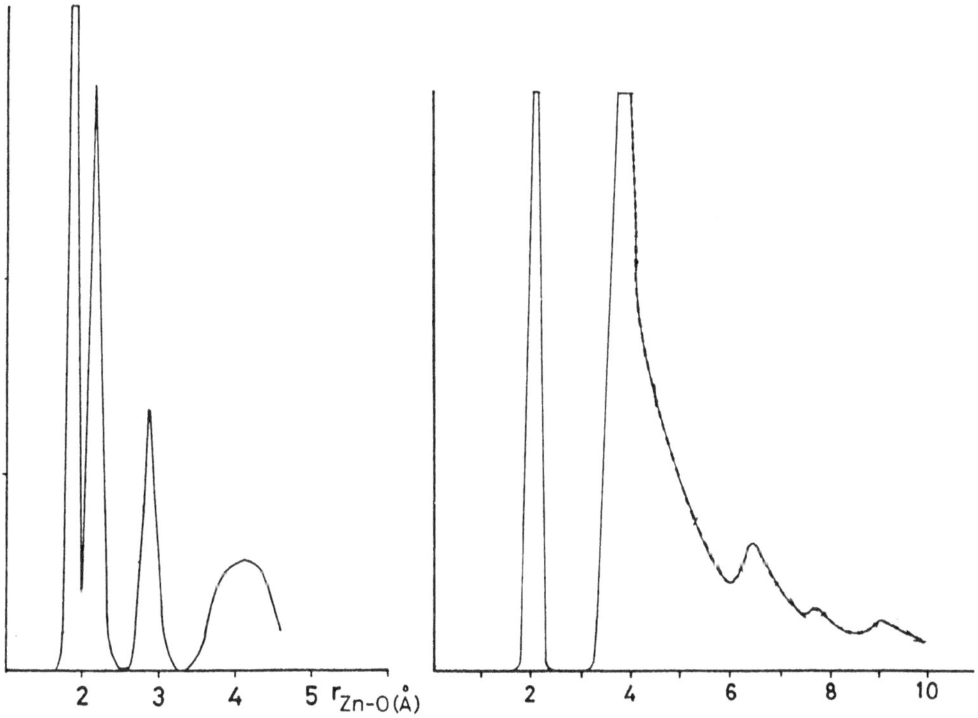

Figure 3. a) Radial distribution function for water oxygen atoms around Zn^{2+} in the enzyme cavity. b) The same quantity for a Zn^{2+} ion in a water cluster.

to one water molecule each, while the fourth broad peak corresponds to another four water molecules. By comparing with the radial distribution function for water oxygen around a zinc ion only (figure 3b) it is seen that the first three peaks in the zinc enzyme radial distribution function are all within the radius of the second shell for the zinc ion in a cluster.

It is also possible to analyze the results in a somewhat different way: in figure 4 are shown oxygen probability densities in successive

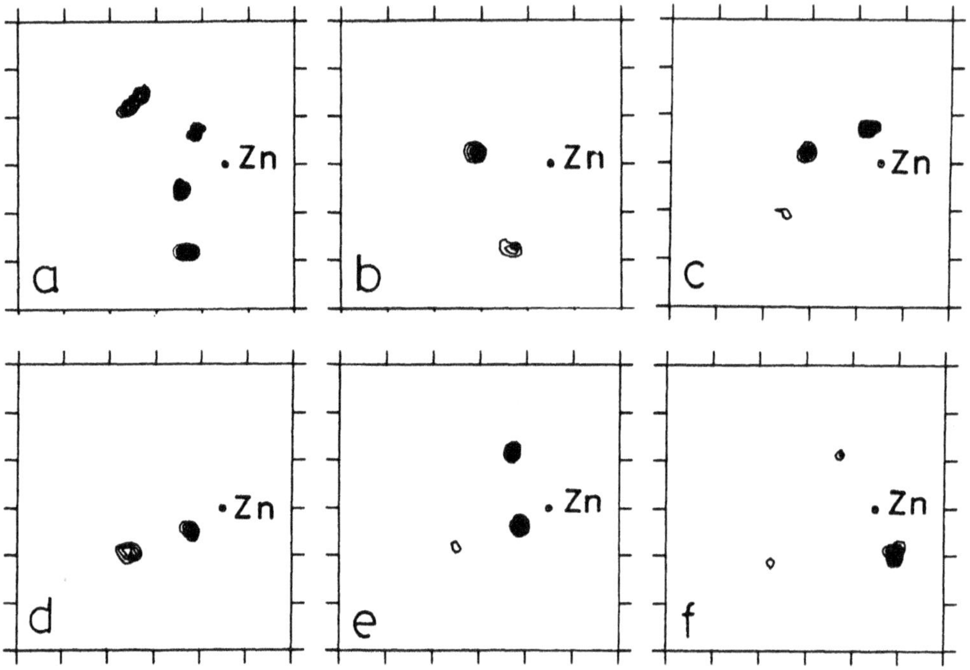

Figure 4. Probability density maps for water oxygens. The maps correspond to cross sections perpendicular to the approximate axes (see figure 1) of the cavity, taken at different heights above the Zn^{2+} ion. The thickness of the layers were 1 Å and the different heights were: a=0.5, b=1.5, c=2.5, d=3.5, e=4.5 and f=5.5 Å above the Zn^{2+} ion. The interval between the marks in the perimeter is 2 Å.

layers (of thickness 1 Å) cut out perpendicular to the approximate cone constituted by the active cleft . As may be seen the densities are very sharp, indicating a rather well-ordered water structure in the cleft as compared to that in bulk water. The average water-water interaction energy in the cleft was found to be -10.5 kJ/mol, which can be compared to the same quantity for bulk water, -35.6 kJ/mol[22].

THE REACTION $OH^- + CO_2 \rightleftarrows HCO_3^-$

One of the suggested reaction mechanisms for carbonic anhydrase involves a Zn-bound OH^- attack on CO_2[3,23]. Several questions can be

$$E\text{-}Zn\text{-}OH^- + CO_2 \rightleftarrows E\text{-}Zn + HCO_3^- \qquad \text{(II)}$$

raised in connection with this mechanism: it is known that the uncatalyzed reaction in aqueous solution proceeds with a rather high activation energy, 55 kJ/mol[24], and thus it is not apparent how a Zn-bound OH^- could be a better nucleophile. In this mechanistic model, the pH-dependence of the enzymatic reaction rate (see figure 1b), is explained by the equilibrium

$$EH_2O \rightleftarrows EOH^- + H^+. \qquad \text{(III)}$$

A quantum mechanical _ab initio_ calculation of a potential energy surface for the enzymatic reaction involving even a small part of the active site is impossible today. Thus if one intends to approach such biochemical problems via _ab initio_ methods, it must be done using either very simplified model systems or by an indirect approach.

In this work the problem has been approached by considering the uncatalyzed reaction

$$OH^- + CO_2 \rightleftarrows HCO_3^-. \qquad \text{(IV)}$$

However, it is nonetheless still not feasible to compute a full surface including all internal coordinates, instead, a number of approximations have been introduced in order to simplify the computational task:

a) The system has been assumed to be planar.
b) The O-H distance was not varied.
c) The C-O(H) distance was chosen as "reaction coordinate", whereas the other remaining degrees of freedom were assumed not to be involved in the reaction coordinate.

The procedure was then to minimize the energy with respect to the dependent variables for each value of the approximate reaction coordinate, $R_{C-O(H)}$. This procedure is not without objections, as has been discussed in the literature[8,25], but it is probably a good approximation in this case.

All calculations were done within the Hartree-Fock approximation. The optimization of the dependent variables was done with a small primitive basis set contracted to double zeta and the final reaction profile (figure 5) was obtained by performing calculations with a larger basis, using the opitimized geometries previously obtained. The latter basis set consisted of a double zeta basis augmented with diffuse functions in order to achieve a better description of the negative ions[9]. As is seen in figure 5 the reaction proceeds without any barrier, in

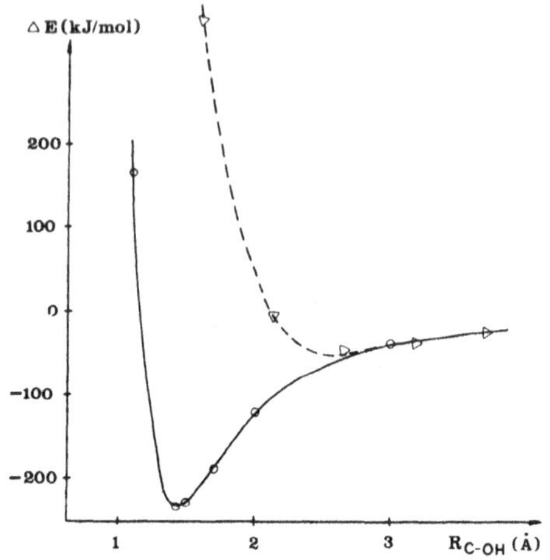

Figure 5. Variation of the interaction energy along the reaction coordinate $R_{C-O(H)}$. The dashed line represents an approach of OH^- and CO_2 under C_{2v} symmetry without relaxation of the dependent parameters.

contrast to what has been found in aqueous solutions. Thus we conclude that the barrier present in aqueous solution is in all likelihood due to the strong solvation of the OH^- ion.

An interesting feature to note is that the calculated barrier[8] for the reaction,

$$H_2O + CO_2 \rightleftarrows H_2CO_3 \qquad (V)$$

is much larger than that experimentally determined[26], thus in this case the effect of the solvent is to <u>lower</u> the barrier.

CONCLUSIONS

From the Monte Carlo simulation of the water structure in the active cleft of carbonic anhydrase, it is evident that there are considerable differences from the structure in bulk water. Thus an OH^- bound to the zinc ion in the active cleft, would obviously be much less solvated than in bulk water. For the formation of HCO_3^-, this might mean that the barrier found for the uncatalyzed reaction is lowered, since it was found in the theoretical calculations that in the gas phase there is no barrier at all. Finally, such a model giving $HCO_3^- + H^+$ as the reaction products for the rate determining step is in agreement with the results of the diffusion study, in which it was concluded that the overall products for the enzymatic hydration of CO_2 are

$HCO_3^- + H^+$.

ACKNOWLEDGEMENTS

This work was initiated by Sture Forsén, and various sections have been done in cooperation with Jan Almlöf, Enrico Clementi, Giorgina Corongiu, Gunnar Karlström, Silvano Romano, Björn Roos and Håkan Wennerström.

REFERENCES

1. S. Lindskog, L.E. Henderson, K.K. Kannan, A. Liljas, P.O. Nyman and B. Strandberg, 1971, Enzymes, 5, pp. 587.

2. K.K. Kannan, B. Notstrand, K. Fridborg, S. Lövgren, A. Ohlsson and M. Petef, 1975, Proc.Natl.Acad.Sci. USA, 72, pp. 51.

3. J.E. Coleman, 1976, J.Biol.Chem., 242, pp. 5212.

4. R.G. Khalifah, 1971, J.Biol.Chem., 246, pp. 2561.

5. B. Jönsson, G. Karlström and H. Wennerström, 1975, Chem.Phys. Letters, 30, pp. 58.

6. B. Jönsson and B. Nelander, Chem.Phys., 1977, 25, pp. 263.

7. B. Jönsson, G. Karlström, H. Wennerström and B. Roos, 1976, Chem.Phys. Letters, 41, pp. 317.

8. B. Jönsson, G. Karlström, H. Wennerström, S. Forsén, B. Roos and J. Almlöf, 1977, J.Am.Chem.Soc., 99, pp. 4628.

9. B. Jönsson, G. Karlström and H. Wennerström, 1978, J.Am.Chem.Soc., 100, pp. 1658.

10. E. Clementi, G. Corongiu, B. Jönsson and S. Romano; Monte Carlo simulation of the water structure around Zn^{++}. To be published.

11. B. Jönsson and H. Wennerström, 1978, Biophys.Chem., 7, pp. 285.

12. E. Clementi, G. Corongiu, B. Jönsson and S. Romano, FEBS Letters, in press.

13. S.H. Koenig and R.D. Brown, 1973, Biochim.Biophys.Res.Commun., 53, pp. 624.

14. R.G. Khalifah, 1973, Proc.Natl.Acad.Sci. US, 70, pp. 1986.

15. S. Lindskog and J.E. Coleman, 1973, Proc.Natl.Acad.Sci. USA, 70, pp. 2505.

16. M. v. Smulochowski, 1917, Z.Physik.Chem., 92, pp. 129.

17. B-H. Jonsson, H. Steiner and S. Lindskog, 1976, FEBS Letters, 64, pp. 310.

18. Gmelin, 1956, "Handbuch der anorganischen Chemie", 8 auflage, 32, pp. 720.

19. H. Ohtaki, T. Yamaguchi and M. Maeda, 1977, Essays Anal.Chem., pp. 163.

20. O. Matsouka, E. Clementi and M. Yoshimine, 1976, J.Chem.Phys., 64, pp. 1351. H_2O-H_2O interaction potential.

21. E. Clementi, F. Cavallone and R. Scordamaglia, 1977, J.Am.Chem. Soc,. 99, pp. 5531; G. Bolis and E. Clementi, 1977 ibid, 99, pp. 5550; G. Ranghino and E. Clementi, to be published. H_2O-amino acid potentials.

22. G.C. Lie, E. Clementi and M. Yoshimine, 1976, J.Chem.Phys., 64, pp. 2314.

23. P. Woolley, 1975, Nature, 258, pp. 677.

24. B.R.W. Pinsent, L. Pearson and F.J.W. Roughton, 1956, Trans. Faraday Soc., 52, pp. 1512.

25. C.P. Baskin, C.F. Bender, C.W. Bauschlicher Jr and H.F. Schaefer III, 1974, J.Am.Chem.Soc., 96, pp. 2709.

26. E. Magid and B.O. Turbeck, 1968, Biochim.Biophys. Acta, 165, pp. 515.

A COMPARISON OF THE ENERGETICS OF PROTON TRANSFER IN THE SERINE AND CYSTEINE "CHARGE RELAY" SYSTEMS AND THE ROLE OF THE PROTEIN ELECTROSTATIC POTENTIAL ON THE PROTON TRANSFER ENERGETICS

David M. Hayes and Peter A. Kollman
Department of Pharmaceutical Chemistry, UC, San Francisco, 94143. and Department of Chemistry, Union College, Schenectady, N.Y.

ABSTRACT

We present *ab initio* calculations on models of the "charge relay" triads Serine-Histidine-Aspartate, Cysteine-Histidine-Asparagine and Cysteine-Histidine-Aspartate. The direct calculations on double proton transfer indicate that it is significantly more favourable than single proton transfer in both serine and cysteine protease models and Asp is considerably more effective than Asn as the third member of the triad. However, correcting these proton transfer energies for known deficiencies in calculated proton affinities makes single proton transfer more likely than double proton transfer. The electrostatic potential in α-chymotrypsin makes the proton transfer more difficult, whereas that in subtilisin facilitates it. In papain, the potential favours single proton transfer, but not double proton-transfer.

INTRODUCTION

One of the best characterized enzyme mechanisms is that in the serine proteases, where, as pointed out by Blow *et al.* (1969), the "charge relay triad" Aspartate-Histidine-Serine (Figure 1) is apparently crucial for effective catalytic action. This triad and model substrates have been the subject of numerous theoretical studies by Umeyama *et al.* (1973, 1974, 1975), Nakagawa and Umeyama (1977), Scheiner and Lipscomb (1975, 1976), Demoulin and Allen (1976) and Amidon (1974). These studies all found that serine → histidine and accompanying histidine → aspartate was energetically much more favourable than only serine → histidine proton transfer and was essential to the catalytic action of the serine. In fact, Scheiner and Lipscomb (1975) pointed out that 80% of the proton transfer step took place before there was significant movement of the serine nucleophilic oxygen toward the substrate. Since all of the above calculations involved only those residues directly involved in the catalytic action, we decided to evaluate the effect of other residues on the proton transfer process using an approach we applied to

B. Pullman (Ed.): Catalysis in Chemistry and Biochemistry. Theory and Experiment. 77–90.

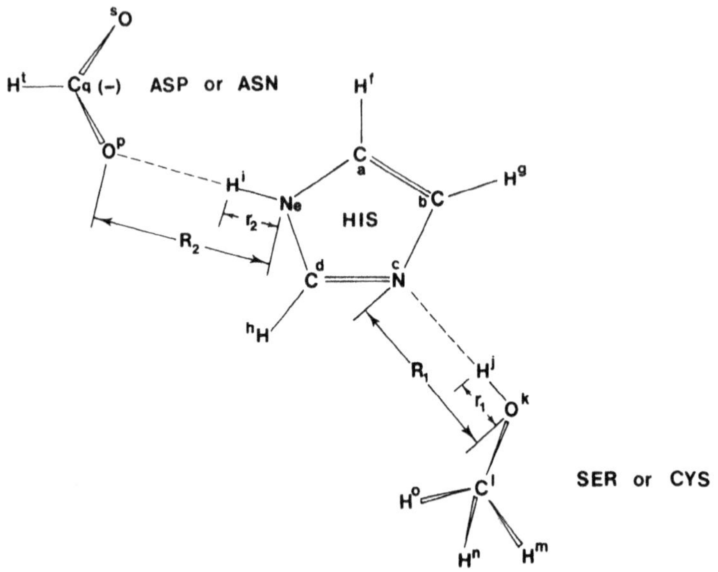

Figure 1. Schematic model of charge relay triad

carboxypeptidase A (Hayes and Kollman, 1976b).

There have been many physical chemical studies of the Asp-His-Ser triad with the goal of elucidating its acid base properties. Although there are conflicting reports in the literature, the most recent studies suggest a normal pKa near 7 for the Histidine and a lowered pKa of about 2 for the Aspartate (Bachovin and Roberts, 1978).

Since subtilisin and α-chymotrypsin apparently have similar mechanisms of catalytic action despite their very different primary structures, we thought it would be of interest to compare the electrostatic potential in the active site region of these enzymes.

Sulfur is generally considered to be a better nucleophile than oxygen, so the cysteine protease papain has been previously thought to involve only histidine-cysteine as a "charge relay" diad, with an asparagine H-bonding to the histidine N-H. Demoulin and Allen (1977) have considered possible proton transfer to the asparagine in a model papain calculation and Broer *et al.* (1976) have modelled the cysteine S-H to the Histidine proton transfer with and without a formaldehyde C=O...H-N Histidine H bond and noted that this H-bond did facilitate single proton transfer. Recently, however, Angelides and Fink (1979) pointed out that a rotation around the CαCβ of the histidine in papain would allow Asp 158 to H-bond with the imidazole N-H of this essential histidine. Thus, papain may contain a Cys-His-Asp "charge relay" triad. We thus thought it would be of interest to compare the electronic structure and proton transfer energies of the serine and cysteine protease "charge relay" triads.

METHOD OF CALCULATION

Because of the size of the systems considered, we were forced to use a minimal, STO-3G, basis set in our *ab initio* calculations (Hehre *et al.*, 1969). We used the programme GAUSSIAN 70 (QCPE # 236), modified to incorporate the partial charges of protein groups, as we have previously described (Hayes and Kollman, 1976b).

Our choice of molecular geometry for the *ab initio* calculations on the charge relay triad (Figure 1) came from the X-ray structure of α-chymotrypsin determined by Blow *et al.*[1]. We started with the X-ray co-ordinates for histidine 57 Cα, Cβ and Cγ to generate the locations of Nδ and Cδ, using X-ray co-ordinate dihedral angles, but the internal co-ordinates (R and θ) from the neutron diffraction work of Lehmann *et al.* (1972). The Asp 102 was located using the dihedral angles from the crystal structures, the internal geometry for neutral and anionic Asp from Lehmann *et al.* (1972) and the Nδ-Oγ distance fixed at 2.8Å (the X-ray value is 2.88Å). The serine 195 was located by using the dihedral angles from the X-ray structure, the internal co-ordinates from the experimental structure of methanol and $R(N_{\varepsilon}-O_{\gamma})$ = 2.8Å (this value was 3.6Å in the Blow and Birktoft co-ordinates.) Asp, His and Ser were modelled by formate, imidazole and CH_3OH, respectively in the quantum mechanical calculations. For calculations to model Cys 25 in papain, we used the same structure with S replacing O, $R(N_{\varepsilon}-S_{\gamma})$ = 3.4Å, and the internal geometry of methanethiol. When we wished to model asparagine of papain, we used formamide, with its carbonyl O located in the same position as the proton accepting oxygen of Asp, its C=O bond along the same line as in Asp (R(C=O)= 1.21Å) and the NH_2 replacing the other oxygen (see Table I).

In the electrostatic potential calculations, we used the X-ray structure of Blow and Birktoft, generated hydrogens in standard locations and placed partial charges on all the atoms, as previously described (Hayes and Kollman, 1976a), with the charges of Asp 102, His 57 and Ser 195 deleted from the electrostatic potential calculations.

Our *ab initio* calculations were carried out with the above model system as well as (a) including all the residues with a formal charge (+1 or -1) located at the centroid of the charged "group" (Lysines on the nitrogen, Arginines on the guanidium carbon, Asp or Glu on the carboxyl carbon) (b) including charged residues <u>and</u> all the partial charges of atoms within 5Å of any atom in the charge relay system in the one-electron Hamiltonian.

For subtilisin (Kraut *et al.*, 1971), we merely evaluated the electrostatic potential in the location of the charge relay triad after

atoms	R(Å)	atoms	θ(°)	atoms	φ(°)
f-a	1.062	f-a-e	121.1	b-a-e-d	0.0
a-e	1.376	f-a-b	129.1	e-a-b-c	0.0
a-b	1.372	a-e-d	105.4	e-a-b-g	180.0
e-d	1.309	a-b-c	105.7	b-c-d-h	180.0
b-c	1.376	a-b-g	131.4	c-d-e-i	180.0
b-g	1.062	c-d-h	123.1	e-d-c-j	180.0
d-h	1.075	d-e-i	127.3	e-d-c-k	180.0
e-i	1.038→1.84	d-c-j	126.6	d-c-k-l	335.0
c-j	1.84→1.038	d-c-k	126.6	c-k-l-m	148.5
c-k	2.80	c-k-l	109.5	c-k-l-n	268.5
k-l	1.43	k-l-m	109.5	c-k-l-o	28.5
l-m	1.09	k-l-n	109.5	c-d-e-p	180.0
l-n	1.09	k-l-o	109.5	d-e-p-q	156.6
l-o	1.09	d-e-p	127.3	e-p-q-s	312.7
e-p	2.80	e-p-q	120.0	e-p-q-t	132.7
p-q	1.25→1.31	p-q-s	124.0		
q-s	1.25	p-q-t	118.0		
q-t	1.09				

[a] figure 1. [b] changes for Asp→Asn. R(q-s) = 1.33Å, s becomes N. Added two hydrogens to s at R(N-H) = 1.01Å, θ = 120°, φ (pqsH) = 0, 180°. [c] changes for Ser→Cys. k becomes S. R(c-k) = 3.4Å.

Table I. Internal Geometry for Ser-Im-Asp[a,b,c]

deleting the partial charges of the Asp-His-Ser residues.

To model the papain "charge relay triad", we evaluated the proton transfer processes *ab initio* for isolated Asn-His-Cys and Asp-His-Cys, as well as evaluating the electrostatic potential, using the X-ray coordinates of Drenth *et al.* (1971) and deleting the partial charges of Cys 25, His 159 and Asn 175.

RESULTS AND DISCUSSION

Single vs. Double Proton Transfer

As one can see in Table II, the calculations find double proton transfer in the charge relay system more favourable than single proton transfer, by a substantial amount, similar to what was found in previous studies employing semi-empirical molecular orbital theory (Umeyama *et al.*, Scheiner and Lipscomb, Demoulin and Allen, Amidon). The absolute values of the endothermicities of proton transfer are exaggerated, because of the tendency of the minimal basis set to underestimate the stabilities of the anions, CH_3O^- and $HCOO^-$, compared to the neutral molecules CH_3OH and HCOOH. The calculated proton affinity of imidazole, like those of CH_3O^- and $HCOO^-$ is also too high, but is no over-estimated as much as the anions. We can attempt to correct for these defects, as did Scheiner and Lipscomb, by using the proton affinities of CH_3O^-, $HCOO^-$, CH_3S^- and imidazole. The calculated and experimental proton affinities are presented in Table III. The "corrected" values for the proton transfer energies can now be easily estimated. For example, the energy of Ser-His → $Ser^{(-)}$-$His^{(+)}$ of Table II (114 kcal/mole) becomes 14 kcal/mole because the error in the calculated proton affinity of $Ser^{(-)}$ exceeds the error for His by 100 kcal/mole. Table IV now contains these "corrected" energies from Table II. The fact that the basis set error for the proton affinity of CH_3O^- and CH_3S^- is almost identical (158 vs. 155 kcal/mole) encourages us that a comparison of the proton transfer process in cysteine and serine protease systems will be valid.

Molecule	isolated	Relative Energy[a] with point charges	Point charges and partial charges
His-Ser	0	-	-
$His^{(+)}$-$Ser^{(-)}$	114.3	-	-
$Asp^{(-)}$-His-Ser	0	0(-60.8)[b]	0(-63.6)[b]
$Asp^{(-)}$-$His^{(+)}$-$Ser^{(-)}$	92.9	105.0	103.5
Asp-His-$Ser^{(-)}$	38.7	60.5	63.4
His-Cys	0	-	-
$His^{(+)}$-$Cys^{(-)}$	102.3	-	-
$Asp^{(-)}$-His-Cys	0	-	-
$Asp^{(-)}$-$His^{(+)}$-$Cys^{(-)}$	74.5	-	-
Asp-His-$Cys^{(-)}$	13.8	-	-
Asn-His-Cys	0	-	-
Asn-$His^{(+)}$-$Cys^{(-)}$	96.0	-	-
$Asn^{(+)}$-His-$Cys^{(-)}$	-	-	-

[a] kcal/mole; [b] in parentheses are the stabilization energies of the triad by the protein.

Table II. Calculated Relative Energies for Protease Models $Asp^{(-)}$-His-Ser, $Asp^{(-)}$-His-Cys, Asn-His-Cys, His-Cys and His-Ser.

Molecule	Calculated PA	Experimental PA
CH_3O^-	535	377[a]
CH_3S^-	492	337[b]
$HCOO^-$	466	342[c]
Imidazole	278	220[d]

[a] McIver and Miller, 1974
[b] estimated[2]
[c] Yamdagni and Kebarle, 1973
[d] estimated [3]

Table III. Calculated and Experimental Proton Affinities (kcal/mole).

Molecule	Relative Energies[b]
His-Ser	(0)
$His^{(+)}$-$Ser^{(-)}$	14
$Asp^{(-)}$-His-Ser	(0)
$Asp^{(-)}$-$His^{(+)}$-$Ser^{(-)}$	-7
Asp-His-$Ser^{(-)}$	5
His-Cys	(0)
$His^{(+)}$-$Cys^{(-)}$	5
$Asp^{(-)}$-His-Cys	(0)
$Asp^{(-)}$-$His^{(+)}$-$Cys^{(-)}$	-22
Asp-His-$Cys^{(-)}$	-17
Asn-His-Cys	(0)
Asn-$His^{(+)}$-$Cys^{(-)}$	-4
$Asn^{(+)}$-His-$Cys^{(-)}$	- [c]

[a] using the calculated and experimental proton affinities in Table III.
[b] kcal/mole.
[c] did not converge, but should be less favourable than single proton transfer by at least 15 kcal/mole.

Table IV. Relative Energies for Proton Transfer Corrected for Calculated Errors in Proton Affinities[a]

The results in Table IV are consistent with the greater stability of CH_3S^- than CH_3O^- both in the gas phase and in solution in that the $Asp^{(-)}$-$His^{(+)}$-$Cys^{(-)}$ system is more stable than $Asp^{(-)}$-His-Cys by 22 kcal/mole and the corresponding quantity with Ser replacing Cys is only 7 kcal/mole. The results also suggest, in disagreement with previous quantum mechanical calculations and Table II, that single proton transfer is favoured over double proton transfer.

It is likely that we are overcorrecting the basis set error by using the full error between the calculated and experimental proton affinities to alter the energies in Table II to those in Table III. This is because the presence of the imidazole orbitals can cause artificial stabilization (Johansson *et al.*, 1973) of the serine anion. In addition, there is probably some real stabilization due to the fact that the serine oxygen bears a smaller negative charge when $\overset{+}{His}$ is nearby. Based on the experimental proton affinities of CH_3O^-, Imidazole and $HCOO^-$, one can calculate that $Ser^{(-)}$-His-Asp is more stable than $Ser^{(-)}$-$His^{(+)}$-$Asp^{(-)}$ by about 120 kcal/mole at infinite separation of these species, but electrostatic stabilization will greatly reduce this difference in stability and make these two states comparable in stability. In addition, Warshel (1978) has suggested that one might have important stabilization of ionic triads such as - + - by enzymes, due to polarization effects not included in these calculations.

Based on the calculations by Umeyama *et al.* and the findings by

Scheiner and Lipscomb that most of the proton transfer takes place prior to nucleophilic attack, one might conclude that the presence of substrate will not alter the above arguments. Whether the proximity of a His^+ (as opposed to neutral His) would interact favourably or unfavourably with the substrate might depend, e.g., on whether the substrate carbonyl carbon, oxygen or other polar groups are closest to the His^+ positive charge.

Even though the calculations reported here are not definitive in locating the proton positions of the charge relay system, we can attempt to make a more precise estimate of these positions using more accurate calculations and experiment on model compounds. Table V summarizes the results on the complexes $H_3N...HCl$ and $H_3N...HF$ and the fragments which make up these complexes.

Molecule	Calculated PA	Experimental PA
NH_3	224	207
Cl^-	349	334
F^-	401	370
Complex	Calculated ΔE (proton transfer)	Experimental ΔE (proton transfer)
$H_3N...HCl$	~0	~0
$H_3N...HF$	26	-

Table V. Calculated and Experimental Properties of H_3NHF, H_3NHCl, and appropriate fragments.

For $NH_3...HCl$, calculations by Clementi (1967) and ourselves (Kollman *et al.*, 1975) have indicated a "half-proton transferred" structure, in excellent agreement with matrix IR experiments (Ault and Pimentel, 1975). Thus, the energy difference between a neutral $H_3N...H\text{-}Cl$ and ionic $H_3NH^+...Cl^-$ structure is calculated and found to be near zero. As one can see in Table V, the probable reason that the calculations reproduce the experiment so well is that the error in the calculated proton affinity of Cl^- (15 kcal/mole) is so similar to the error in the calculated proton affinity of NH_3 (17 kcal/mole). In $H_3N...HF$, the 431G calculated energy to form the ion-pair structure is 40 kcal/mole (Kollman and Kuntz, 1976); correcting this for the difference (14 kcal/mole) between the error in proton affinities of NH_3 and F^- leads to a ΔE for proton transfer of 26 kcal/mole.

Using this data and the data in Table III, we now turn to an analysis of the proton positions in imidazole...$HSCH_3$, imidazole... $HOCH_3$ and imidazole...HOOCH: The experimental proton affinity of Cl^- is only 3 kcal/mole less than that of CH_3S^-, whereas the proton affinity of imidazole is ~ 13 kcal/mole more than that of NH_3. Thus, in the gas phase, the energy of the ion-pair imidazole...$HSCH_3$ complex is likely to be close to that of the neutral complex. Our calculations (Table IV) suggest that the ion-pair structure is less stable by only 5 kcal/mole. More accurate *ab initio* calculations by van Duijnen *et al.* (1979a, 1979b) find the ion-pair structure less stable by 14 kcal/mole (corrected for errors in the proton affinities). Because van Duijnen *et al.* did not optimize the N-S distance for the ionic structure, it is likely that 14 kcal/mole is an upper bound for the energy difference with their basis set. In aqueous solution, the similarity in pKa's of His and Cys certainly argues that, if these two molecules are held in proximity, they will be most stable as a zwitterionic pair. In papain, the presence of Asn 175 (or Asp 158) H bonded to the His and the electrostatic potential (see below) reinforces this zwitterionic tendency and leaves little doubt that the ion-pair structure is likely to be the form that actually exists in the enzyme even in the absence of substrate.

In the serine proteases, we are dealing with an anion (CH_3O^-) that is less intrinsically stable by ~ 40 kcal/mole. Using calculations on H_3N HF as a guide, the ion-pair structure is intrinsically less stable by about 26 kcal/mole. The $Asp^{(-)}$ stabilizes this ionic structure (Table II) by about 21 kcal/mole[4], leaving an activation barrier of ~ 5 kcal/mole for proton transfer. Again, by the same arguments as used for above CH_3S^-, $HCOO^-$ has only an 8 kcal/mole greater proton affinity than $C\ell^-$; taken together with the loss of stabilization of CH_3O^- upon neutralization of $His^{(+)}$-$Asp^{(-)}$; we argue that there is no tendency for $His^{(+)}$ to give up its proton to $Asp^{(-)}$ in the serine proteases even when the substrate binds. In particular, in α-chymotrypsin, the presence of H-bonds to Asp 102 and the unfavourable electrostatic potential gradient for proton transfer to Asp 102 (see below) all argue that Asp 102 is functioning to stabilize the His-Ser ion pair.

One must emphasize that the above analysis on the model systems is based on experimental proton affinities of CH_3O^-, $HCOO^-$, F^-, $C\ell^-$, "estimated" proton affinities for CH_3S^- and imidazole[5], the experimental structure of $H_3N...HC\ell$ and calculations in the estimate of the energy of proton transfer for $H_3N...HF$. However, we do feel that the evidence we present here for an intrinsically stable zwitterionic $Cys^-...His^+$ system and the operation of the Serine "charge relay system" involving only single proton transfer is as reasonable as those just employing uncorrected quantum mechanical calculations[6].

The Role of the Protein Electrostatic Potential

It is interesting that the presence of the electric field of charged surrounding residues in α-chymotrypsin (Table II) does not facilitate proton transfer, but rather hinders both single and double proton transfer. The charged and partial charges within 5Å effect is similar, with the partial charges making the first proton transfer somewhat easier, but inhibiting the second. Even though these calculations do not allow polarization or conformational changes in the enzyme atoms, the results of the calculations suggest that in α-chymotrypsin, the protein groups actually inhibit charge relay proton transfer. To see if this effect could be "predicted" without detailed *ab initio* studies, we evaluated the electrostatic potential at the locations of the serine Oγ, the histidine N_ε and N_γ and the asp $O_{\delta 1}$ and $O_{\delta 2}$ (Table VI) (points k, c, e, s and p, see Figure 1). As one can see, the electrostatic potential gradient across the "charge relay" triad foreshadows the *ab initio* result. Based on this point charge model, we expect single proton transfer to be inhibited by ~ 15 kcal/mole and double proton transfer to be inhibited by ~ 40 kcal/mole; these can be compared with the quantum mechanically calculated values (10.6 and 24.7) in Table II. The simple electrostatic predictions exaggerate the magnitude of the endothermicities of proton transfer because they do not consider the charge redistribution and polarization effects, but they do give us a correct qualitative picture.

When we turn (Table VI) to the potential gradient in subtilisin, we find it is almost the reverse of that in α-CT; here, both single and double proton transfers are made more favourable by the presence of the enzyme surrounding. This surprising result seems to suggest that either the proton transfer step is not rate limiting (assuming similar mechanisms for the two enzymes) or if proton transfer is at least partially rate limiting, that subtilisin should be a more effective catalyst than α-CT. Deuterium isotope effects, (Bender and Kilheffer (1973)), suggest that proton transfer plays a role in the rate limiting steps of acylation and deacylation, so it will be of interest to compare in detail these steps for chymotrypsin and subtilisin; our calculated results suggest that the latter should have a lower activation barrier for proton transfer.

Protein	Point			
	k	c	e	s,p[b]
α-CT[c]	0.08	0.10	0.15	0.235
Subtilisin[c]	-0.09	-0.15	-0.18	-0.195
Papain[d]	0.28	0.23	0.26	0.33

[a] partial charge in atomic units/Å; conversion factor to kcal/mole = 332.
[b] average of potential at $O\delta_1$ and $O\delta_2$ ($O\delta$ and N_δ if Asn).
[c] charges of Asp, His, Ser removed.
[d] charges of Asn, His, Ser removed.

Table VI. Electrostatic Potentials[a] at Points in Active Site.

If we now consider papain, our *ab initio* calculations on Asp-His-Cys indicate that single and double proton transfer are easier than found in the corresponding Ser triad, consistent with the greater stability in the gas phase of CH_3S^- than CH_3O^-. The charge relay triad Asn-His-Cys has a higher energy requirement for proton transfer than Asp-His-Cys. Our SCF calculations did not converge for $Asn^{(+)}$-His-$Cys^{(-)}$, but calculations on the corresponding "diads" Asn-His, $Asp^{(-)}$-His, His-Ser and His-Cys indicate that one can estimate the energetics of formation of $Asn^{(+)}$-His-$Cys^{(-)}$ compared with Asp-His-$Cys^{(-)}$ as follows: The calculated proton transfer energy going from $Asp^{(-)}$-$His^{(+)}$ to Asp-His is -60 kcal/mole; going from Asn-$His^{(+)}$ to $Asn^{(+)}$-His, it is 15 kcal/mole. Thus, the calculated tendency for double compared to single proton transfer will be about 75 kcal/mole greater for the $Asp^{(-)}$ than the Asn triad. This rules out Asn proton acceptance. Thus, unlike the suggestion by Demoulin and Allen (1977), we find no tendency of the amide to accept a proton in the Asn-His-Cys triad. The asparagine does contribute to the proton transfer in that the calculated transition Asn-His-Cys → Asn-$His^{(+)}$-$Cys^{(-)}$ is 6 kcal/mole less endothermic than His-Cys → $His^{(+)}$-$Cys^{(-)}$. The C=O...H-N hydrogen bond is thus about 6 kcal/mole stronger with cationic than with neutral His.

As noted in Table VI, the electrostatic potential gradient in the papain "charge relay" system Asn-His-Cys favours single proton transfer. If one moves the Asp to hydrogen bond with His, (the catalytically active conformation suggested by Angelides and Fink (1979)) the electrostatic potential gradient in the papain charge relay Cys-His increases from -.035 au/Å (the difference between the potential at point k and the average of the potential at the two histidine nitrogens c and e) to about -.15 au/Å. Based on the rough relationship between the *ab initio* calculated effect of potential on proton transfer in α-CT and the change in the potential calculated directly (Table III) it is clear that the protein environment in papain significantly stabilizes the proton transferred structure relative to isolated Cys-His.

Hol *et al.* (1978) have recently noted that helix dipole effects can stabilize charges and have pointed out that Serine 195 in subtilisin and Cys 25 in papain both lie near the N-terminal (positive) end of an α-helix. They also note that there is no such helix in chymotrypsin, but that the groups nearby in chymotrypsin would tend to stabilize a negative charge. Our electrostatic potential calculations include the effects of the helix dipole as well as any other electrostatic effects, so our results are inconsistent with theirs for chymotrypsin. It may be, however, that this inconsistency is probably due to the different choice of dielectric constant. We used $\varepsilon = 1$ throughout and Hol *et al.* used a distance dependent dielectric constant. Note that our quantum mechanical calculations in Table II suggest that the nearby partial charges stabilize single but not double proton transfer for α-chymotrypsin. Also, we have not explicitly included the potential due to Asp $102^{(-)}$ for α-chymotrypsin (or the corresponding Asp in subtilisin) in Table IV. Such an inclusion would change the gradient between point k and the average of c and e from 0.045 to -.055 in chymotrypsin and from -.065 to -.165 in subtilisin. In any case, our suggestion of an expected difference in rate constants for proton transfer in these two enzymes still is consistent with both our data and that of Hol *et al.*

Conclusions and Implications for Catalytic Mechanisms

We have calculated the effect of protein environment on the charge relay triads Ser-His-Asp$^{(-)}$, Cys-His-Asp$^{(-)}$ and Cys-His-Asn, involved in the mechanism of action of the proteases α-chymotrypsin, subtilisin and papain. In the triads involving Asp$^{(-)}$, our calculations suggest that the question of single vs. double proton transfer is not as unambiguously answered as previous calculations suggested. In the triad involving Asn, there is no doubt that only single proton transfer occurs.

In fact, our best "estimates" for proton location suggest that in papain, the zwitterionic Cys^{-}...His^{+} ion pair exists in the absence of substrate and that proton transfer should not be a rate limiting step in forming the tetrahedral intermediate (it may well be rate limiting for the subsequent breakdown of the peptide bond). In α-chymotrypsin, we estimate a barrier of 5 kcal/mole for the intrinsic energy Asp$^{(-)}$-His$^{(+)}$-Ser$^{(-)}$ relative to Asp$^{(-)}$-His-Ser. This value is qualitatively similar to the activation energy of 11 kcal/mole observed for acylation of a specific ester substrate by chymotrypsin and is consistent with the fact that deuterium isotope effects suggest rate limiting proton transfer for acylation. However, the approximate numerical agreement is fortuitous because of the uncertainties in our energy estimates.

We calculate that the effect of protein environment is to facilitate proton transfer in subtilisin and papain, but to make it more difficult in α-chymotrypsin. This is not consistent with the expected similarity of the mechanism of α-chymotrypsin and subtilisin and the role of proton transfer in their rate limiting steps (Bender and Kilheffer). This could be due to: the simplicity of our model for the electrostatic potential in the active site; or, if real, would suggest some difference in rate constant for proton transfer in subtilisin and chymotrypsin.

ACKNOWLEDGMENTS

We are pleased to acknowledge the support of the NSF (CHE-7681718) in this research. P.A.K. is a recipient of a Research Career Development Award (GM-70718). Crucial to this work were a receipt of a preprint from K. Angelides and A. Fink and the suggestion by our colleague George Kenyon that a comparison between subtilisin and chymotrypsin would be of interest; we are very grateful to these three for their help. We are also grateful to P. van Duijnen and Lee Allen for the receipt of preprints on their calculations.

REFERENCES

Amidon, G. L.: 1974, J. Theor. Biol. 46, pp. 101.

Angelides, K. and Fink, A. L.: *Cryoenzymology of Papain: Reaction Mechanism with an Ester Substrate* and *The Mechanism of Action of Papain with a Specific Anilide Substrate*, 1979, Biochem. (in press).

Ault, B. S. and Pimentel, G. C.: 1975, J. Phys. Chem. 77, pp. 1649.
Bachovin, W. and Roberts, J. D.: 1978, J. Amer. Chem. Soc. 100, pp.8041
Bender, M. L. and Kilheffer, J.: 1973, CRC Crit. Rev. Biochem. 1, pp. 149.
Blow, D. M., Birkhoft, J. J. and Hartley, B. S.: 1969, Nature 221, pp. 337.
Broer, R., van Duijnen, P. and Nieuwpoort, W. C.: 1976, Chem. Phys. Lett. 42, pp. 525.
Clementi, E.: 1967, J. Chem. Phys. 46, pp. 3851.
Demoulin, D. and Allen, L. C.: 1976, preprint.
Demoulin, D. and Allen, L. C.: 1977, preprint.
Drenth, J., Jansonius, J., Koekock, R. and Wolthers, B.: 1971, Adv. Prot. Chem. 25, pp. 79.
Hayes, D. M. and Kollman, P. A.: 1976a, J. Amer. Chem. Soc. 98, pp. 3335.
Hayes, D. M. and Kollman, P. A.: 1976b, J. Amer. Chem. Soc. 98, pp. 7811.
Hehre, W. J., Stewart, R. F. and Pople, J. A.: 1969, J. Chem. Phys. 51, pp. 2657.
Johansson, A., Kollman, P. and Rothenberg, S.: 1973, Theor. Chim. Acta 29, pp. 167.
Kollman, P., Johansson, A., McKelvey, J. and Rothenberg, S.: 1975, J. Amer. Chem. Soc. 97, pp. 955.
Kollman, P. and Kuntz, I. D.: 1976, J. Amer. Chem. Soc. 98, pp. 6820.
Kraut, J., Robertus, J., Birktoft, J., Alden, B., Wilcox, R. and Powers, J.: 1971, Cold Spring Harbor Symp. Quant. Biol. 36, pp. 117.
Lehmann, M., Koetzle, T. and Hamilton, W.: 1972, Int. J. Protein Pept. Res. 4, pp. 229.
McIver, R. and Miller, J.: 1974, J. Amer. Chem. Soc. 96, pp. 4323.
Nakagawa, S. and Umeyama, H.: 1977, Chem. Pharm. Bull. 25, pp. 909.
Taagepara, M., Henderson, W., Brownlee, R., Beauchamp, J., Holtz, D. and Taft, R.: 1972, J. Amer. Chem. Soc. 94, pp. 1369.
Umeyama, H., Imamura, A., Nagata, C. and Hanano, M.: 1973, J. Theor. Biol. 41, pp. 485.
Umeyama, H.: 1974, Chem. Pharm. Bull. 22, pp. 2518.
Umeyama, H., Imamura, A. and Nagata, C.: 1975, Chem. Pharm. Bull. 23, pp. 3075.
van Duijnen, P., Thole, B. and Hol, W.: *On the Role of the Active Site Helix in Papain: An ab initio Molecular Orbital Study*, Biophys. J. in press, 1979a.
van Duijnen, P., Thole, B., Broer, B. and Nieuwpoort, W. C.: *The active site α helix in papain and the stability of the ion pair RS^--ImH^+. An ab initio molecular orbital study*, Int. J. Quant. Chem., submitted 1979b.
Warshel, A.: 1978, Proc. Nat. Acad. Sci. 75, pp. 5250.
Yamdagni, R. and Kebarle, P.: 1973, J. Amer. Chem. Soc. 95, pp. 4050.

NOTES

1. We used the α-chymotrypsin co-ordinates of Blow *et al.* deposited in

the Brookhaven protein data bank.

2. We estimated this from the following reactions:
   $De(H_2S)$ = 90 kcal/mole (Handbook of Chemistry and Physics); electron affinity HS = 2.32 ev in J. Lowe (J. Amer. Chem. Soc., 99, 5557 (1977). This leads to a proton affinity of 350 kcal/mole for SH^-. By comparing the proton affinity of OH^- (390 kcal/mole; D. R. Stull and H. Prophet, JANAF Thermochemical Tables, 2nd ed., US. Depts. Commerce, NSRDS-NBS, 37 (1971) with that of CH_3O^- (McIver and Miller, 1974), we estimate a proton affinity of CH_3S^- of 337 kcal/mole.
3. We estimate that the proton affinity of imidazole is about that of dimethylamine (222 kcal/mole; see J. Beauchamp, Ann. Rev. Phys. Chem., 22, 527 (1971), since the sp^2 hybridization of N makes it a weaker base by about 6 kcal/mole (comparing the calculated proton affinities of CH_3NH_2 (P. Kollman and S. Rothenberg, J. Amer. Chem. Soc., 99, 1313 (1977)) and CH_2NH (P. Kollman, P. Gund and S. Rothenberg, ibid, 97, 1640 (1975)) and the resonance effect will make it a stronger base by about 4 kcal/mole (L. Pauling, The Nature of the Chemical Bond, Cornell Univ. Press, 1960, estimates that the resonance effect on the basicity of guanidium (3 equivalent resonance structures) is 6-8 kcal/mole).
4. Comparing the proton transfer in the "diads" His-Cys and His-Ser with the triads Asp-His-Cys and Asp-His-Ser, we find that the $Asp^{(-)}$ stabilizes the ionic states $His^{(+)}$-$Ser^{(-)}$ and $His^{(+)}$-$Cys^{(-)}$ by 21 and 28 kcal/mole respectively. The greater stabilization of the His-Cys diad is probably due to the greater N...S (3.4Å) than N...O (2.8Å) distance, thus keeping the anionic $Cys^{(-)}$ further from $Asp^{(-)}$ than the $Ser^{(-)}$.
5. The estimated proton affinity of CH_3S^- assumes a similar "methyl substituent effect" in stabilizing OH^- and SH^-. Our estimated proton affinity of imidazole is likely to be a lower bound because the proton affinity of pyridine (Taagepara *et al.*, 1972) (no resonance effect on protonation) is 225 kcal/mole.
6. Scheiner and Lipscomb did correct their energies for "minimal basis set errors" using more accurate 431G energies. It may be that they came to different conclusions than we concerning single vs. double proton transfer because they assumed a similar minimal basis set error for CH_3O^- and $HCOO^-$. Using more recent experimental values (our Table II) suggests a larger error for CH_3O^- than $HCOO^-$ by 17 kcal/mole for their PRDDO energies. Also, they apparently decided to carry out double rather than single proton transfer based on the results of uncorrected calculations of the two alternatives (p.2607).

DISCUSSION

GREADY : Theoretical results for charged systems commonly show delocalization rather than formal charges as written. Therefore, to what extent is the term "charge-relay mechanism" an appropriate description for the processes described here which involve mostly movements of protons, i.e. do the systems $Asp^{\ominus}$-His-Ser, $Asp^{\ominus}$-$His^{\oplus}$- $Ser^{\ominus}$ and Asp - $His^{\ominus}$-Ser display significant differences in charge distributions.

KOLLMAN : There is relatively little intermolecular charge transfer in all these species, e.g. $Asp^{\ominus}$- $His^{\oplus}$- $Ser^{\ominus}$ is a reasonable description of the charge distribution in this complex. Thus, comparing $Asp^{\ominus}$- His - Ser with Asp - His - $Ser^{\ominus}$, one can reasonably say the charge is "relayed" from Asp to Ser. However, the relative energies we calculate suggest that this charge relay does not occur, i.e. $Asp^{\ominus}$- $His^{\oplus}$- $Ser^{\ominus}$ is more stable than Asp - His - $Ser^{\ominus}$.

SABLE : One has to be cautious about the usual formulation of the charge-relay system. In a recent issue of the Journal of the American Chemical Society, Prof. J.D. Roberts reported a study in which he cultured bacteria in a medium supplemented with ^{15}N-enriched histidine. A serine protease was isolated from this organism and was studied by ^{15}N-NMR. It was found that the pK for protonation - deprotonation of the imidazole ring was out of the region that would allow it to be considered to participate in the "classical" charge-relay system.

KOLLMAN : The results of Roberts (7) suggest apK of His 57 of 7 and that of Asp 102 of 2, reinforcing our conclusion that there is no tendency of a protonated His to transfer its proton to the Asp. Thus, both our results and Roberts suggest that the role of Asp 102 is to stabilize the His cation and not to be the recipient of a proton. The His does not "relay" charge from Asp to Ser, but enables Ser to be a more nucleophilic group than it would be in aqueous solution.

HYDROLYSIS BY ACETYLCHOLINESTERASE. TRIMETHYL AND METHYL SUBSITES

Fariza B. Hasan and Saul G. Cohen
Department of Chemistry, Brandeis University
Jonathan B. Cohen
Department of Pharmacology, Harvard Medical School

Abstract. A study was made of kinetics of hydrolysis by acetylcholinesterase and by hydroxide of acetate esters $RCH_2CH_2OCOCH_3$: R = $(CH_3)_3N^+$, $(CH_3)_3C$-, $(CH_3)_2NH^+$, $(CH_3)_2CH$-, $CH_3NH_2^+$-, CH_3CH_2-, NH_3^+-, CH_3-, H-, HO-, CH_3O-, Cl-, Br-, N≡C-. Comparative values of K_s indicate that positive charge of the β-substituent, R, makes little if any contribution to substrate binding. Acylation rate constants, k_2, and enzymic reactivities, k_2/K_s, are normalized for effects of β-substituents on intrinsic reactivity to hydroxide. A linear relationship is found between log of normalized enzymic reactivity and apparent molal volume of the β-substituent. Cogent evidence is not found for a negative charge in the "anionic" site, which is better considered a Trimethyl site. Enzymic reactivity is determined predominantly by precision of fit of the β-substituent in the Trimethyl site, the acetyl methyl in its Methyl site, and the ester grouping at the serine-OH.

The active site of acetylcholinesterase (E.C. 3.1.1.7) is generally accepted to have an anionic site,[1-5] complementary to the trimethylammonium group of acetylcholine, $(CH_3)_3N^+CH_2CH_2OCOCH_3$ I, and an esteratic site, containing serine[6] which is acetylated and deacetylated in the hydrolysis. The negative charge of the anionic site is considered to facilitate hydrolysis by binding and orienting cationic substrates[7]. The three methyl groups are important for the binding of cationic inhibitors[7] and for the binding and reactivity of cationic[8] and uncharged[9-12] substrates. Neohexyl acetate, $(CH_3)_3CCH_2CH_2OCOCH_3$ II, shows high reactivity, without positive charge. Doubt has been expressed[13] about a negative charge in the "anionic" site, but its presence remains widely accepted.[3 5,14] We wish to report that study of hydrolysis of a group of esters by acetylcholinesterase and by hydroxide reveals little if any evidence that interaction of a positive charge in a substrate with a negative charge in the "anionic" binding site substantially affects binding or enzymic reactivity.

The esters are ethyl acetates, $RCH_2CH_2OCOCH_3$: I R = $(CH_3)_3N^+$, II R = $(CH_3)_3C$-, III R = $(CH_3)_2N^+H$-, IV R = $(CH_3)_2CH$-, V R = $CH_3N^+H_2$-,

B. Pullman (Ed.): Catalysis in Chemistry and Biochemistry. Theory and Experiment. 91–98.

		10^{-2} k_{cat} s^{-1}	K_m mM	10^{-4} k_{cat}/K_m $M^{-1}s^{-1}$	$k_{(OH)}$ $M^{-1}s^{-1}$	10^{-2} $k_2(n)$ s^{-1}	10^{-4} $k_2(n)/$ $M^{-1}s^{-1}$
I	$(CH_3)_3N^+$-	160[a]	0.33[b]	4800	2.7	1100	4800
II	$(CH_3)_3C$-	31	2.6[c]	120	0.11	950	3100
III	$(CH_3)_2N^+H$-	48	1.1	440	2.7	67	470
IV	$(CH_3)_2CH$-	12	2.6	46	0.11	330	1200
V	$CH_3N^+H_2$-	22	20.	11	2.7	26	12
VI	CH_3CH_2-	13	12.	11	0.11	350	270
VII	N^+H_3-	19	24.	8	2.7	23	8
VIII	CH_3-	3	33.	0.8	0.11	67	20
IX	H-	4	220.	0.2	0.11	97	4
X	HO-	2	22.	0.7	0.46	10	4
XI	CH_3O-	8	5.	18.	0.38	64	130
XII	Cl-	15	32.	5.	0.31	150	43
XIII	Br-	10	10.	10.	0.42	68	65
XIV	N≡C-	11	17	6.	1.3	25	14

[a] $k_2 = 1100 \times 10^2\ s^{-1}$. [b] $K_s \cong 1.2$ mM. [c] $K_s = 3.1$ mM.

Table 1. Hydrolysis of $RCH_2CH_2OCOCH_3$ by Acetylcholinesteras pH 7.8, 25°C, 0.18 M NaCl

VI R = CH_3CH_2-, VII R = N^+H_3-, VIII R = CH_3, IX R = H-, X R = HO-, XI R = CH_3O-, XII R = Cl-, XIII R = Br-, XIV R = N≡C-. Enzyme was fro Worthington Biochemical and from Aldrich, and was assayed by hydrolysi of acetylcholine with $k_{cat} = 1.6 \times 10^4\ s^{-1}$.[5] Enzymic hydrolysis was studied at pH 7.8, 0.18 M NaCl, 25° C, with a Radiometer Titrator, Servograph Recorder and Autoburette. Initial linear zero order rates were obtained and corrected for non-enzymic hydrolysis. Slopes and intercepts of plots of 1/V vs. 1/S were obtained by least square analy Rate constants for hydrolysis by alkali, $k_{(OH)}$, were obtained at pH 10 or pH 7.8. The results are summarized in Table 1.

The value of k_2 for acylation of the enzyme by acetylcholine was taken as six[5,8] times the deacylation constant, k_3, and values of k_2 and k_3 for acetylcholine were calculated from the value of k_{cat}, above and equation 1[15]: $k_2 = 1.1 \times 10^5\ s^{-1}$, $k_3 = 1.9 \times 10^4\ s^{-1}$.

$$k_{cat} = k_2k_3/(k_2+k_3) \quad (1$$

Values of k_2 for the other substrates were calculated from their values of k_{cat}, the common value of k_3, and equation 1, and were similar to the values of k_{cat}. Normalized values of k_2, $k_2(n)$, are k_2 multiplied by the ratio of the bimolecular rate constant for hydrolysis of acetylcholine by hydroxide to that for hydrolysis of the substrate by hydroxide, $k_{(OH)}$. Normalized enzymic reactivities are $k_2(n)/K_s$

Values of the equilibrium binding constant, K_s, were calculated from equation 2[15] for all the substrates except acetylcholine, I, and are similar to values of K_m.

$$k_{cat}/K_m = k_2/K_s \qquad (2)$$

Because of the high value of k_2 for I, the steady state expression, equation 3, is better applied, and, for I, this

$$k_{cat}/K_m = k_2/(k_{-1}+ k_2)/k_1 \qquad (3)$$

leads to 2.3 m$\underline{M}$ = $(k_{-1}+ k_2)/k_1$. The initial enzyme-substrate association of I must have rate constant $k_1 \geq k_{cat}/K_m$, $k_1 \geq 4.8 \times 10^7$ $\underline{M}^{-1}s^{-1}$. If the value of k_1 is $\sim 10^8$ $\underline{M}^{-1}s^{-1}$, or if half of all associations lead to acylation, i.e. $k_2 \cong k_{-1}$, the value of K_s for I is $\sim$ 1.2 mM. This is more favorable than that for the uncharged analogue II by a factor of 2.5, corresponding to < 0.6 kcal/mol more favorable binding energy. Values of K_s are similar within each of the other charged and uncharged pairs, III and IV, V and VI, VII and VIII. Values of K_s for compounds XI-XIV, with uncharged polar β-substituents, are also no less favorable than for the charged analogues V and VII. Less favorable K_s of the unsubstituted ethyl acetate, IX, indicates that all the β-substituents may be involved in binding. Binding of substrates, not improved substantially by positive charge, gives no cogent evidence for a negative charge in the "anionic" site. This site, complementary to the $(CH_3)_3N^+$- and $(CH_3)_3C$- groups of I and II respectively would hold $\sim$ 5 molecules of water, which may be fully displaced by these large β-substituents and only partially displaced by the small ones. The N^+H_3- and CH_3- groups of VII and VIII may lead to similar binding because their nearest neighbors are water molecules both in solution and in the binding site. The large difference in solvation energy caused by three N-methyl groups in comparison of N^+H_4 and $(CH_3)_3N^+H$ ions, 12 kcal/mol,[16] may not be reflected in the small difference, 2 kcal/mol, in binding energies of VII and I, since the NH_3^+- need not be desolvated as it lies in the binding site.

Normalization of k_2 and k_2/K_s for intrinsic reactivity toward hydroxide has been applied to chymotrypsin,[17] which is also a serine enzyme, and to acetylcholinesterase.[18] A $\rho^*\sigma^*$ relationship, $\rho^* \sim 3$, similar to that in base catalyzed hydrolysis, has been reported between the polar σ^* constants of substituents in the alkoxyl group of a limited set of uncharged acetate esters and their reactivity toward acetylcholinesterase.[12] Normalization by use of values of $k_{(OH)}$ appears appropriate. This procedure raises $k_2(n)$ and $k_2(n)/K_s$ for the uncharged substrates to values similar to or greater than those for the analogous positively charged compounds, with the sole exception of the strongly hydrophilic β-hydroxyethyl acetate, X. The higher intrinsic hydrolytic reactivity of the positively charged substrates accounts for their higher values of k_{cat} and k_2. Thus, reactivity in acylation, like binding, also does not call for presence of a negative charge in the "anionic" site. Further, positive charge is absent in the deacetylation which is very rapid, almost as rapid as acetylation by the natural substrate I.[5,8] We conclude that the positive charge is not a specificity endowing feature, and the "anionic" binding site is better considered an uncharged Trimethyl site.

A plot of log $(k_2(n)/K_s)$ against hydrophobicity[19] of the β-substituent is linear for the seven β-alkyl, β-halo and β-H compounds, Fig. 1 line A. The less hydrophobic substituents, HO-, CH_3O- and -C≡N, are much more reactive than their hydrophobicity would suggest, and this criterion may also not be applied to the positively charged compounds, I, III, V and VII. A progressive increase in reactivity with increasing size of the β-substituent is observed in both sets of compounds, Fig. 1. A fair linear relationship is obtained, Fig. 2, between the apparent molal volumes, $\overline{V}°_{25}$, of (the substrates or) the β-substituents and log $(k_2(n)/K_s)$ values for all the substrates -- positively charged and uncharged, with β-alkyl and polar β-substituents, both hydrophilic and hydrophobic. Values of $\overline{V}°$ have also been related to the activity of quaternary ammonium salts toward acetylcholine receptors.[20]

We propose that reactivity to acetylcholinesterase of ethyl acetates related to I is predominantly determined by: (i) the intrinsic reactivity of the ester in hydrolysis by hydroxide, which is increased by positively charged and electronegative β-substituents; and (ii) the shape and volume as they affect the capacity of the β-substituents to fill the Trimethyl site.

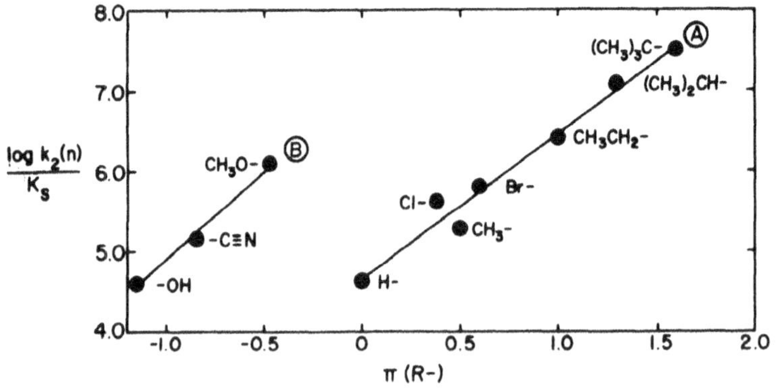

Figure 1. Log normalized enzymic reactivity of $RCH_2CH_2OCOCH_3$, log $k_2(n)/K_s$, vs. hydrophobicity, π(R-).

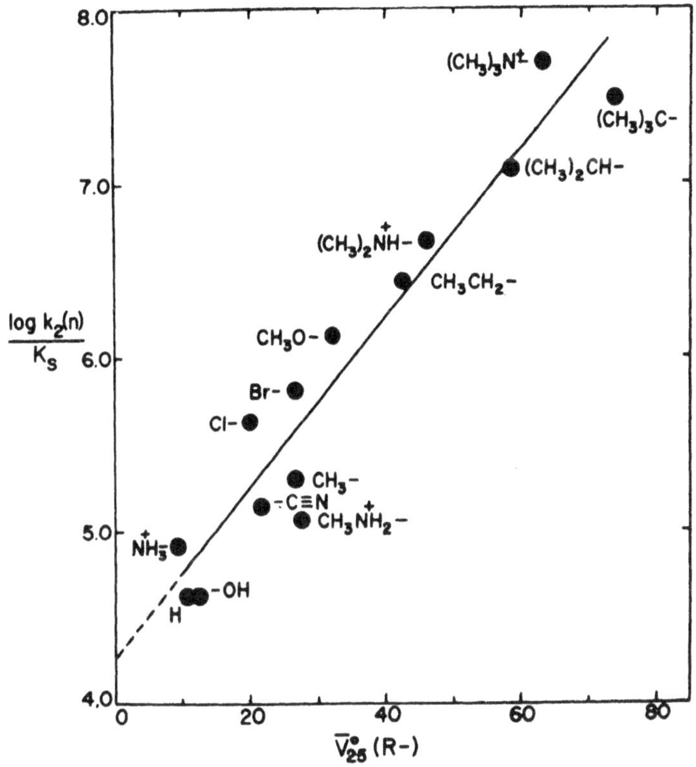

Figure 2. Log normalized enzymic reactivity of $RCH_2CH_2OCOCH_3$, $\log k_2(n)/K_s$, vs. apparent molal volume, $\overline{V}^{\circ}_{25}(R-)$.

The intercept of the least square line, Figure 2, 4.3, corresponds to the log of the normalized reactivity that may be ascribed to the unsubstituted $-CH_2CH_2OCOCH_3$ moiety, and reflects the importance to specificity of the acetyl group. The deviations from the line indicate that hydrophobicity may increase and hydrophilicity decrease rates incrementally.

Precise fit of the β-substituent in the Trimethyl site and the acetyl methyl in the Methyl site, Figure 3, may restrict motion of the substrate in the active site and direct the ester group properly -- alkoxyl-O toward H and carbonyl-C toward O of the serine-O-H. Precise fit displaces water from the enzyme-substrate interface, and part of the binding energy may be used to this end. Desolvation of the ester raises its reactivity, and hydrogen-bonding at the alkoxyl-O activates it for attack at the carbonyl.[21] After the alkoxyl is replaced by the serine, this covalent bond and the fit at the Methyl site may provide

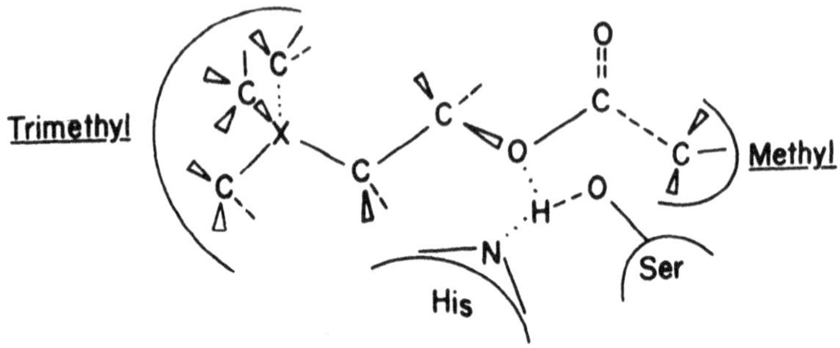

Figure 3. Substrate and the Trimethyl and Methyl subsites. X = N^+, Acetylcholine I. X = C, Neohexyl acetate II.

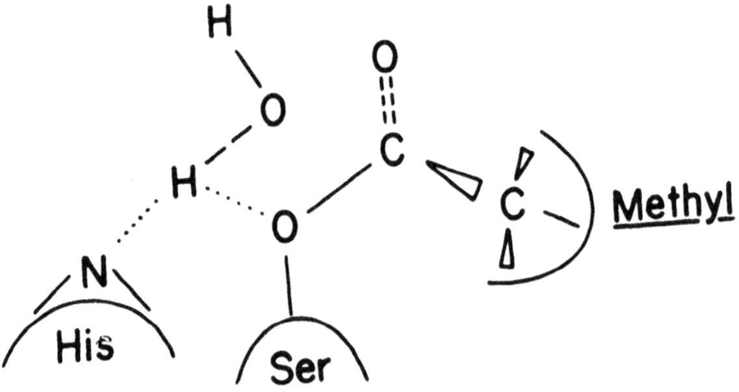

Figure 4. Methyl subsite, O-Acetyl Acetylcholinesterase.

precise orientation for protonation of the serine-O and attack by water, Figure 4. In acetyl acetylcholinesterase the acetyl group provides specificity and is hydrolyzed off rapidly. In hydrolysis by chymotrypsin an aromatic or other specific acyl group provides precise orientation in both acylation and deacylation, but acetylchymotrypsin is hydrolyzed very slowly.[17a] Non-enzymic ester hydrolysis has low activation energy and is entropy controlled,[22] $\Delta S^{\pm} \sim -30$ cal/deg. Hydrolysis of I by acetylcholinesterase has $\Delta H^{\pm}$ similar to non-enzymic hydrolysis, but much more favorable $\Delta S^{\pm}$, > +16 cal/deg, consistent with the proposal that precise fit and juxtaposition of reacting groups lead to high reactivity.

These conclusions are based on the premise that the charged and uncharged substrates are hydrolyzed at the same active site and that Compounds I and II occupy this site in the same way. This view is supported by work in progress in which hydrolysis of Compounds I and II by acetylcholinesterase responds in quite the same way to commonly used and newly designed reversible inhibitors. The view that the binding site for the β-substituent is uncharged is leading to the development of novel reversible and irreversible inhibitors and labelling agents which possess a tertiary-butyl group. These compounds may not be affected by peripheral anionic groups and may allow more specific study of the active site. Further, the positive charge of acetylcholine, brought into an uncharged site by the three N-methyl groups, may not neutralize a negative charge or displace a positive gegenion, but may increase the positive charge. The implications of this for the mode of action of acetylcholine with the enzyme, and possibly also with receptors are being explored.

Acknowledgment: This work was initiated under grant GM 04584 from the National Institute of Health and supported by the Rowland Foundation.

References

1. Adams, D.H., and Whittaker, V.P.: 1950, Biochim. Biophys. Acta 4, pp. 543-558.
2. Wilson, I.B., and Bergmann, F.: 1950, J. Biol. Chem. 185, pp. 479-489.
3. Michelson, M.J., and Zeimal, E.V., "Acetylcholine" Pergamon Press, New York: 1973, p. 13.
4. Froede, H.C., and Wilson, I.B. in "The Enzymes" 3rd ed., P.B. Boyer, Ed., Academic Press, New York, N.Y.: 1972, Ch. 5, pp. 87-114.
5. Rosenberry, T.L., "Advances in Enzymology and Related Areas of Molecular Biology", A. Meister, Ed.: 1975, Vol. 43, pp. 103-218.
6. Krupka, R.M.: 1964, Can. J. Biochem. 42, pp. 677-693.
7. Wilson, I.B.: 1952, J. Biol. Chem. 197, pp. 215-225.
8. Wilson, I.B., and Cabib, E.: 1956, J. Am. Chem. Soc. 78, pp. 202-207.
9. Adams, D.H.: 1949, Biochim. Biophys. Acta 3, pp. 1-14.
10. Adams, D.H., and Whittaker, V.P.: 1949, Biochim. Biophys. Acta 3, pp. 358-366.
11. Mounter, L.A., and Cheatham, R.M.: 1963, Enzymologia 25, pp. 215-214.
12. Järv, J., Kesvatera, T., and Aaviksaar, A.: 1976, Eur. J. Biochem. 67, pp. 315-322.
13. O'Brien, R.D., "Drug Design" E.J. Ariens, Ed., Academic Press, New York and London: 1971, Vol. II, Ch. 3, pp. 161-212.
14. Barnett, P., and Rosenberry, T.L.: 1977, J. Biol. Chem. 252, pp. 7200-7206.
15. Zerner, B., and Bender, M.L.: 1964, J. Am. Chem. Soc. 86, pp. 3669-3674.
16. Arnett, E.M., Jones, III, F.M., Taagepera, M., Henderson, W.G., Beauchamp, J.L., Holtz, D., and Taft, R.W.: 1972, J. Am. Chem. Soc. 94, pp. 4724-4726.
17. a. Bender, M.L., Kezdy, F.J., and Gunter, C.R.: 1964, J. Am.

Chem. Soc. 86, pp. 3714-3721; b. Ingles, D.W., and Knowles, J.R.: 1967, Biochem. J. 104, pp. 369-377.
18. Schowen, K.B., Smissman, E.E., and Stephen, W.F. Jr.: 1975, J. Med. Chem. 18, pp. 292-300.
19. Leo, A., Hansch, C., and Elkins, D.: 1971, Chem. Rev. 71, pp. 525-616.
20. Barlow, R.B., Lowe, B.M., Pearson, J.D.M., Rendall, H.M., and Thompson, G.M.: 1971, Molec. Pharmac. 7, pp. 357-366.
21. Cohen, S.G., Vaidya, V.M., and Schultz, R.M.: 1970, Proc. Natl. Acad. Sci. 66, pp. 249-256.
22. Humphreys, H.N., and Hammett, L.P.: 1956, J. Am. Chem. Soc. 78, pp. 521-524.

CREATINE KINASE: AN ASYMMETRIC DIMER OPERATING BY A "ONE-AT-A-TIME" MECHANISM

Yair Degani and Chemda Degani
Departments of Organic Chemistry and Isotope Research
The Weizmann Institute of Science, Rehovot, Israel

INTRODUCTION

It has become increasingly evident that most intracellular enzymes exist and act as multisubunit structures. Yet, despite the great progress that has been made in determining quarternary structure for multisubunit enzymes, the functional significance of subunit assembly and intersubunit interaction in oligomeric enzymes has remained unclear. In virtually all of the oligomeric enzymes whose quarternary structure has been determined by X-ray crystallography, the protomers were found to be arranged in a strictly symmetrical manner, as was expected by Monod et al. [1]. In one case, dimeric yeast hexokinase [2,3], asymmetry in the subunit arrangement has been observed in the crystal state. Thus far, however, there has been no evidence that identical protomers can associate in an asymmetric manner in solution.

In this paper we present evidence that the subunits of the dimeric enzyme creatine kinase from rabbit muscle are associated asymmetrically in solution. The enzyme (ATP: creatine phosphotransferase) has a molecular weight of 82,600 [4]. It contains two substrate-binding sites and two reactive thiol groups located in proximity to the active sites [5]. In denaturing media the enzyme dissociates into two subunits of identical molecular weight. Peptide mapping in a variety of different systems points to an identity of the two polypeptide chains. In particular, the two reactive thiol groups were shown to be identical in that they both formed part of a unique 25-residue tryptic peptide sequence [6,7]. In the following we report that these two sequentially identical thiol groups react non-identically with the reagent 2-nitro-5-thiocyanobenzoic acid (NTCB) to form a catalytically inactive enzyme derivative, differentially modified at its two subunits. By demonstrating the formation of the highly active isomer of this derivative, we provide conclusive evidence for preexisting asymmetry in the subunit arrangement of the enzyme. From the properties of these isomers we conclude that the subunits of the enzyme are associated heterologously and that intersubunit conformational cooperativity is mechanistically essential in the catalytic activity of the enzyme. Taken together with

B. Pullman (Ed.): Catalysis in Chemistry and Biochemistry. Theory and Experiment. 99–111.

earlier reported observations, our results lead us to suggest that creatine kinase operates by a "one-at-a-time" mechanism.

RESULTS AND DISCUSSION

Preexisting Asymmetry in Creatine Kinase.

The thiol groups of peptides and proteins can be converted quantitatively to their S-cyano derivatives by the cyanylating reagent NTCB, with release of thionitrobenzoate (TNB) [8,9].

$$\text{R-SH} + \text{ArSC*N} \longrightarrow \text{R-SC*N} + \text{ArS}^-$$

Under denaturing conditions, all the titratable thiol groups of rabbit muscle creatine kinase (CK) reacted with NTCB in the normal way, undergoing essentially quantitative cyanylation. In the native state, however, the two reactive thiol groups in the dimeric enzyme reacted with [^{14}C]NTCB in a non-identical manner. While one thiol group was cyanylated (with equivalent release of TNB), the other thiol abnormally formed a TNB mixed disulfide group. The isolated derivative, designated S-CN-S'-TNB-CK(I), was catalytically inactive. When this derivative was treated with 0.11 M[^{14}C]KCN at pH 9.5, 1.0 mole TNB per mole enzyme was displaced, with a second order rate constant of 3.0 M^{-1} $min.^{-1}$. The resulting dicyano enzyme, designated S,S'-di-CN-CK, was found to possess 75% of the original enzymic activity. These results mean that the catalytically inactive product formed by NTCB does not contain any dicyano enzyme. Hence the cyanylation and arylthiolation occurring during NTCB modification of the enzyme are not random, but subunit-selective reactions reflecting a differential chemical behavior of the two subunits.

The 75% active S,S'-di-CN-CK is also formed by total cyanolysis of the inactive derivative S,S'-di-TNB-CK, produced by a previous reaction of the enzyme with 5,5'-dithio-bis-(2-nitrobenzoic acid) (DTNB). Fig.1 presents the time course of this reaction. During the reactivating cyanolysis, one S-TNB group is displaced much more rapidly than the other. Analysis of the semi-logarithmic plot for the rate of TNB release showed that 0.85±0.05 S-TNB groups per mole enzyme were involved in the fast phase of the reaction ($k=7.8\ M^{-1}min^{-1}$), and 0.9±0.05 S-TNB groups in the slow phase ($k=0.3\ M^{-1}\ min^{-1}$). The regeneration of enzymic activity coincided solely with the faster of the two cyanolysis reactions. The slower reaction did not contribute further to the maximal activity of 75% possessed by the half-cyanolysis product of S,S'-di-TNB-CK. This monocyano-mono-TNB derivative of the enzyme is designated S-TNB-S'-CN-CK(II). A summary of the reactions described above is presented in Fig. 2.

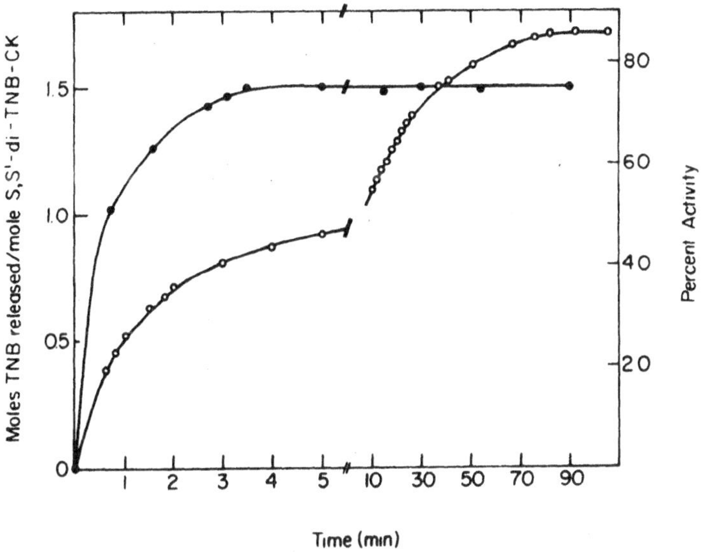

Figure 1. Cyanolysis of S,S'-di-TNB-CK. The modified enzyme (containing 1.8 S-TNB groups per mole) was treated with 0.11 M KCN in 0.125 M potassium phosphate buffer at pH 9.5. The reaction was followed by the release of TNB (o—o) and by the regeneration of enzymic activity (o—o).

Thus, unlike the inactive S-CN-S'-TNB-CK(I) formed by NTCB, the half-cyanolysis product S-TNB-S'-CN-CK(II) is highly active. Furthermore, its cyanolysis rate constant is ten times smaller than that of S-CN-S'-TNB-CK(I) under the same conditions. Therefore, these two derivatives of creatine kinase, possessing such different enzymic and chemical properties, cannot be identical. Since both derivatives have the same composition, and both are derived by selective chemical modifications of the same pair of sequentially identical cysteinyl residues (as indicated by the fact that the reactions with NTCB and DTNB are mutually exclusive), they must be isomers. In isomer I the cyano group is linked to subunit A and the TNB group to subunit B, whereas in isomer II the cyano group is linked to subunit B, and the TNB group to subunit A. Hence, the subunits of the enzyme are inherently and initially non-equivalent. Initial non-equivalence of covalently identical promoters can only arise from preexistant asymmetry in the dimeric structure.

Heterologous Association and Intersubunit Conformational Cooperativity.

Asymmetric arrangement of identical subunits has been previously reported for hexokinase B from yeast [3,10]. In the crystal state, the subunits of this dimeric enzyme are associated heterologously, being

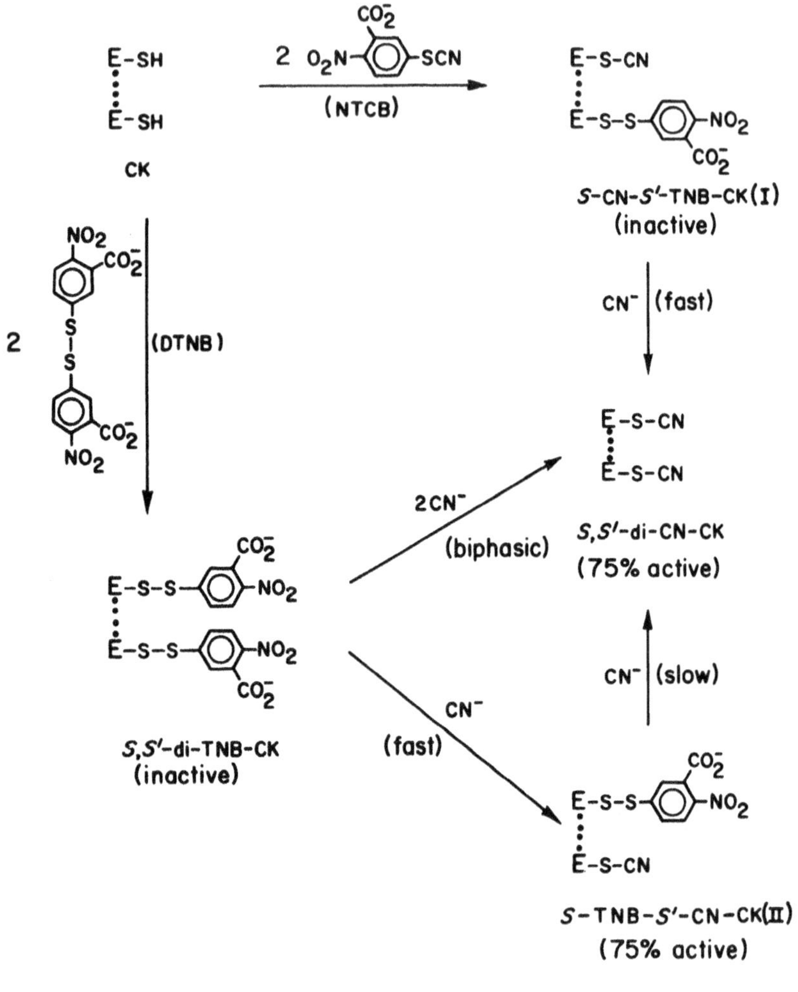

Figure 2. Derivatizations of creatine kinase. "E" designates enzyme subunit.

related by a 156° rotation and a 13.8 Å translation along a molecular screw axis. Consequently, the two subunits are not in identical environments. Our evidence for preexisting asymmetry in CK strongly suggests that a similar situation exists in this enzyme, in solution. Indeed, a heterologous structure of the dimer could well account for the nonidentical chemical environments existing at the two subunits, as sensed by NTCB. Furthermore, as shown below, a heterologous association of the subunits could also provide the only plausible explanation for the finding that the bulky and charged TNB group, when present at subunit A, has virtually no effect on the catalytic activity of the enzyme beyond a small inhibitory effect caused by the tiny and unchanged cyano group at subunit B, but when present at subunit B, the single TNB block completely prevents catalysis in both subunits. The fact that the 75% residual activity of S-TNB-S'-CN-CK(II) is not increased whatsoever by displacing its bulky TNB group with the small cyano group, clearly shows that the catalytic efficiency of subunit A is not significantly affected by the blocking of its own thiol group, even with a TNB group. Therefore, the inactivity of this cyanylated subunit A in the isomer S-CN-S'-TNB-CK(I), indicates that the catalytic reaction at subunit A requires a conformational change at the adjacent subunit, which in this derivative is prevented by the TNB block at thiol B. Thus, the subunits in the dimer are incapable of acting independently. This conclusion is supported by previous evidence of cooperative interactions in nucleotide binding under conditions simulating the transition-state complex of the enzyme, as first reported by McLaughlin [11], and further referred to below. The insensitivity of the dimer to S-substitution at subunit A can be explained by the following analysis of the structural implications of intersubunit conformational cooperativity in a heterologous dimer.

Conformational cooperativity of the subunits requires the existence of an intersubunit "communication line" which connects the two catalytic sites and transmits induced conformational changes from one site to the other during catalysis. In analogy with the term "binding set" as defined by Monod et al. [1], we shall use here a similar term, "communication set", to mean, "the conformationally organized collection of all groups and residues of one enzyme-subunit which are dynamically involved in its communication with one other subunit during catalysis". (A "communication set" could be a polypeptide chain segment extending between a substrate-binding site and a subunit contact site). Together, two communication sets form the intersubunit communication line which mediates conformational interactions between the catalytic sites. In a heterologous dimer, the two binding sets are, by definition, non-identical, and therefore, the communication sets must also be non-identical. Furthermore, since the separate subunits are individually identical, each subunit of a heterologous dimer must possess two potential communication sets, one of which remains unconnected (both to the active site and to an opposite communication set) after dimerization, and hence non-operative during catalysis. This is illustrated in the schematic model of native CK presented in

Fig. 3, where the operative communication sets are designated a and b and the corresponding non-functional communication sets are designated a' and b'.

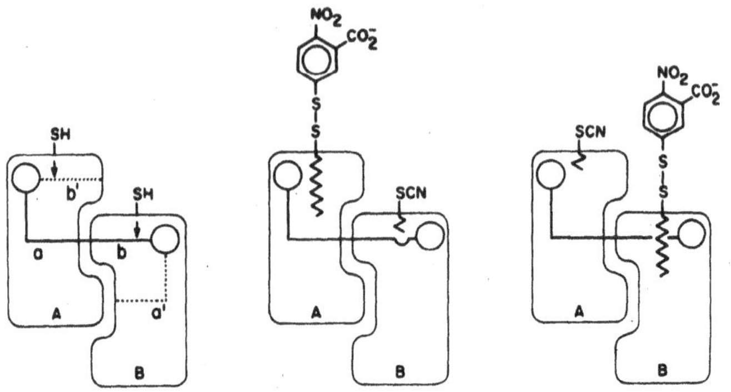

Figure 3. Schematic representation of models describing the heterologously-associated dimer of creatine kinase and its two isomeric derivatives. The two subunits, A and B, are individually identical but non-equivalent in the dimer. The two identical catalytic sites (circles) are connected by an inter-subunit communication line, consisting of two different "communication sets", a and b. The two non-operative communication sets, a' and b' (dotted lines) are shown only in the CK model. As indicated by the arrows in the CK model, the cysteine in subunit B is part of the communication set b, whereas the cysteine in subunit A is correspondingly part of the non-functional communication set b'. The steric requirements of the S-substituents are symbolized by the zig-zag lines. As a consequence of the asymmetry of the inter-subunit communication line, S-substitution affects intersubunit communication during catalysis only at subunit B.

Thus, it is the asymmetry of the path of communication between the active sites in a heterologously-bound dimer, which can well account for the striking difference in the sensitivity of creatine kinase to S-substitution at two subunits. As indicated by our results, the cysteine at subunit B appears to be part of the communication set b (Fig. 3). Under the conformational constraints existing at this cysteine during catalysis, its thiol can barely accommodate the tiny cyano group, without greatly affecting the efficiency of the intersubunit conformation interactions during catalysis. Blocking thiol B with the bulky TNB group completely prevents intersubunit communication and abolishes catalysis at both subunits. At subunit A, the cysteine must correspondingly belong to the non-functional communication set b', which cannot be involved in intersubunit conformational interactions in the heterologous dimer. It is therefore sufficiently free to carry the bulky TNB group without interfering with the propagation of the intersubunit conformational changes during catalysis.

To gain further insight into the role of intersubunit conformational changes in the catalytic activity of creatine kinase, the ligand binding properties of S,S'-di-CN-CK and S-CN-S'-TNB-CK have been studied by means of difference spectroscopy. In the presence of the ligand mixture MgADP plus creatine plus nitrate ions, the native enzyme is known to form a stable quarternary complex E•MgADP•NO_3•creatine. This complex is considered to be a transition-state analog (t.s.a.), in which the planar trigonal nitrate ion simulates and occupies the vacant site of the planar metaphosphate ion intermediate which migrates between MgADP and creatine during the enzyme-catalysed transphosphorylation reaction [12-18] (Fig. 4).

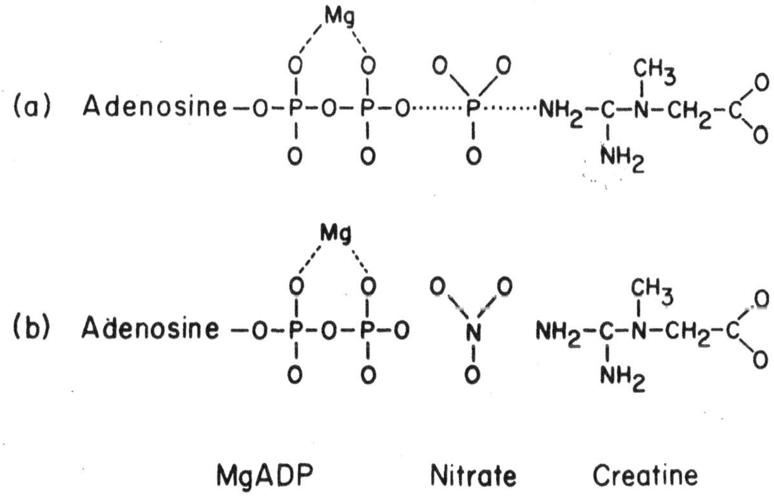

Figure 4. Diagramatic representation of (a) the transition state complex of the CK reaction and (b) the postulated t.s.a. quarternary complex.

Indeed, this complex possesses all the properties expected of a t.s.a., in accordance with the "strain" and "transition-state-affinity" theories of enzymic catalysis [19,20]. These properties include (1) exceptionally tight binding; (2) strong competitive inhibition and (3) effective protection of the active site region against chemical modifications.

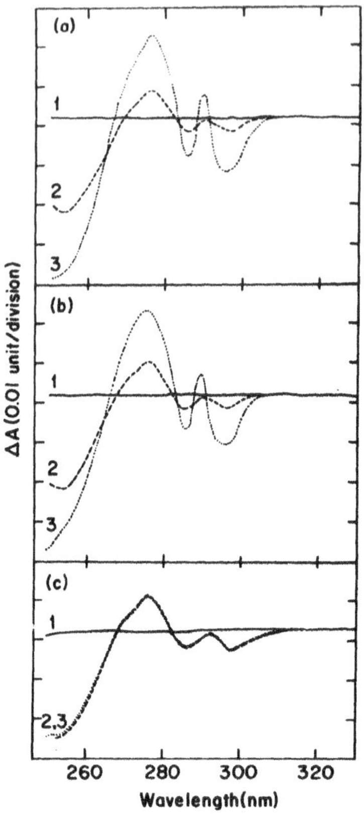

Figure 5. Difference-spectra measurements on creatine kinase and its S-substituted derivatives. (a) Native enzyme. (1)——, baseline; (2)---, interaction of 0.014 mM CK with 0.09 mM MgADP and 30 mM creatine in 0.05 M Tris-acetate buffer, pH 8.5; (3)..., effect of 50 mM $NaNO_3$ on the difference spectrum. (b) S,S'-di-CN-CK. The same experiment was carried out with 0.014 mM of the 75% active dicyano enzyme. (3) S-CN-S'-TNB-CK. The same experiment was carried out with 0.014 mM of the inactive S-CN-S'-TNB-CK.

Furthermore, the incorporation of NO_3^- into the ternary E·MgADP·creatine complex induces prominent changes of conformation in the entire active site region, consistently revealed by a large variety of spectrophotometric and magnetic resonance effects [12-18]. As indicated by the typical difference spectrum for the E·MgADP·creatine produced by S-CN-S'TNB-CK (Fig. 5c), this inactive enzyme derivative is capable of forming a normal ternary complex at both of its subunits, but remains inert to the subsequent addition of nitrate ions, indicating failure to form the quarternary complex at either subunit. However, after displacing its single TNB block at subunit B by the small cyano group, the resulting 75% active S,S'-di-CN-CK regains the enzyme's normal ability to form the quarternary t.s.a. complex at each of the subunits (compare Figs. 5a and 5b). Therefore, the failure of the complexed S-CN-S'-TNB-CK to incorporate NO_3^- indicates that the active-site conformation at its cyanylated subunit A cannot be sufficiently stretched to accommodate the nitrate ion between the bound MgADP and creatine, because this conformational state is unattainable in the dimer without an intersubunit conformational change which in this derivative is prevented by the S-TNB block near the active site of subunit B. These results strongly suggest that it is the formation of the activated transition-state complex during catalysis which requires intersubunit conformational cooperativity in the mechanism of action of this enzyme. Our results corroborate and explain earlier observations indicating that the binding of MgADP to the native enzyme exhibited negative cooperativity (or two affinity constants) only in the presence of creatine plus nitrate, but not in the absence of nitrate [11,21,22]. Since the formation of the t.s.a. complex (and correspondingly, catalysis) is unattainable at either subunit without an intersubunit conformational rearrangement, the active-site conformations in the two subunits cannot be strained simultaneously to optimally bind the quarternary complex. We propose, therefore, that CK operates catalytically by a "one-at-a-time" mechanism defined as follows: At any instant of time during catalysis, irrespective of substrate concentration level and degree of site occupancy, not more than one subunit per oligomer forms - and is conformationally capable of forming - a productive transition state. In other words, we propose that a single molecular turnover in the catalytic cycle of a saturated CK molecule is rate-limited by the propagation of the intersubunit conformational change which permits the "second" site to reach the productive transition state after the "first" site had formed product(s).

Besides the half-of-the-sites inactivation and the negative cooperativity in the t.s.a. binding, the proposed "one-at-a-time" mechanism is also consistent with earlier reports [23,24] indicating that CK from rabbit muscle does not obey Michaelis-Menten kinetics. According to this mechanism, the Michaelis-Menten equation would apply only for single-site turnovers occurring at half-saturated CK molecules, where the formation of the transition state is the sole rate limiting step. The "one-at-a-time" mechanism can also account for substrate inhibition by high concentration of substrates. Note that simultaneous binding of the productive substrate complex at the two active sites would be catalytically unfavorable, inasmuch as mutual intersubunit conformational interference and anticooperativity in attaining the productive transition state at either subunit, should retard the turnover at the "first" site, and thus decrease the rate of the molecular turnover in this molecule. Given that the average turnover number per site for the forward CK reaction at pH 8 under the concentration conditions of the quenched-flow experiments described by Engelborghs et al. [25] and Barman et al. [26], is of the order of

30-40 sec^{-1} (corresponding to an average lifetime of ∿30 msec for a single turnover per site), the actual magnitude of the intersubunit interference effect may well depend on the exact timing of the co-saturation in a given CK molecule. Statistical analysis indicates that the magnitude of this interference effect would be of minor functional significance during continuous steady-state catalysis at moderate substrate concentration levels, maintained at their likely normal physiological values. However, at increasingly high concentration levels, the intersubunit interference effect would become important. Indeed, Morrison and Cleland [27] who investigated the kinetics of the reaction catalyzed by CK over a wide range of substrate concentrations, consistently observed a gradual depression in the initial rate of both steady-state catalysis and equilibrium isotope exchange, with increase of substrate concentrations in the range 5-30 mM. Although these authors suggested that the substrate inhibition was not due to the substrate itself but to the presence of NaCl which was introduced when MgATP was formed from $MgCl_2$ and the sodium salt of ATP, their explanation seems rather unlikely considering that all the kinetic measurements in both the initial velocity and equilibrium isotope exchange studies were carried out over a background of 100 mM chloride ions introduced into all reaction mixtures in the form of triethanolamine•HCl buffer (pH 8).

It should be noted that neither the negative cooperativity in the binding of the t.s.a. complex, nor the postulated "one-at-a-time" mechanism for CK, are related to the preexisting asymmetry of the subunit arrangement of the enzyme, since notwithstanding the <u>structural</u> nonequivalence of the subunits in the heterologous dimer (Fig. 3), the two subunits are <u>functionally</u> equivalent. In other words, subunit A and B in the enzyme population have equal probability of binding the t.s.a. complex tightly or loosely, or of being "first" or "second" site to form the productive transition state during catalysis in a saturated CK molecule. The catalytic equivalence of the two subunits has been demonstrated by the recent study of Bickerstaff and Price [28] on the properties of matrix-bound dimer and monomer derivatives of immobilized CK from rabbit muscle. As reported by these authors, the matrix-bound monomer possessed essentially the same specific activity (in μmol/min per mg protein) as the matrix bound native dimer and the matrix-bound reassociated dimer (under non-saturating conditions).

Lazdunski [29,30] has proposed a different type of mechanism to account for half-of-the-sites reactivity and negative cooperativity in ligand binding by oligomeric enzymes. In contrast to the present "one-at-a-time" proposal, which assumes <u>conformational</u> (but not catalytic or thermodynamic) interdependence of the subunits in catalysis and postulates intersubunit catalytic anticooperativity, Lazdunski's "flip-flop" model postulates <u>catalytic</u> interdependence and <u>catalytic cooperativity</u> of the subunits to enhance catalytic efficiency through intersubunit thermodynamic coupling between successive steps of the catalytic cycle, (e.g. substrate binding at one subunit promoting product formation or desorption at the other), and implies pure Michaelian [29] or pseudo-Michaelian [31] steady-state kinetics. Note that the steady-state cycle of a flip-flop-type mechanism does not contain the free enzyme, and implies, therefore, continuous total re-saturation of the entire enzyme population as a prerequisite for effective steady-state product formation or product release at any level of substrate concentration. Note also, that while the "flip-flop" model can plausibly account for negative cooperativity in substrate binding by dimeric enzymes, it cannot account for negative cooperativity in the binding of a t.s.a., or for negative cooperativity in substrate binding by oligomeric enzymes other than dimers (involving more than two different binding constants). Furthermore, the findings of Bickerstaff and Price [28] that dimerization is not required for the expression of the catalytic activity of the monomer argues against intersubunit catalytic cooperativity in creatine kinase, but is in harmony with conformational interdependence and catalytic anticooperativity of the subunits.

CONCLUSION

The present study has demonstrated two fundamental features of creatine kinase: (1) The subunits of the enzyme are associated asymmetrically. (2) The catalytic activity of the dimer requires intersubunit conformational cooperativity, consistent with a "one-at-a-time" mechanism. Re-examination of the literature reveals a large body of evidence to suggest that many other oligomeric enzymes operate by a "one-at-a-time" mechanism. The case of creatine kinase may end the controversy over the question whether negative cooperativity in ligand binding and half-of-the-sites reactivity of catalytic sites are generated by ligand-induced or preexisting subunit asymmetry [32,33]. These widespread phenomena are characteristic features of the "one-at-a-time" mechanism which appears to operate in oligomeric enzymes, independent of the mode of association of the subunits.

REFERENCES

1. Monod, J., Wyman, J., and Changeux, J.P. (1965) J. Mol. Biol. 12, 88-118.

2. Steitz, T.A., Fletterick, R.J., and Hwang, K.J. (1973) J. Mol. Biol. 78, 551-561.

3. Anderson, W.F., Fletterick, R.J., and Steitz, T.A. (1974) J. Mol. Biol. 86, 261-269.

4. Watts, D.C. (1973) Enzymes, 3rd Ed. 8, 383-455.

5. Cohn, M., Diefenbach, H. and Taylor, J.S. (1971) J. Biol. Chem. 246, 6037-6042.

6. Thomson, A.R., Eveleigh, J.W., and Miles, B.J. (1964) Nature (London) 203, 267-269.

7. Atherton, R.S., Laws, J.F., Miles, B.J. and Thomson, A.R. (1970) Biochem. J. 120, 589-600.

8. Degani, Y., Neumann, H., and Patchornik, A. (1970) J. Amer. Chem. Soc. 92, 6969-6971.

9. Degani, Y., and Patchornik, A. (1974) Biochemistry 13, 1-11.

10. Steitz, T.A., Fletterick, R.J., Anderson, W.F., and Anderson, C.M. (1976) J. Mol. Biol. 104, 197-222.

11. McLaughlin, A.C. (1974) J. Biol. Chem. 249, 1445-1452.

12. Milner-White, E.J., and Watts, D.C. (1971) Biochem. J. 122, 727-740.

13. McLaughlin, A.C., Cohn, M., and Kenyon, G.L. (1972) J. Biol. Chem. 247, 4382-4388.

14. Reed, G.H., and Cohn, M. (1972) J. Biol. Chem. 247, 3073-3081.

15. Schimerlik, M.I., and Cleland, W.W. (1973) J. Biol. Chem. 248, 8418-8423.

16. Focant, B., and Watts, D.C. (1973) Biochem. J. 135, 265-276.

17. Milner-White, E.J. and Kelley, I.D. (1976) Biochem. J. 157, 23-31.

18. Reed, G.H., Barlow, C.H., and Burns, R.A., Jr. (1978) J. Biol. Chem. 253, 4153-4158.

19. Jencks, W.P. (1969), "Catalysis in Chemistry and Enzymology", Chapter 5, McGraw-Hill, New York.

20. Wolfenden, R. (1976) Annu. Rev. Biophys. Bioeng. 5, 271-306.

21. Price, N.C., and Hunter, M.G. (1976) Biochim. Biophys. Acta 445, 364-378.

22. Milner-White, E.J., and Rycroft, D.S. (1977) Biochem. J. 167, 827-829.

23. Chetverikova, E.P., Krinskaya, A.V. Rozanova, N.A., Rybina, V.V., and Alievskaya, L.L. (1970) Biochemistry (USSR) 35, 827-833.

24. Chetverikova, E.P., and Rozanova, N.A. (1977) Biochemistry (USSR) 42, 369-375.

25. Engelborghs, Y., Marsh, A., and Gutfreund, H. (1975), Biochem J. 151, 47-50.

26. Barman, E., Travers, F., Bertrand, R., and Roseau, G. (1978) Eur. J. Biochem. 89, 243-249.

27. Morrison, J.F. and Cleland, W.W. (1966) J. Biol. Chem. 241, 673-683.

28. Bickerstaff, G.F., and Price, N.C. (1978) Biochem. J. 173, 85-93.

29. Lazdunski, M., Petitclerc, C., Chappelet, D., and Lazdunski, C. (1971) Eur. J. Biochem. 20, 124-139.

30. Lazdunski, M. (1974) Prog. Bioorg. Chem. 3, 81-140.

31. Gache, C. (1974) FEBS Lett. 49, 5-9.

32. Levitzki, A., and Koshland, D.E., Jr. (1976) Curr. Top. Cell. Regul. 10, 1-40.

33. Seydoux, F., Malhorta, O.P. and Bernhard, S.A. (1974) CRC Crit. Revs. Biochem. 2, 227-257.

STUDIES ON THE MECHANISM OF CATALYSIS BY THIAMIN: PROGRESS AND PROBLEMS

Henry Z. Sable and Debra J. Meschke
Case Western Reserve University, Cleveland, Ohio USA

ABSTRACT

The mechanism of catalysis by thiamin is discussed with respect to the function of the NH_2 group. Hypotheses that assume that the NH_2 group behaves as an intramolecular general base are untenable because such amino groups are not basic. On the contrary, many lines of evidence show that the NH_2 groups of aminopurines and aminopyrimidines should be considered as Lewis acids. The evidence includes the following: the π-electronic charge density is low; in H-bond formation these NH_2 groups are H-donors and never H-acceptors; rate constants for specific acid- and specific base-catalyzed exchange of the NH_2 protons with solvent water are $10^6 - 10^8$ and 10^2 times smaller, respectively, than those for the corresponding diffusion-controlled reactions. A catalytic mechanism involving a Lewis acid is proposed. The exchange of the thiazolium (C-2) proton with solvent has been reinvestigated by two NMR methods: (a) disappearance of the signal of (C-2)H in D_2O due to H—D exchange; (b) broadening of the (C-2)H signal in H_2O solution due to H—H exchange. Rates calculated from the latter method are approximately two orders of magnitude larger than those from H--D exchange.

INTRODUCTION

Thiamin was one of the first vitamins to be identified and its physiological role, as "cocarboxylase", also was recognized early, but many incorrect theories of its mechanism of catalysis were proposed before Breslow (1) discovered that the proton attached to C-2 of the thiazolium ring exchanged rapidly with D_2O. It had been known that both thiamin and cyanide could catalyze acyloin condensations; Breslow recognized the analogy between CN^- and the thiamin C-2 carbanion and adapted an earlier mechanism for catalysis of the benzoin condensation by CN^- to the acyloin condensation (2,3). (See (4) for a recent, detailed discussion of the mechanism). Breslow proposed that the condensation would proceed through an intermediate in which an

B. Pullman (Ed.): Catalysis in Chemistry and Biochemistry. Theory and Experiment. 113–123.

α-hydroxyalkyl group was attached to C-2, and he postulated that this intermediate, too, would have carbanionic character. Krampitz et al.

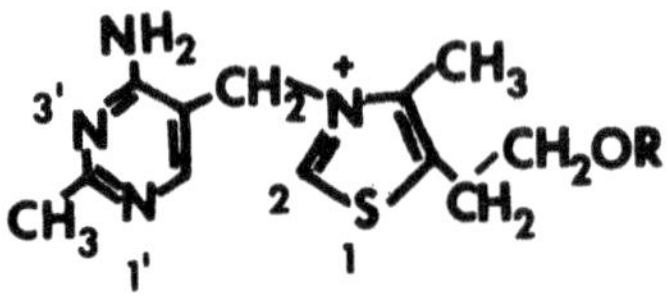

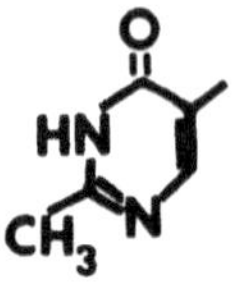

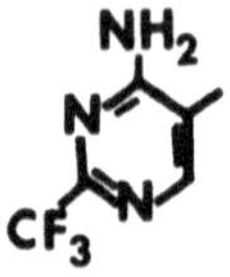

Ia Thiamin, R=H II III

Ib TPP, $R = P_2O_6^{3-}$

(5,6) proved that 2-(α-hydroxyethyl)thiamin-PP was indeed an intermediate in the enzymic synthesis of acetoin and acetolacetate; Holzer et al. (7) found that pyruvate dehydrogenase complex formed (^{14}C)acetyl-CoA from (^{14}C)HETPP[1], and HETPP was isolated from preparations of pyruvate decarboxylase that were producing acetaldehyde from pyruvate (8,9). Proof of the existence of the postulated second carbanion was obtained in our laboratory (4,10,11).

POSSIBLE FUNCTION OF THE AMINO GROUP

A large number of thiazolium salts in addition to thiamin can catalyze the model acyloin condensation in the absence of enzyme (1-3, 12,13), but thiamin is the most efficient. Breslow proposed that the function of the aminopyrimidine ring was to produce inductive stabilization of the C-2 carbanion. He suggested, very cautiously, that the NH_2 group might participate as an intramolecular general base, but pointed out that this was unlikely for theoretical reasons: even in the unprotonated state there is significant delocalization of the π electrons of the NH_2 group into the conjugated ring system (14); in addition protonation of such compounds occurs on the ring, rather than on the NH_2 group (15-17). When we began to study the problem 15 years ago, we felt that the notion of general basic catalysis, though unproved, needed to be considered. We found that the rate of exchange of the (C-2)H of thiamin was independent of concentration, and the addition of thiamin did not increase the slow rate of exchange of the same proton of oxythiamin (18); this eliminated the possibility that any group of the pyrimidine moiety was acting intermolecularly as a basic catalyst, but left unanswered the question of an intramolecular role. We wanted to use analogues, but we have had great difficulties in the synthesis of certain key compounds, which we are just overcoming. Schellenberger (19) has been the strongest proponent of a mechanism involving intramolecular basic catalysis, and he has postulated that when TPP is bound in the holoenzyme, the basicity of the NH_2 group is increased so that intramolecular proton transfer to that group is enhanced. We hope to use model systems to test such a mechanism when the desired analogues become available. In the absence of those compounds it is still possible to make a strong case against an intramolecular basic function for the NH_2 group. All physical evidence, which we shall cite, points to the definition of the amino groups of aminopurines and aminopyrimidines as being non-basic, and we prefer to designate them as Lewis acids.

The delocalization of the π-electrons noted above is great, the charge density (14) on the N-atom of 4-aminopyrimidine being only 1.66, and the exocyclic C—N bonds are correspondingly much shorter than those of ordinary amines (Table I). When the rings are protonated, these bonds become even shorter. The charge density on the N-atom of aniline, which is a true aromatic amine, is 1.80 (14); in such

Table I. Length of Exocyclic C—N Bonds, A°

Type or Compound	Unprotonated	Protonated	References
CH_3NH_2	1.426		
Cyclohexylamine		1.488	(31)
Aromatic	1.377-1.380		(32,33)
"		1.461-1.464	(34,35)
Adenine and Guanine compounds*	1.341		(36)
		1.315	(36)
Cytosine compounds*	1.34	1.32	
Thiamin and derivatives*	1.34		(37)
		1.32	(38)

* For these categories average values are given, including data from studies not specifically cited.

compounds protonation does occur on the N-atom and the C—N bond lengthens. The lengthening is due partly to the increase in single-bond character, and possibly also to the change from pyramidal to tetrahedral structure of the N-atom. In the case of aliphatic amines the C—N bond lengths are still greater. In this case the lengthening on protonation is due exclusively to the change from the time-averaged pyramidal to the tetrahedral structure. The weak basicity of the NH_2 groups in purine and pyrimidine compounds is easily inferred from the well-known hydrogen bonding in the Watson-Crick structure of nucleic acids (20), in which the H-bonds in G-C and A-T pairs involve the donation of an H-atom of the NH_2 group or of a (ring)N-H to a suitable base, structure IV, and never the donation of a proton from a (ring)N-H to the unshared pair of the NH_2 group, structure V. The

```
        H
      --NH····:O═C    or    --NH····:N(ring)      IV

         H
      --N:····H-N(ring)                           V
         H
```

specificity of this hydrogen bonding is not a consequence of the structure of the nucleic acid polymer. The same specificity is seen in crystallographic studies of G-C or A-T pairs (21). Similar H-bonding causes the formation of planar tetramers of guanosine-5-P (22) in aqueous solutions, and the NH_2 protons of flavin adenine dinucleotide form an H-bond with the isoalloxazine moiety (23).

Such behavior is not characteristic of bases, but of Lewis acids. In agreement with such a formulation the amino protons of such compounds exchange very slowly with the protons of solvent water. The exchange is so slow that proton NMR spectra of aqueous solutions of thiamin (18) show a signal for the NH_2 protons over a range of 9 pH units (Fig. 1). Integration of the NH_2 signal gives a maximum value, in the acidic region, of 1.8 protons, showing that even when $[H^+]$ = 1.0 M, the amino group remains unprotonated. Solutions in concentrated H_2SO_4 at -20° show two signals (24), ascribed to protonation of both N-atoms of the pyrimidine ring, and still only two amino protons. Fig. 1 shows also that the line width of the NH_2 proton signal is pH dependent; the values are shown in Fig. 2, and the lifetime of a proton in the NH_2 group can be calculated from these data (Fig. 3). From the pH-dependence of τ^{-1} the catalytic rate constants can be calculated: at 30°C $k_{H_3O^+} = 7 \times 10^2\ M^{-1}sec^{-1}$ and $k_{OH^-} = 2 \times 10^8\ M^{-1}sec^{-1}$. These values are 10^8 and 10^2 times slower, respectively, than the corresponding diffusion-controlled rates characteristic of aromatic amines (25). McConnell et al. obtained comparable rates for adenine and guanine mononucleotides and for polyadenylic acid (26-29). Similar H-bonding in a model system is reported by Lancelot (30). We consider that all these observations point to the non-basic nature of such NH_2 groups, and justify our considering them as Lewis acids.

While emphasizing that there is no proof that the NH_2 group participates directly in the catalytic process, one may speculate on how such participation might occur if it involved the H-bonding ability already described. Since TPP-requiring enzymes use substrates as varied as α-ketoacids and ketose-ω-phosphates, it is unlikely that the function would involve the recognition of anions such as carboxylate or phosphate: such a function should be assigned to groups on the apoprotein. Once such recognition of substrate has occurred, hydrogen bonding with the carbonyl group could assist in orienting the substrate for the nucleophilic attack of the (C-2)-carbanion (Y^- in structure VI).

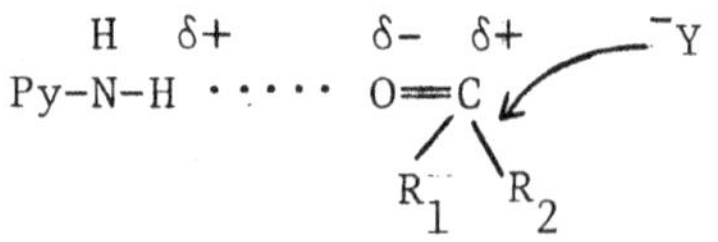

VI

In addition, as a result of the attack of the carbanion, the C=O bond becomes polarized and a full negative charge develops on the oxygen atom. The hydrogen bond indicated in VI would help to decrease the density of that charge; this would lower the free energy of formation of the transition state and should have a favorable effect on the

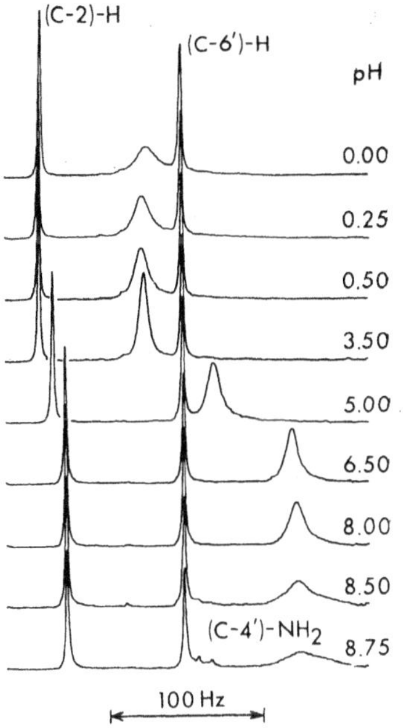

Fig. 1. 60 MHz proton NMR spectra of thiamin in H_2O at various pH values. The (C-6')H signal is 491 ± 3 Hz from that of 2,2-dimethyl-2-silapentane-5-sulfonate. The spectra are aligned with respect to the (C-6')H signal. Solutions of pH 0.00, 0.25 and 0.50 contained aqueous H_2SO_4 to give the indicated pH, on the basis of the Hammett acidity function for such solutions.

(The figures on this page are reproduced from (18) by permission of the American Society of Biological Chemists and the authors)

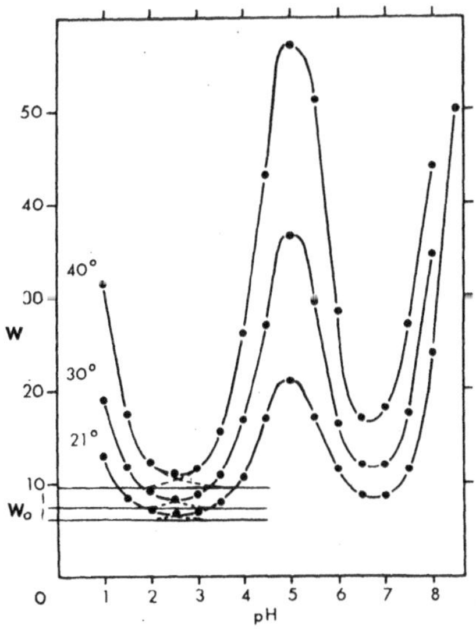

Fig. 2. pH dependence of the line-width of the (C-4')NH_2 signal; 1 M solutions; W is in Hz; W_o is the line-width in the absence of exchange.

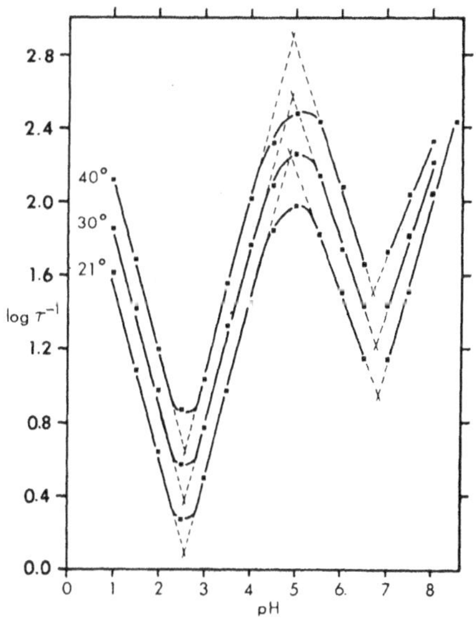

Fig. 3. pH dependence of log τ^{-1} for the NH_2 protons of thiamin; τ^{-1} in sec^{-1} refers to $2\pi(W - W_o)$.

progress of the reaction. In the ring-protonated form of thiamin the NH_2 protons will be better Lewis acids. TPP-requiring enzymes function in a pH range in which the ring would not be protonated. If the NH_2 group does participate as suggested, the acidity could be enhanced by ligation of N-1' to a positively charged group in the active site.

MODEL SYSTEMS

Early studies on the mechanism of action of thiamin (2,12,13) showed that many simpler thiazolium salts could catalyze acyloin condensations, but at a slower rate than thiamin. We wished to compare a large series of analogues of thiamin in an attempt to relate the structure of the substituent attached to N-3 of the thiazolium moiety to the rate of H—D exchange at C-2 and to the rate of formation of the α-hydroxyalkyl derivatives (and related adducts) at C-2. As noted, over the range of pH 3—5, log $t_{1/2}$ for the exchange rate is linear with pH (18) and seems therefore to depend only on the concentration of OH^-; in a repetition of that study we found[2] that in the 2'-trifluoromethyl analogue III, the (C-2)H exchanged almost half as rapidly as thiamin. Eight analogues[3,4] were then tested under identical conditions (Table II). The nitrobenzyl analogues all exchange at

Table II. Rate of H—D Exchange of Thiamin Analogues

Two series of thiazolium salts were studied. Type A has the same thiazolium group of thiamin I; type B has an H at C-5 in place of the side chain. Compounds were 0.5 M in a buffer 0.15 M acetate, 0.15 M formate in D_2O, pH obs. 4.06; T=40°. The numbers represent the half-time for disappearance of the (C-2)H signal, in minutes.

A		B	
4'-deaminothiamin	22	CH_3-thiazolium	201
o-NO_2-Ph-CH_2-	28	o-NO_2-Ph-CH_2-	24
m-NO_2-Ph-CH_2-	28	Ph-CH_2-	40
p-NO_2-Ph-CH_2-	24		

comparable rates, and it is clear that the β-hydroxyethyl side chain does not influence the rate of exchange. The benzyl analogue exchanges about 2/3 as rapidly as the methyl analogue about 1/7 as rapidly as the nitrobenzyl analogues, in agreement with the suggestion that the substituent on the N-atom of the thiazolium ring stabilizes the C-2 carbanion. Under these experimental conditions $t_{1/2}$ for thiamin is about 8 min., about 3 times faster than deaminothiamin. This can be explained by the greater electron withdrawing ability of the aminopyrimidine ring.

In our earlier studies, both published (18) and unpublished[2], we had difficulty in extending our studies of H—D exchange above pH 5. At that pH, thiamin is about 50% in the protonated form. It seemed important to know whether the linear dependence of $t_{1/2}$ on pH (18) would continue when the pyrimidine ring was unprotonated. Chauvet-Monges et al. (39) reported that the (C-2)H of TPP exchanges more rapidly than that of thiamin, in disagreement with unpublished work[6] in which we found that the rate of exchange for TPP was the same as that of mixtures of thiamin and an organic pyrophosphate. Recently, Petzold (40) studied the H—H exchange of (C-2)H in H_2O, and assumed that the two methods were continuous with one another. We then reinvestigated the problem by both methods.[7] It was possible to measure the rate of H--D exchange up to pH 6.5 at 42°C. Fig. 4 shows that when the ionic strength is constant, the rate of H—D exchange is the same for

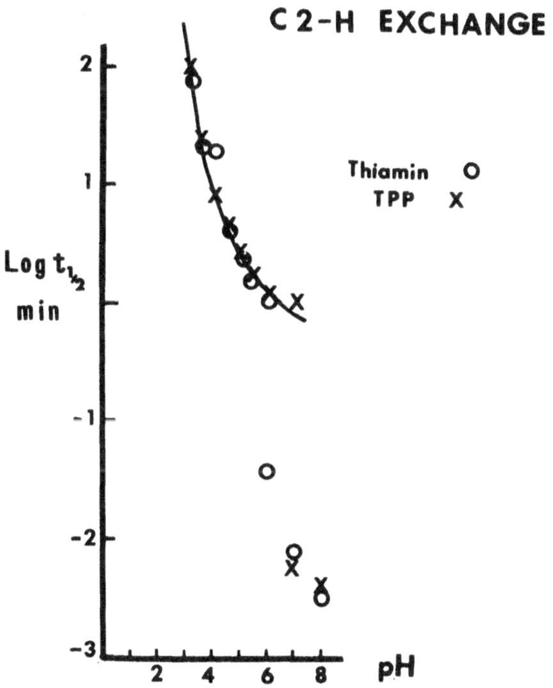

Fig. 4. pH-dependence of log $t_{1/2}$ for exchange of thiazolium protons; conc. 0.5 M. The contribution to the ionic strength of TPP was calculated on the basis of pK=6 for the secondary phosphoryl OH, and the ionic strength of all solutions was adjusted to 3.5 with NaCl. Points in the upper portion of the figure refer to H—D exchange. The five values in the lower portion refer to H—H exchange. $t_{1/2} = \ln 2/k$.

thiamin and TPP over the range of pH 3.0 to 6.5. In addition, the dependence on pH is not linear above pH 5, suggesting that there are at least two separate rates of exchange, characteristic of protonated and unprotonated thiamin respectively. Petzold (40) did not extend his studies to low enough pH to determine whether the rate of H—H exchange, measured by line-broadening, was the same as the rate of H—D exchange. When we attempted to make such measurements, we found that at 42°C, the temperature of the probe of the NMR spectrometer, the line-width, W, was too small to permit accurate measurements of the variation of k (the rate constant for H—H exchange) with pH. Consequently, we measured the line-widths and computed k for several temperatures over the range 42° - 80°C and determined the value of k at 42°C from Arrhenius plots. $k = \tau^{-1} = 2\pi\ (W/2 - Wo/2)$ where W is the width at half height of the (C-2)H signal, and Wo is that for (C-6')H (41). The results are shown in the lower part of Fig. 4: clearly the rates of exchange determined in this manner are nearly two orders of magnitude larger than those measured by H—D exchange. This was a wholly unexpected result, and we have consulted colleagues in an effort to find alternative explanations. If no such explanation is available, we will be forced to conclude that the measurement of line-width describes a process that is not the same as that described by the measurement of H—D exchange. We are unable to offer any suggestions as to what such a process might be.

ACKNOWLEDGEMENTS

This work was supported by research grants from the National Science Foundation and the National Institutes of Health. We thank Professor T.J. Swift for valuable discussion concerning the interpretation of the results obtained in the study of NMR line-widths. Figs. 1,2 and 3 are reproduced from reference (18) by permission of the Journal of Biological Chemistry.

NOTES

This publication is paper XII in the series "Coenzyme Interactions" for paper XI, see reference (4).

1. The following abbreviations are used in the text and formulas: HETPP, 2(α-hydroxyethyl)thiamin pyrophosphate; Hz, hertz; NMR, nuclear magnetic resonance; Ph, C_6H_5; Py, pyrimidine; TPP, thiamin pyrophosphate.

2. G.A. Bantle and H.Z. Sable, unpublished experiments.

3. We thank Professor D.E. Metzler for supplying the benzyl and nitrobenzyl analogues used in the experiments reported in Table II.

4. We thank Dr. Akira Takamizawa of Shionogi and Co., Ltd.,

Osaka, Japan, for a sample of deaminothiamin.

5. We thank Cynthia F. Bearer for assistance with these experiments.

6. I.L. Hansen and H.Z. Sable, unpublished experiments.

7. D.J. Meschke, Dissertation in preparation, Cleveland, 1979.

REFERENCES

1. Breslow, R.: 1958, J. Amer. Chem. Soc. 80, pp. 3719-3726.

2. Breslow, R. and McNelis, E.: 1959, J. Amer. Chem. Soc. 81, pp. 3080-3082.

3. Breslow, R.: 1962, Ann. N.Y. Acad. Sci. 98, pp. 445-452.

4. Gallo, A.A., Mieyal, J.J., and Sable, H.Z.: 1978, in Bioorganic Chemistry, Vol. IV, pp. 147-177, E.E. van Tamelen, ed., Academic Press.

5. Krampitz, L.O., Greull, G., Miller, C.S., Bicking, J.B., Skeggs, H.R., and Sprague, J.M.: 1957, J. Amer. Chem. Soc. 80, pp. 5893-5894.

6. Krampitz, L.O., Suzuki, I., and Greull, G.: 1962, Brookhaven Symp. Biol. 15, pp. 282-292.

7. Goedde, H.W., Inouye, H., and Holzer, H.: 1961, Biochim. Biophys. Acta 50, pp. 41-44.

8. Holzer, H. and Beaucamp, K.: 1959, Angew, Chem. 71, p. 776.

9. Carlson, G.L. and Brown, G.M.: 1961, J. Biol. Chem. 236, pp. 2099-2108.

10. Mieyal, J.J., Votaw, R.G., Krampitz, L.O., and Sable, H.Z.: 1967 Biochim. Biophys. Acta 141, pp. 205-208.

11. Mieyal, J.J., Bantle, G., Votaw, R.G., Rosner, I.A., and Sable, H.Z.: 1971, J. Biol. Chem. 246, pp. 5213-5219.

12. Yatco-Manzo, E., Roddy, F., Yount, R.G., and Metzler, D.E.: 1959, J. Biol. Chem. 234, pp. 733-737.

13. Yount, R.G. and Metzler, D.E.: 1959, J. Biol. Chem. 234, pp. 738-741.

14. Mason, S.F.: 1958, J. Chem. Soc., pp. 3619-3627.

15. Metzler, D.E.: 1960, in The Enzymes, 2nd ed., Vol. 2, Ch. 9, pp. 295-337, P.D. Boyer, H. Lardy, and K. Myrbäck, eds., Academic Press.

16. Mizukami, S. and Hirai, E.: 1966, J. Org. Chem. 31, pp. 1199-1202.

17. Brown, D.J., Hoerger, E., and Mason, S.F.: 1955, J. Chem. Soc., pp. 4035-4040.

18. Suchy, J., Mieyal, J.J., Bantle, G., and Sable, H.Z.: 1972, J. Biol. Chem. 247, pp. 5905-5912.

19. Schellenberger, A.: 1967, Angew. Chem., Int. Ed. Engl. 6, pp. 1024-1035.

20. Steiner, R.F. and Beers, R.F., Jr.: 1961, The Polynucleotides, Elsevier.

21. Pauling, L. and Corey, R.B.: 1956, Arch. Biochem. Biophys. 65, pp. 164-181.

22. Pinnavaia, T.J., Miles, H.T., and Becker, E.D.: 1975, J. Amer. Chem. Soc. 97, pp. 7198-7200.

23. Raszka, M. and Kaplan, N.O.: 1974, Proc. Nat. Acad. Sci. U.S.A. 71, pp. 4546-4550.

24. Chauvet-Monges, A.M., Martin-Borret, Y., Crevat, A., and Fournier, J.: 1974, Biochemie 56, pp. 1269-1277.

25. Grunwald, E., Loewenstein, A., and Meiboom, S.: 1957, J. Chem. Phys. 27, pp. 630-640.

26. McConnell, B., Raszka, M., and Mandel, M.: 1972, Biochim. Biophys. Res. Commun. 47, pp. 692-698.

27. McConnell, B. and Seawell, P.C.: 1972, Biochemistry 11, pp. 4382-4392.

28. McConnell, B. and Seawell, P.C.: 1973, Biochemistry 12, pp. 4426-4434.

29. McConnell, B.: 1974, Biochemistry 13, pp. 4516-4523.

30. Lancelot, G.: 1977, J. Amer. Chem. Soc. 99, pp. 7037-7042.

31. Rao, S.T. and Sundaralingam, M.: 1969, Acta Cryst. B25, pp. 2509-2515.

32. Sax, M., Pletcher, J., and Gustafsson, B.: 1970, Acta Cryst. B26, pp. 114-124.

33. Lai, T.F. and Marsh, R.E.: 1967, Acta Cryst. 22, pp. 885-893.

34. Pletcher, J., Sax, M., and Yoo, C.S.: 1972, Acta Cryst. B28, pp. 378-387.

35. Arora, S.K., Sundaralingam, M., Dancz, J.S., Stanford, R.H., and Marsh, R.E.: 1973, Acta Cryst. B29, pp. 1849-1855.

36. Watson, D.G., Sutor, D.J., and Tollin, P.: 1965, Acta Cryst. 19, pp. 111-124.

37. Pletcher, J., Sax, M., Sengupta, S., Chu, J., and Yoo, C.S.: 1972, Acta Cryst. B28, pp. 2928-2935.

38. Pletcher, J., Wood, M., Blank, G., Shin, W., and Sax, M.: 1977, Acta Cryst. B33, pp. 3349-3359.

39. Chauvet-Monges, A.M., Rogeret, C., Briand, C., and Crevat, A.: 1973, Biochim. Biophys. Acta 304, pp. 748-752.

40. Petzold, D.R.: 1976, Stud. Biophys. 54, pp. 159-162.

41. Jackman, L.M. and Sternhell, S.: 1969, Applications of Nuclear Magnetic Resonance Spectroscopy in Organic Chemistry, 2nd Ed., pp. 55-58, Pergamon Press.

THE CONFORMATION OF COCARBOXYLASE (THIAMINEPYROPHOSPHATE)

Gavriella Gabor
Israel Institute for Biological Research
Department of Chemistry
P.O. Box 19
Ness-Ziona, Israel

The role of the Pyrophosphate group in Cocarboxylase, as a conformation determining factor, is suggested.

Comparison of fluorescence intensities of thiamine and cocarboxylase - thiaminepyrophosphate (TPP) - implies the rigidity of the TPP molecule, resulting from an internal H-bond between the pyrophosphate oxygen and the 4'-amino proton.

Cocarboxylase acts as a coenzyme to pyruvate decarboxylase [1]. Its chemical structure as thiaminepyrophosphate (TPP) has been established [2]. The enzymatic degradation of α-ketoacids proceeds via the binding of the substrate, such as pyruvic acid, at the 2 positions of TPP followed by irreversible release of CO_2 yielding "active acetaldehyde" [3], which functions as an acetaldehyde transfer agent [4]. Through all these processes the pyrophosphate group (PP) is bound to the thiamine (T) moiety; it remains a part of the TPP molecule; no phosphate bonds are broken or produced, in order to release or absorb energy, as in other biochemical systems. In TPP catalyzed reactions the pyrophosphate does not function as an energy source in any known biological system [5].

In the present study a connection between the tertiary structure of TPP and the presence of the PP group is suggested. Electrostatic attraction results in the folding back of the negatively charged PP side chain over the positive centers of thiamine in TPP [5, 6] (Fig. 1). According to CPK models, the P-2 oxygen (of the PP group) is hydrogen bonded by one of the 4'-amino protons, forming the 4'NH... OP-2 hydrogen bond. As a result the molecule obtains a completely rigid conformation in which its "active sites", 2CH, 4'NH and 1'N, involved in the enzymatic reaction are exposed (Fig. 2). Simultaneously, interaction - a non-linear H-bond - between the 2C-proton, or any other atom (X) bound to it, and the 4'-amino nitrogen [2C-X...N4'] is permitted in this conformation.

Further contributions of the [4'NH...OP-2] H-bond may be: The amine nitrogen sharing one of its protons in this H-bond, acquires larger electronegativity, and, thus, enhanced attraction for the 2C-

B. Pullman (Ed.): Catalysis in Chemistry and Biochemistry. Theory and Experiment. 125–130.

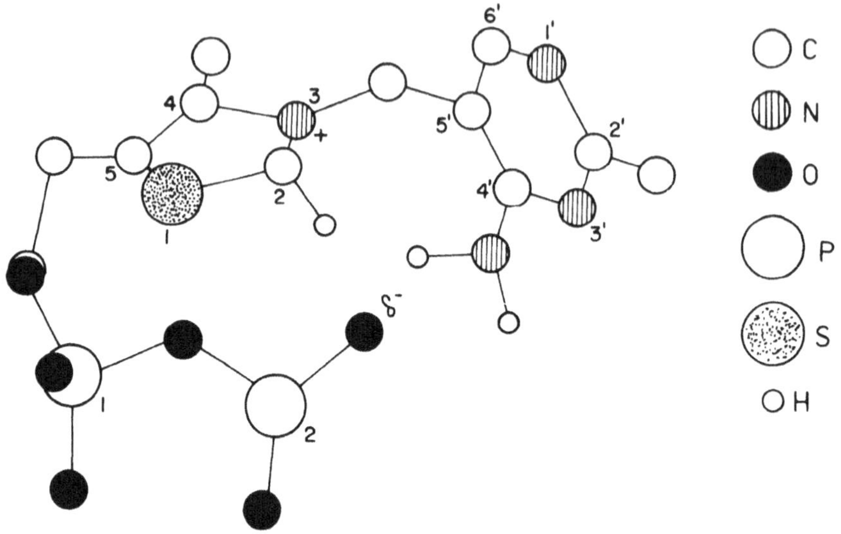

Fig. 1. Folded structure of TPP.

Fig. 2. CPK model of the H-bonded TPP molecule.

proton, or the OH proton, or the positively charged acetyl carbon atom of the substrate. A prerequisite for such an interaction is a twist of the 4'-NH_2 group out of its best resonance position relative to the aromatic pyrimidine ring [1]. Schellenberger suggested that this is achieved by formation of an H-bond, between the 4'-amino-proton and an appropriate binding site of the apoenzyme. However, according to CPK models (Fig. 2), the 4'NH...OP-2 H-bond would cause the appropriate twist. Changing the 4'-NH_2 group to an OH group reduces the coenzymatic activity by about 52% [7]. 4-OH-pyrimidines are known to exist mainly in the amidine form

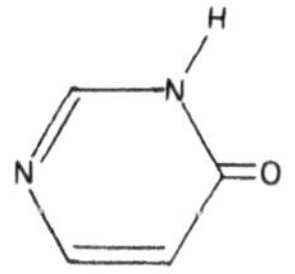

[8]. As a result there is no available proton on the 4' substituent to form the necessary H-bond with the 2P oxygen, for obtaining the suitable "coenzymic" conformation. Thiamine, or TPP, is converted to thiol in basic solutions (Scheme 1) [9]. Due to the 4'-NH...OP-2 bond, TPP

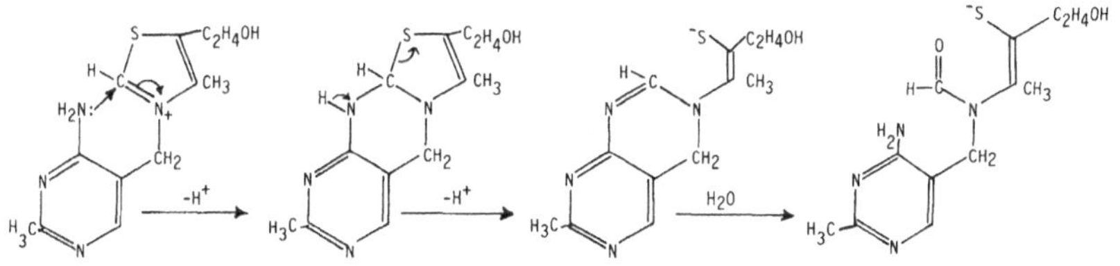

I II III IV

Scheme 1.

would retain its conformation in space through this transformation, and, thus, possibly its coenzymic activity. The possibility of other molecular forms of T being involved in the enzymatic reaction was considered previously [10].

Rigidity of a molecule is known to increase the probability of its being fluorescent. The fluorescence of T and TPP was compared. At neutral pH (∿7), a fresh solution of T or TPP fluoresces at 390-400 nm. The TPP fluorescence is about twice as intense as that of T (Table 1). Lowering the pH to pH 2-3, results in a small increase of the ratio of the fluorescence intensities of the respective compounds, $\frac{A[TPP]}{A[T]}$ reaches the value of 4. The relative increase of fluorescence intensity results from the 4'NH...OP-2 H-bond becoming stronger, due to the partial protonation of the 4'NH_2 group at this pH. The protonation of the thiamine pyrimidine occurs mainly at N1 with a pKa of 5.0 [10]. The unusually high pKa of 4-amino-pyrimidines results from the resonance stabilization of the protonated species:

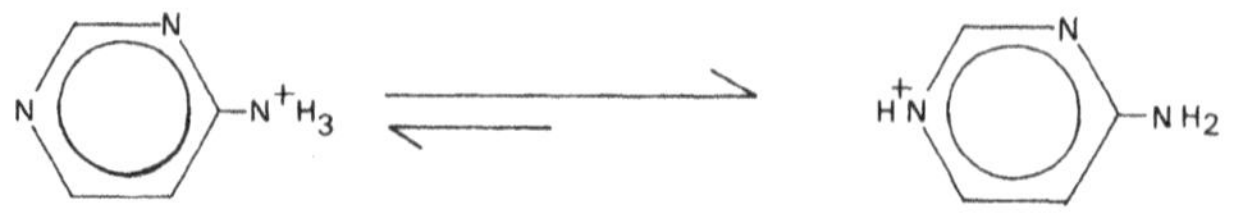

(a) (b)

Although the 4-amino group itself is not protonated, even in highly acidic solutions [11], the partial positive charge on the amino nitrogen, as in structure (a), increases slightly the strength of the 4'NH...OP-2 H-bond, i.e. the rigidity of the molecule. The small increase of fluorescence obtained in acid solutions implies that the contribution of species (a) is rather small.

The fluorescence maximum (λ_{MAX}) of a fresh $2{,}5.10^{-2}$ M solution at neutral pH of T or TPP is at 385 nm, while λ_{MAX} of excitation is at 330-340 nm. "Aging" of such a solution at room temperature yields a λ_{MAX} of fluorescence at 430 nm, and λ_{MAX} of excitation at 370 nm. Simultaneously the absorption above 320 nm is considerably increased. The "aging" is accelerated by a) heating or b) increasing pH or c) adding less polar solvent, like ethanol. The slow aging at biological conditions (temperature, and pH in aqueous solutions), resembles the faster process in basic solutions of T [9], and the product is most probably thiol III (or IV). The absorbancies at 370 nm (above 320 nm), as well as the fluorescence intensities of derivatives III, for both T and TPP, are highly increased as shown in Table 1. The larger increase for TPP is assigned to the rigidity of TPP-1 (due to the internal 4'NH...OP-2 H-bond) promoting conversion I → III (IV).

The preservation of the conformation of TPP during the I → III (IV) conversion, which occurs at biological pH, may have a functional importance, implying the retention of the coenzymatic activity, throughout these conversions. The stabilization of III (or IV) in the enzymatic process, by binding the thiol (III or IV) to lipoic acid, should be considered.

Acknowledgments

The skillful technical assistance of Mr. Z. Samoha is kindly acknowledged. I am grateful to Dr. C. Lachman for helpful discussions.

Table 1.

Absorbancies (D_{370}) and Fluorescent Intensities (A) of T and TPP, 0.0125 M solutions, before (I) and after (III) heating to 70 C for 12 hrs.

	D_{370}	λ_{MAX}, nm		A, arbitrary units
		ex.	flu.	
T-I	0.03	330-340	385	6.0
T-III	0.05	365	420	96.0
TPP-I	0.03	330-340	385	13.0
TPP-III	1.60	370	430	1800

A - fluorescence intensity - measured and excited at the respective ϕ - relative to fluorescence intensity of quinine sulfate.
D_{370} - optical density at 370 nm.
All fluorescence measurements have been carried out in dilute solutions having optical density ≤ 0.1 at the exciting wavelength.

References

1. Schellenberger, A. (1967) Angew. Chem. Intern. Edit. 6, 1024-1035 and other references quoted therein.
2. Lohman, K. & Schuster, P. (1937) Biochem. Z. 294, 188; Weilard, J. & Tauber, H. (1938) J.A.C.S. 60, 2263-2264.
3. Holzer, H. & Beaucamp, K. (1959) Angew. Chem. 71, 776 (1961) Biochim. Biophys. Acta 46, 225-24[illegible]
4. Breslow, R. (1962) in Unsolved Problems of Thiamine. Ann. N.Y. Acad. Sci. 98(2), pp. 445-452.
5. Pletcher, J. & Sax, M. (1966) Science 154, 1331-1333.
6. Gallo, A.A., Hausen, I.L., Sable, H.Z. & Swift, T.J. (1972) J. Biol. Chem. 247, 5913-5920.
7. Schellenberger, A., Muller, V., Winter, K. & Hubner, G. (1966) Hoppe-Seylers Z. Physiol. Chem. 346, 123.
8. Albert, A. & Phillips, J.N. (1956) J. Chem. Soc. 1294.
9. Maier, G.D. & Metzler, D.E. (1957) J.A.C.S. 79, 4386-4391.
10. Metzler, D.C. & Maier, G.D. (1962) Ref. 4. pp. 495-498.
11. Suchy, J., Mieyal, J.J., Bantle, G. & Sable, H.Z. (1972) J. Biol. Chem. 247, 5905-5912 and other references quoted therein.
12. A. Albert (1969) in "Synthetic Procedures in Nucleic Acid Chemistry" (Zorbach, W. & Tipson, R., ed.), Vol. 2, pp. 6-12, Wiley, N.Y.

PROBING THE MECHANISM OF BINDING OF THE THIAMIN PYROPHOSPHATE CO-ENZYME ON YEAST PYRUVATE DECARBOXYLASE

Frank Jordan[1], Bijan Farzami[2], and Olufemi B. Akinyosoye with the assistance of Robert Aslanian
Department of Chemistry, Rutgers, The State University of New Jersey, Newark, New Jersey 07102, U.S.A.

ABSTRACT. Several aspects of the co-enzyme function of thiamin pyrophosphate and Mg(II) in the action of brewer's yeast pyruvate decarboxylase were examined. All experiments to be reported were performed by first resolving then reconstituting holoenzyme. The principal findings are the following: (1) An aminopyrimidine derivative and the 2,4-dinitrophenylether of thiamin were inhibitory; (2) upon reconstitution of apoenzyme with either pyrophosphate or Mg(II) there is an enhancement in the UV absorption between 300-400 nm and this enhancement develops with a $t_{\frac{1}{2}}$ of ca 2 min. There is a lagtime in the regaining of the activity as well and it is demonstrated that the formation of the enhanced absorption is a necessary but not a sufficient condition for regaining of the activity; (3) reconstitution with Eu(III) in place of Mg(II) affords ca 50% of the activity compared to Mg(II) and leads to a large enhancement in the fluorescence emission spectrum which enhancement can be subsequently quenched by the addition of thiamin pyrophosphate. The fluorescence data for the apoenzyme-Eu(III) binary complex indicate the existence of two types of binding sites with K_d's of 200 μm and (6$\pm$2) μm respectively. A qualitative assessment of the quenching curve by thiamin pyrophosphate indicates that there is an apparent K_d of the apoenzyme-Eu(III) complex of ca 1 μm.

[1]Author to whom correpondence should be addressed.

[2]Current address: Department of Biochemistry, University of Theran, Tehran, Iran.

B. Pullman (Ed.): Catalysis in Chemistry and Biochemistry. Theory and Experiment. 131–146.

INTRODUCTION

Of the enzymes requiring thiamine pyrophosphate (TPP, the vitamin B_1 co-enzyme) as a cofactor, yeast pyruvate decarboxylase (PDCase) has been the most widely studied (Krampitz, 1969):

$$\underset{\text{pyruvate}}{CH_3COCO_2^-} + TPP + Mg(II) \xrightarrow{\text{PDCase}} \underset{\text{acetaldehyde}}{CH_3CHO} + CO_2 + OH^- \quad (1)$$

The role of the various portions of the coenzyme have been studied by several groups. The most important advance in our understanding of the mechanism was made

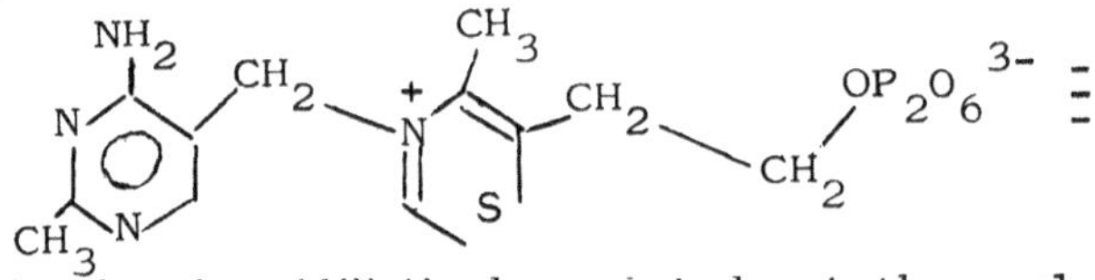

by Breslow (1957) who pointed out the nucleophilic activity of the C^2 atom in the conjugate base of the thiazolium ring. a further impetus was given to studies on thiamin catalysed reactions by model studies that indicated that all reactions, i.e. binding of pyruvate to thiamin, decarboxylation, loss of acetaldehyde, as well as ylide formation in the proposed sequence were accelerated by transferral of the model reaction from an aqueous to a lower dielectic environment (Crosby et al., 1970; Crosby and Lienhard, 1970; Kemp and O'Brien 1970). These findings led to the proposal that such solvent effects, if created by hydrophobic environments on the enzyme site, could explain part of the enormous catalytic rate factors enzymic reactions are capable of exhibiting. Sable and his associates demonstrated that tryptophan derivatives formed weak complexes in aqueous solution with TPP and especially with the thiazolium portion of the ring (Sable and Bigalow, 1965; Bigalow et al., 1969; Mieyal et al., 1969). Subsequently it was shown that the pyrimidine ring also quenches fluorescence of tryptophan and solvent effects were established on TPP-tryptophan (presumably stacking-type) interactions (Farzami et al., 1977). It was also demonstrated that the thiazolium ring can give rise to an ionic association with negative charges (e.g. I^-) and such interactions are stronger in media of lower dielectric constant (Jordan and Farzami, 1977 There arose the suggestion that a tryptophan residue on the enzyme and/o a carboxylate side chain may act in the coenzyme binding mechanism.

The role of the Mg(II) ion has also been studied in model systems. Since Mg(II) is inactive from the point of view of both electronic and magnetic resonance spectroscopy, such studies employed Ni(II) or Co(II) (White and Drago, 1971); Ni(II) and Mn(II) (Gallo et al., 1972; Gallo and Sable, 1975); or Mn(II) binding to TPP (Grande et al., 1973) as well as the binding of both Mn(II) and Ni(II) to TPP and to thiamin (Jordan and Mariam, 1977). The studies employing Mn(II) appear to be most informative and they demonstrated that the solution conformation of TPP-Mn(I is a folded one, i.e. the dimethylene phosphate is folded back onto the thiazolium ring.

The role of the pyrimidine group has also been scrutinized. Aside from the obvious possibility of the participation of this ring in the binding mechanism, Schellenberger (1967) had also provided data indicating the essential nature of the amino group in enzymatic catalysis. Gallo at al. (1978) in a recent review article on thiamin models and Jordan and Mariam (1978) took this suggestion a step further and demonstrated in model systems that attaching an electron withdrawing group to pyrimidine (2'-CF_3 in the former, an N1'-CH_3 induced positive charge in the latter study) could convert the amino group into a weak acid which then could act either as an acid-base or nucleophilic catalyst.

It is our intention to test the validity of some of the findings on model systems in enzymatic processes reqiring TPP and Mg(II). In this communication we report some of our findings on PDCase isolated from brewer's yeast. Some of the techniques we have employed have clear potential in applications to other TPP requiring enzyme systems. Our preliminary findings are reported in three areas: (1) testing of TPP analogues exhibiting inhibitory function; (2) observation of UV spectroscopic changes accompanying coenzyme binding as well as the time development of such changes; and (3) Eu(III) fluorescence enhancement studies (in place of Mg(II)) to study the active site environment and the binary apoenzyme-Eu(III) and tenary apoenzyme-Eu(III)-TPP complexes.

EXPERIMENTAL SECTION

Pyruvate decarboxylase from brewer's yeast was purified according to the method of Ullrich (1970) and further purified according to the method of Gounaris et al. (1975). The procedure taking the protein through a DEAE ion-exchange column gave an enzyme with specific activity of ca 60 units/mg protein. Although this is lower than the specific activity reported for the crystalline enzyme (Ludewig and Schellenberger, 1974) according to slab gel electrophoresis the enzyme was homogeneous. Enzyme assay was performed either by the coupled alcohol dehydrogenase or the pH stat method (Schellenberger et al., 1968). UV measurements were performed on a Beckman Acta (III) spectrophotometer usually at $30 \pm 0.1^\circ C$. Fluorescence measurements were performed on a Farrand MK-1 spectrofluorimeter at $30 \pm 0.1^\circ C$. pH measurements and pH-stat kinetics were performed on a Radiometer Model 26 pH meter equipped with automatic titrator and recorder. Prior to the pH-stat assay all our solutions were thoroughly flushed with N_2. In most of our studies we had to resort to resolution of the holoenzyme prior to kinetic or spectroscopic measurements. The earlier studies followed a protocol of Ullrich (1970), the later that of Gounaris et al. (1971) in the resolution and reconstitution of the enzyme. Protein concentrations were determined by the method of Lowry et al. (1951).

RESULTS AND DISCUSSION

1. Potential Inhibitors of PDCase

No active site labelling studies have been reported to date on this enzyme. For various reasons it was deemed desirable to test the following compounds for possible inhibitory effects. Thiamin (1) was included essentially as a control. It had been tested by several groups before and not found to be inhibitory.

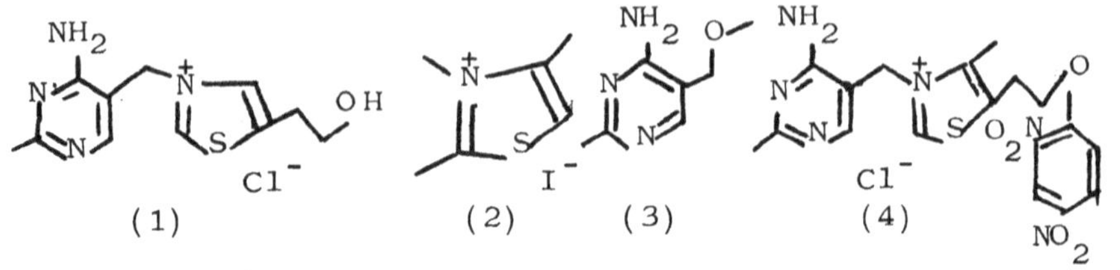

(1) (2) (3) (4)

TPP is reported to be tightly bound to the apoenzyme. Therefore we had resort to the following protocol in testing these compounds. The holoenzyme was first resolved into apoenzyme (Ullrich, 1970 or Gounaris *et a* 1971). Next the apoenzyme was incubated with the particular compound for a period of ca 5 mins and the activity continuously monitored on the pH-stat. After this incubation TPP was added and the initial velocity resulting was measured. Table I lists the results of such studies. As anticipated 1 had minimal effect on the subsequent addition of TPP. Although none of the four compound tested could replace TPP to give an active holoenzyme, 3 and 4 had significant effects in that their presenc produced greatly diminished activity. We do not yet know whether this inhibition is accomplished at the catalytic site or at an allosteric site. Further experiments which will determine steady-state inhibition patterns should establish this point.

Conceivably a highly hydrophobic active site may bind 3 and 4 more strongly than 1 or 2. Establishing the nature of the inhibition provide by 3 and 4 will allow the design of affinity and photoaffinity labels related to these compounds. Functionalization of 3 and 4 should be considerably simpler than of TPP. It is worth noting also that most oth TPP analogues acting as inhibitors possessed the diphosphate side chain. Perhaps the lack of pyrophosphate is compensated for in 3 and 4 by thei enhanced hydrophobic character.

2. UV Spectroscopic Changes Accompanying Reconstitution of Yeast Holopyruvate Decarboxylase

During studies directed towards determination of the inhibition characteristics of some TPP analogues we had noticed that the UV spectrum of freshly reconstituted holoenzyme underwent a time dependent change. It was of interest to determine if such changes could be attributed to one of the cofactors interacting with the apoprotein. Table II summarizes the data employing Mg(II) and Eu(III) in the reconstitution in addition to TPP. Figure 1 represents repetitive scan spectrophotometric measurements performed between 300 and 400 nm and the accompenying reconstitution. Figure 2 represents the time course of such spectral changes. Of greatest importance is the finding that the introduction of either metal ion or TPP or both together cause a spectral change as does a sequential addition of Mg(II) followed by TPP. The approximat $t_{\frac{1}{2}}$ for the apparent first order spectral processes is ca 2 ± 0.5 min at 25°C. If reconstitution is followed by pH-stat methods, the activity is only regained when both Mg(II) and TPP are added (see Table II). If both are added simultaneously, with the recorder running and monitoring continuously the release of OH^-, there is found a lagtime of perhaps 25-30 sec after mixing before zero order kinetics are observed. If the enzyme is reconstituted first and then added to pyruvate no such lagtim

TABLE I - The Effect of Reconstitution with Potential Inhibitors on the Activity of Yeast Pyruvate Decarboxylase[a].

Experiment	Compound	Activity upon[b] reconstitution with the inhibitor (unit/ml).	% Activity regained upon subsequent TPP addition
1	TPP[c]	5	
2	<u>1</u>	0	75-80[d]
3	<u>2</u>	0	70[d]
4	<u>3</u>	0	30[d]
5	<u>4</u>	0	10[d]

[a] Experiments were performed at 30°C at a nominal pH of 6.2. [b] The incubation was performed for ca 5 min with the inhibitor listed under "compound". The release of OH^- during this time was monitored on the pH-stat. All inhibitors were present at the same concentration as TPP, all experiments were performed at 0.001 <u>M</u> Mg(II) concentration.
[c] Reconstitution employing the natural co-enzyme. It is estimated that ca 60-70% of the enzyme activity was lost during the resolution and reconstitution process. [d] % of activity regained upon addition of TPP (same amount as in experiment 1). Calculated from the ratio of initial rates comparing experiments 2,3,4 or 5 to experiment 1.

TABLE II - Ultraviolet Absorption Characteristics of Reconstituted Yeast Holo Pyruvate Decarboxylase[a].

No.	Apoenzyme ml	[Mg(II)] $\times 10^3$	[Eu(III)] $\times 10^3$	[TPP] $\times 10^3$	[Tryptophan] $\times 10^5$	Appearance of excess absorption at 300–400 nm[b]	Activity[c] %
1	0.1	-	-	-	-	no	1
2	0.1	-	-	1.2	-	yes	1
3	0.1	1.2	-	-	-	yes	1.5[d]
4	0.1	1.2	-	1.2	-	yes	100
5	0.1	-	1.2	1.2	-	yes	60
6	-	-	1.2	1.2	-	no	-
7	-	-	1.2	-	7.0	no	-
8	-	1.2	-	-	7.0	no	-
9	-	-	-	1.2	7.0	yes	-

[a]All experiments were performed at 30°C on a Cary 14 spectrophotometer at pH 6.2, 0.05 M phosphate buffer. [b]Recorded after ca 5 seconds of mixing. [c]The activity compared to that found in experiment 4. This residual activity maybe due to a lack of complete removal of TPP during the resolution process.

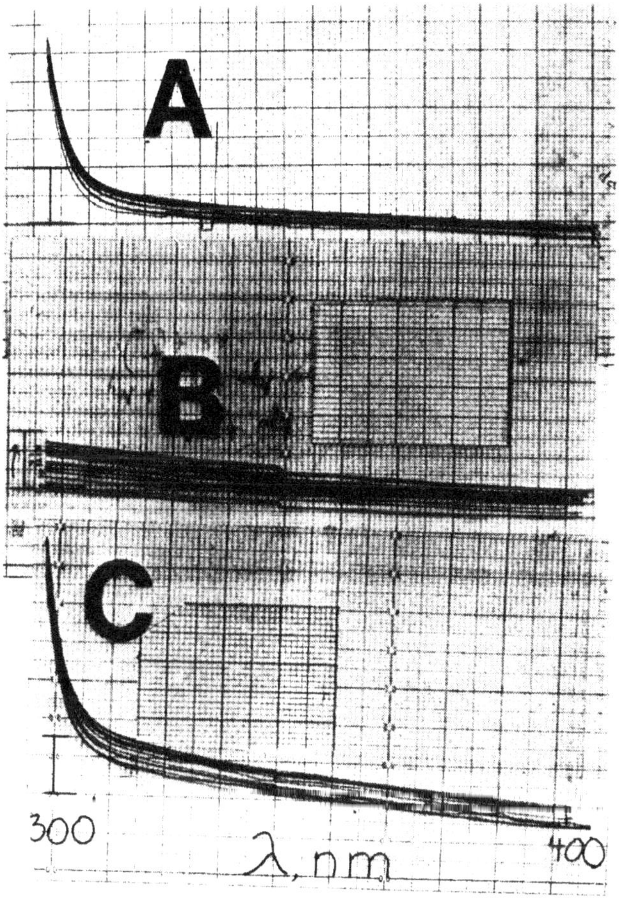

Figure 1 - Repetitive scan UV measurements on reconstitution of apo pyruvate decarboxylase with A. TPP; B. Mg(II); C. Apo + TPP first mixed for five minutes then Mg(II) added. The reference cuvette contained apoenzyme in A and B, and apoenzyme + TPP in C. All measurements were made at 25°C, pH 6.2 between 300 and 400 nm. 0.1 OD is indicated on the ordinate for reference. In all three cases the OD increased with time.

is observed. There clearly appears to be some conformational change induced in the enzyme before full activity is regained. The observed change may be due to recombination of the subunits. The recombination process has previously been studied by Gounaris et al.(1971) on baker's yeast PDCase. The changes in ORD, CD and UV absorption properties of holo transketolase, another TPP requiring enzyme, compared to the apo-protein had been reported by two groups (Kochetov et al., 1970; Koletov and Usmanov, 1970; Heinrich and Schmidt, 1973). They both attributed the observed changes to the formation of a charge transfer complex between a tryptophan residue on the enzymes and TPP. This would be analogous to that observed in the model studies of Sable and Biaglow (1965). In our hands addition of either Mg(II) or TPP to the apoprotein caused changes in the UV absorption, changes similar to those reported in the transketolase spectrum (Kochetov and Usmanov, 1970). It is not clear what the change induced by Mg(II) is due to. Nevertheless, it is tempting associate the spectral change induced by TPP with the formation of a tryptophan-TPP charge transfer complex.

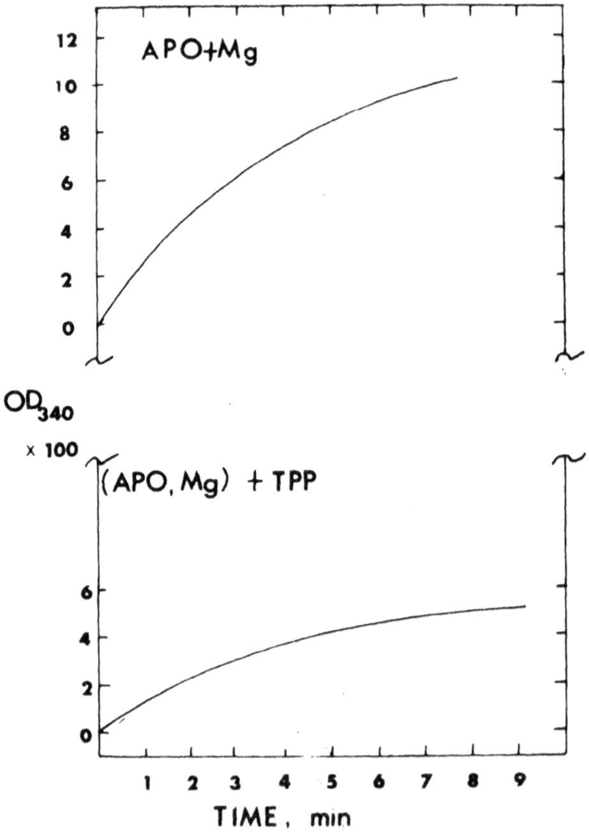

Figure 2 - Time course of the development of the enhanced absorption at 340 nm at 25°C, pH 6.2.

What is clear, looking at the activity vs. appearance of enhanced UV absorption under different conditions (Table II), is that formation of the enhanced absorption is a necessary but not sufficient event for reconstitution of the activity. Our observations allow us to undertake further studies into the kinetics of reconstitution and to probe the molecular events accompanying this process.

3. Fluorescence Spectroscopic Studies on Reconstituted Yeast Holo Pyruvate Decarboxylase

Neither TPP nor Mg(II) possess either electronic or magnetic properties which would be readily discernible when the co-factors are enzyme bound. The binding of TPP to the apoenzyme could be studied employing conventional equilibrium dialysis techniques and radioactive TPP (as was done for transketolase - see Heinrich, 1973). We decided to explore a technique not hitherto applied to TPP requiring enzymes but potentially applicable to any of them, irrespective of their size. The technique employs lanthanide metal ions as a replacement for Mg(II). Such a technique had been employed before for example by Wolfson and Kearns (1974) who replaced Mg(II) by Eu(III) to study the structure of t-RNA in solution.

The procedure we adopted again involved resolution of holo pyruvate decarboxylase followed by reconstitution under different conditions. Table II demonstrates some typical reconstitution data employing Eu(III) instead of Mg(II). Typically, we obtained 30-60% activity when making this substitution. More importantly, the fluorescence emission at 580 nm (excitation at 287 nm) was enhanced very greatly (some 60-fold) when employing Eu(III) compared to Mg(II). While this wavelength region is not a particularly clean one due to interference from second order light scattering, one can subtract such effects and analyse the intensity due to the emission from the apoenzyme Eu(III) binary complex alone.

Table III summarizes some qualitative aspects of the fluorescence emission spectra of the various solutions employed in our studies. Figure 3 indicates the intrinsic protein fluorescence behaviour of apoenzyme upon addition of 1 mM TPP or Mg(II) or both. The emission intensity at 340 nm (excitation into the aromatic band at 287 nm) is decreased by the addition of either Mg(II) or TPP or both. Figure 4 demonstrates the qualitative behaviour of apoenzyme fluorescence upon the addition of 1 mM Eu(III) followed by the addition of 1 mM TPP. This demonstrates the utility of monitoring the behaviour of the long wavelength spectral band. The intensity of the 580 nm emission peak is employed in constructing Figure 5a. The experiment was initiated with ca 2 μM subunit concentration of apopyruvate decarboxylase which was then titrated with μl quantities of Eu(III) at pH 6.5. The data treatment followed that due to Lehrer and Fasman (1966). One first plots $1/\Delta F$ ($F_{apo-Eu(III)} - F_{apo}$) against 1/- Eu(III)) to obtain $F\Delta$. Such a plot exhibited distinct curvature. A simple explanation of this biphasic behaviour would be the presence of two different classes of binding sites. The limiting slopes to the curve can be extrapolated to infinite Eu(III) concentration to yield two values of F_∞ . With these values of F_∞ one can construct log-log plots as illustrated in

TABLE III - Fluorescence Emission Characteristics of Yeast Pyruvate Decarboxylase [a].

No.	Apoenzyme	[Mg(II)]	[Eu(III)]	[TPP]	364_{nm}	388_{nm}	580_{nm}
	ml	$x10^3$	$x10^3$	$x10^3$	(Arbitrary units)		
1	0.1	1.2	-	-	0.54	-	0.18
2	0.1	-	-	1.2	-	0.18	-
3	0.1	1.2	-	-	0.72	-	0.42
4	0.1	1.2	-	1.2	-	0.06	-
5	0.1	-	1.2	0.2	0.54	-	0.24
6	0.1	-	1.2	-	4.05	-	16.5

[a] All experiments were performed at 30°C, pH 6.2, 0.05 M phosphate buffer, excitation at 287 nm. Reconstitution was performed for ca one half hr before recording of the spectrum.

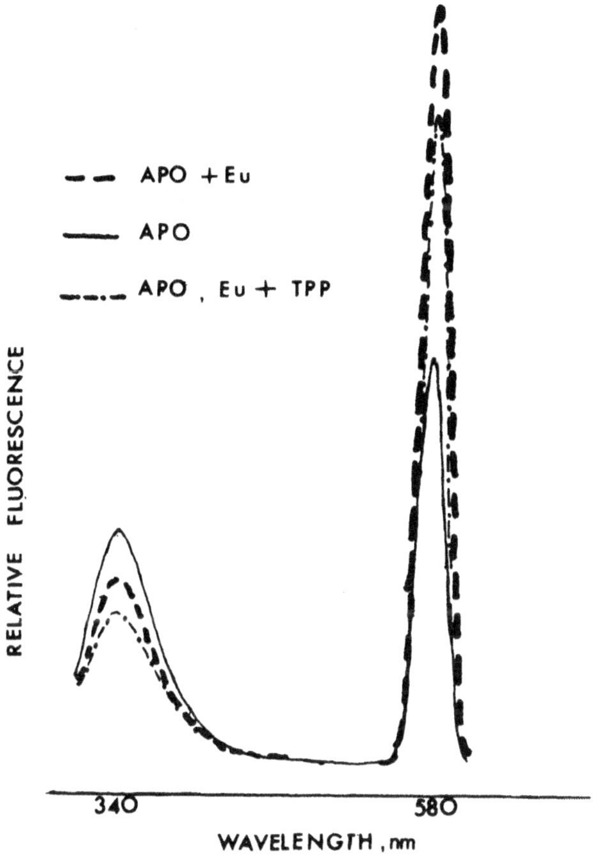

Figure 3 - Fluorescence emission curves for apo pyruvate decarboxylase and reconstituted enzyme under the indicated conditions, excitation at 287 nm.

Figure 5b. The data at low Eu(III) concentration are scarce but can still provide an order of magnitude value. K_d can be interpolated to that Eu(III) concentration at which $\log(F - F_o) / (F_{\infty} - F)$ equals zero (half saturation). In this manner we obtain pK_d's of -3.68 and ca -5.2, or K_d's of 200 M and (6 $\pm$ 2) M for the weaker and stronger metal binding sites respectively. The stronger binding site was not unexpected and was suggested earlier by Schellenberger (1967). The observation of a weaker binding site, in addition to the stronger one, to our best knowledge, is novel and suggests that the metal ion (as well as perhaps TPP) may have two distinct functions. Gounaris et al. (1971) had earlier suggested that the co-factors may have a function in maintaining an active holoenzyme conformation in addition to the well documented catalytic function assigned to TPP. Further substantiation of this observation is currently undertaken and should offer new insight to the mechanism of TPP requiring enzymes. Such studies will also employ other lanthanides. In addition, data will also be collected and analysed by exciting at 395 nm and monitoring emission at 595 - a region characteristic to Eu(III).

Next TPP was added to a highly fluorescence enhanced apoenzyme-Eu(III) solution (2 μM subunit concentration, 0.4 mM Eu(III), pH 6.5). Figure 6 demonstrates the change in fluorescence emission intensity (580 nm) with added TPP. Analysis of this data at higher TPP concentrations (7 - 13 μM) allows only very approximate estimation of an apparent binding constant. Employing a Benesi-Hildebrand approach (for TPP >> apoenzyme) the approximate 1/ ΔF vs. 1/(TPP) plot provides a K_d for the apo-Eu(III) to TPP apparent binary dissociation constant of ca 1 ± 0.5 μM. This value is certainly of the same order of magnitude found in the apotransketolase-Mg(II) to TPP binding determined by the equilibrium dialysis technique (1 μM - see Heinrich, 1973). An important question remains as to whether TPP quenches the enhanced fluorescence by binding near the Eu(III) binding site as well as close to a tryptophan

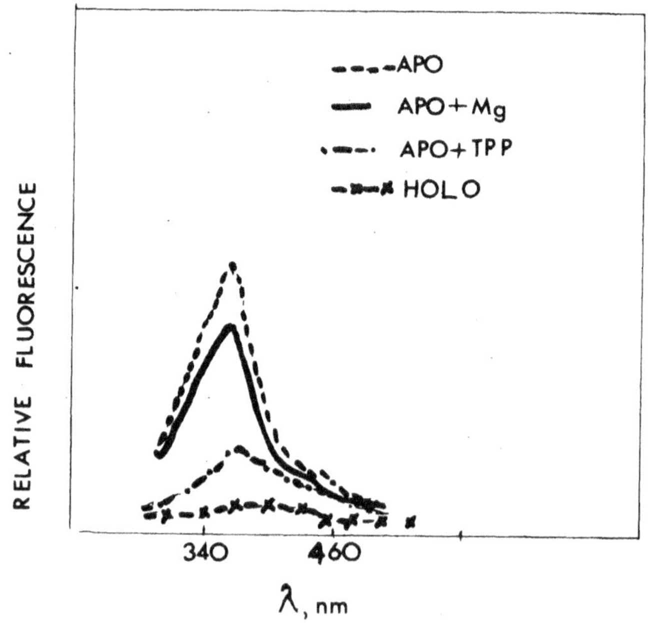

Figure 4 - Relative fluorescence of resolved apoenzyme and enzyme reconstituted with Eu(III) and TPP; excitation at 287 nm.

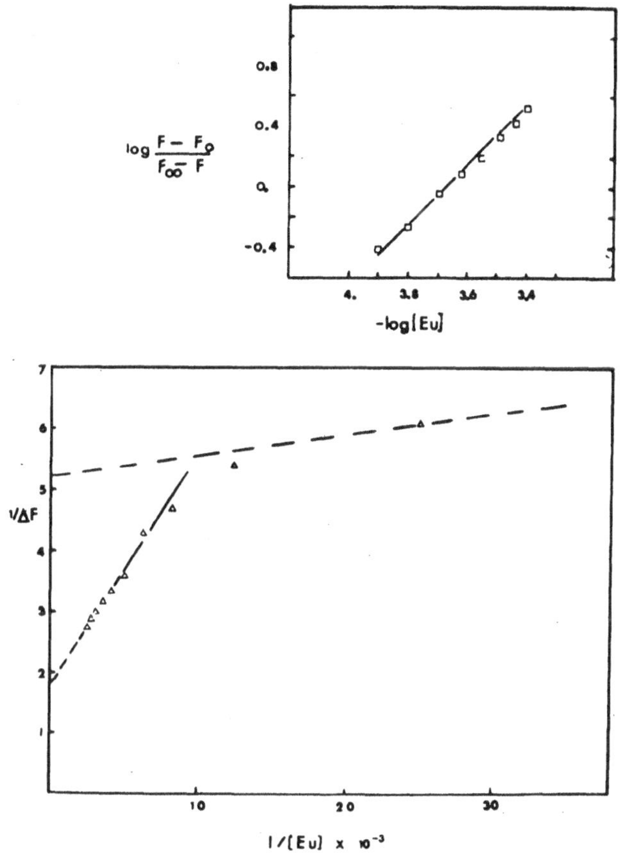

Figure 5 - Estimation of K_d for the apoenzyme-Eu(III)binary complex; fluorescence data collected at 580 nm, excitation at 287nm. Lower figure allows estimation of F which is employed in the calculation of K_d in the upper figure. The solution contained 2 M subunit concentration at pH 6.2, 25°C.

which presumably participates in the observed fluorescence phenomenon. Our preliminary results confirm such an hypothesis. In a control experiment in which 4 μM TPP was first added to 2 μM subunits at pH 6.3 gave, within experimental error, negligible enhancement even upon the addition of 400 μM Eu(III). If this result can be further substantiated it would strongly imply proximity of the two co-factors at least in the stronger binding site.

CONCLUSIONS

The research undertaken in our laboratory has substantiated the suggestions by others concerning rather tight binding of both the metal

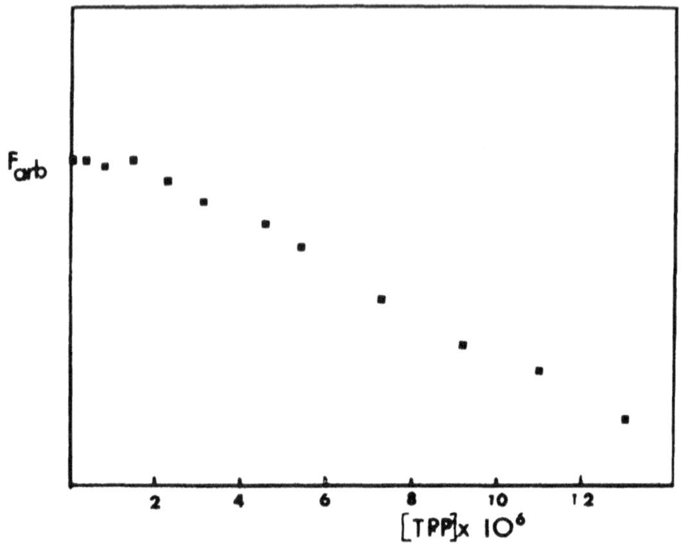

Figure 6 - The quenching of Eu(III) enhanced apoenzyme fluorescence by TPP; excitation at 287 nm, emission measured at 580 nm. The reaction contained ca 2 M subunit concentration 0.4mM Eu(III), pH 6.2, 25°C.

ion and of TPP in the ternary holoenzyme complex. The implication of the existence of two classes of binding sites with a large difference in K_d values raises important questions as to their roles. On account of the tight binding the protocol we have followed, i.e. resolution of holoenzyme followed by reconstitution, will be essential in our future studies. We have developed spectroscopic tools which will be useful in the more detailed investigation of both the reconstitution process and of the catalytic site of the enzyme. This investigation will be particularly aided by our ability to reconstitute holoenzyme with Eu(III) and perhaps in the future with other lanthanide ions.

ACKNOWLEDGEMENTS

We are grateful to the National Institutes of Health (Grant NIH AM 17495) and to the Rutgers University Research Council for continued support of our research and for enabling us to present our results during these Proceedings of the Jerusalem Symposia.

REFERENCES

Biaglow, J. E., Mieyal, J. J., Suchy, J. and Sable, H.Z.: 1969, J. Biol. Chem. 244, pp.4054-4062.

Breslow, R.: 1957, Chem. Ind. pp. 893-894.

Crosby, J., Stone, R. and Lienhard, G. E.: 1970, J. Amer. Chem. Soc. 92, pp. 2891-2900.

Crosby, J. and Lienhard, G. E.: 1970, J. Amer. Chem. Soc. 92, pp. 5707-5716.

Farzami, B., Mariam, Y. H. and Jordan, F.: 1977, Biochemistry, 16, pp. 1105-1110.

Gallo, A. A., Hansen, J. L., Sable, H. Z. and Swift, T. J.: 1972, J. Biol. Chem. 247, pp. 5913-5920.

Gallo, A. A. and Sable, H. Z.: 1975, J. Biol. Chem. 250, pp. 4986-4991.

Gallo, A. A., Mieyal, J. J. and Sable, H. Z.: 1978, Bio-organic Chem. Vol. 4, pp. 147-177, Academic Press, N.Y.

Grande, H. J., Houghten, R. L. and Veeger, C.: 1973, Eur. J. Biochem. 37, pp. 563-569.

Gounaris, A. D., Turkenkopf, I., Buckwald, S. and Young, A.: 1971, J. Biol. Chem. 246, pp. 1302-1309.

Gounaris, A. D., Turkenkopf, I., Civerchia, L. L. and Greenlie, J.: 1975, Biochim. Biophys. Acta, 405, pp. 492-499.

Heinrich, C. P. and Schmidt, D.: 1973, Experientia, 29, pp. 1226-1227.

Heinrich, C. P.: 1973, Experientia, 29, pp. 1227-1228.

Jordan, F. and Farzami, B.: 1977, J. Chem. Soc. Perkin Trans. II, pp. 1484-1486.

Jordan, F. and Mariam, Y. H.: 1977, Microchem. J. 22, pp. 182-200.

Jordan, F. and Mariam, Y. H.: 1978, J. Amer. Chem. Soc. 100, pp. 2534-2541.

Kemp, D. S. and O'Brien, J. T.: 1970, J. Amer. Chem. Soc. 92, pp. 2554-2555.

Kochetov, G. A., Usmanov, R. A. and Merzlov, V. P.: FEBS Lett. 9, pp. 265-266.

Kochetov, G. A. and Usmanov, R. A.: 1970, Biochem. Biophys. Res. Commun. 41, pp. 1134-1140.

Krampitz, L. O.: 1969, Ann. Rev. Biochem. 38, pp. 213-240.

Lehrer, S. S. and Fasman, G. D.: 1966, Biochem. Biophys Res. Commun. 23, pp. 133-138.
Lowry, O. H., Rosebrough, N. J., Farr, A. L. and Randall R. J.: 1951, J. Biol. Chem. 193, pp. 265-275.
Ludewig, R. and Schellenberger, A.: 1974, FEBS Lett. 45, pp. 340-343.
Mieyal, J. J., Suchy, J., Biaglow, J. E. and Sable, H. Z.: 1969, J. Biol. Chem. 244, pp. 4063-4071.
Sable, H. Z. and Biaglow, J. E.: 1965, Proc. Natl. Acad. Sci. U.S.A. 54, pp. 808-814.
Schellenberger, A.: 1967, Angew. Chemie. Int. Ed. 6, pp. 1024-1040.
Schellenberger, A., Hübner, G. and Lehmann, H.: 1968, Angew. Chemie. Int. Ed. 7, pp. 886-887.
Ullrich, J.: 1970, Methods Enzymol. 18A, pp. 109-115.
White, W. D. and Drago, R. A.: 1971, Inorg. Chem. 10, pp. 2727-2735.
Wolfson, J. M. and Kearns, D. R.: 1974, J. Amer. Chem. Soc. 96, pp. 3653-3654.

AB INITIO MOLECULAR ORBITAL STUDIES OF THE FLAVINS

Michael H. Palmer,* and Roelof J. Platenkamp††

* Department of Chemistry, University of Edinburgh, West Mains Road, Edinburgh, EH9 3JJ, Scotland.

†† Gorlaeus Laboratories, Department of Chemistry, The University of Leiden, Leiden, The Netherlands.

The biochemical importance of riboflavin (1a, R = $-CH_2(CHOH)_4CH_2OH$, T = Me) is well known; numerous studies have shown that its role is a redox one, and in particular with net hydride ion transfer from pyridine nucleotides eg NADH to 1a [1-5].

Our present study is concerned with simple molecules related to 1a, which are none-the-less realistic analogs since in-vitro studies show that the simplified molecules do carry out the same reactions. We have performed ab initio molecular orbital studies of simple 10H-iso-alloxazines and 1H-alloxazines namely:- the oxidised forms (1b, 1c), the reduced form (2b), and the radicals (3-5, where R=H) which occur at appropriate levels of pH. Because of the

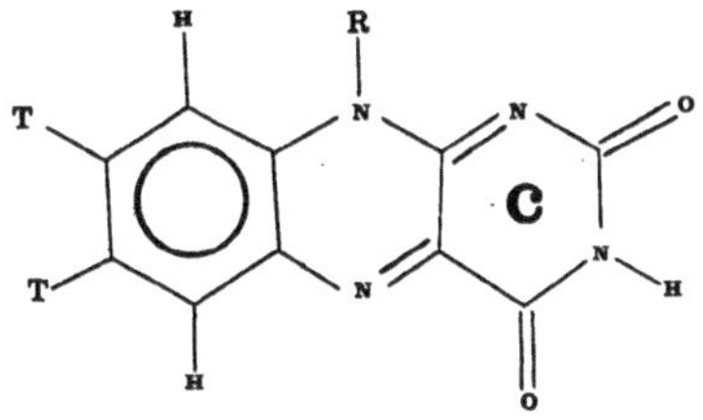

(a) T = Me, R = $-CH_2(CHOH)_4CH_2OH$

(b) T = H, R = Me

(c) T = H, R = H

(d) T = Me, R = Me

B. Pullman (Ed.): Catalysis in Chemistry and Biochemistry. Theory and Experiment. 147–169.

similarity of the right hand ring (C), we have also studied uracil (6), lumazine (7) and the 1H-alloxazine (8). The study is completed with results for the ions formed by addition or removal of $H^{\oplus}$ to the oxidised and reduced forms respectively, namely the cations (9 and 10), anions (11 and 12) and the dianion (13).

There have of course been previous M.O. studies of some members of the flavin series, using Hückel M.O. [6-8], EHM [9,10], CNDO-2 [9] and PPP-π [11,12] methods. These appear to be the first *ab initio* studies of the present range of molecules.

Where possible we have used experimental geometries, for example for uracil [13] and lumazine [14]. 1H-Alloxazine (5) was based on the latter, while the 10H-isomer (1c) was based on lumiflavin (1d) [15] and the reduced

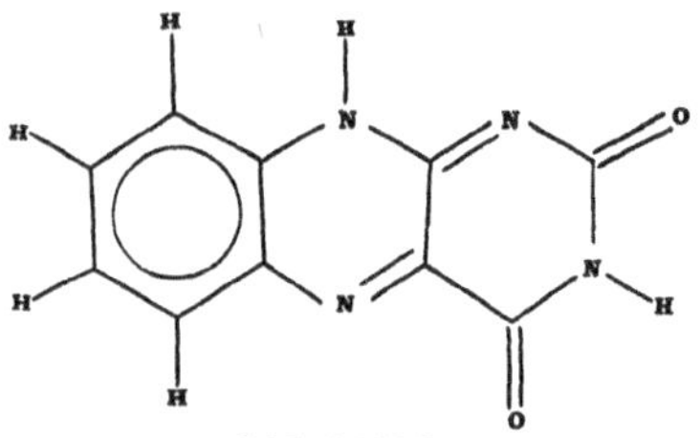

-747.40595
(1c)
10H isoAlloxazine

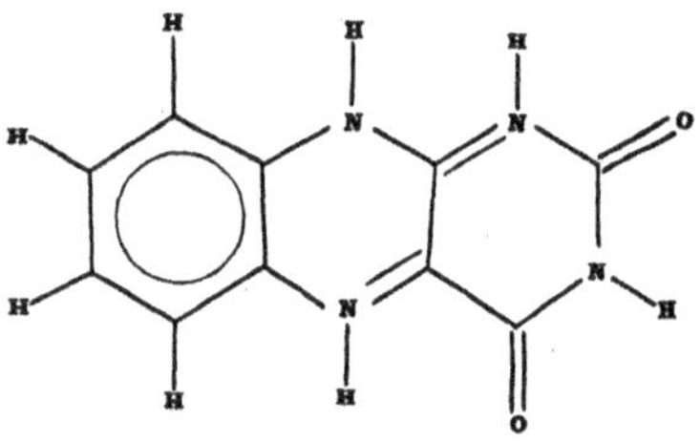

-748.62732
(2b)
Reduced Alloxazine

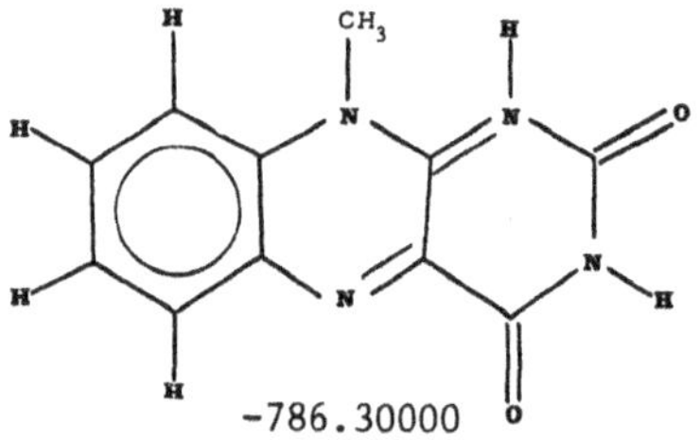

-786.30000
(1d)
10-Methyl isoAlloxazine

MOLECULAR STRUCTURES AND TOTAL ENERGIES (a.u.)

flavin (2b) on the 9-bromo-1,3,7,8,10-pentamethyl derivative [16,17]. The computations used were our previous single zeta Gaussian basis [18] in the ATMOL-3 system on an IBM 370/168 computer. The present results are summarised in Table 1 for total energy, and Table 2 for atomic populations.

MOLECULAR STRUCTURES AND TOTAL ENERGIES (a.u.)

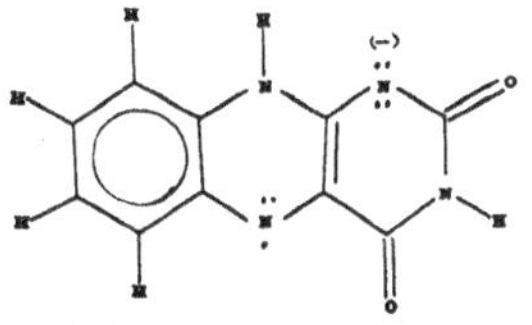

(3) ANION RADICAL
-747.45957

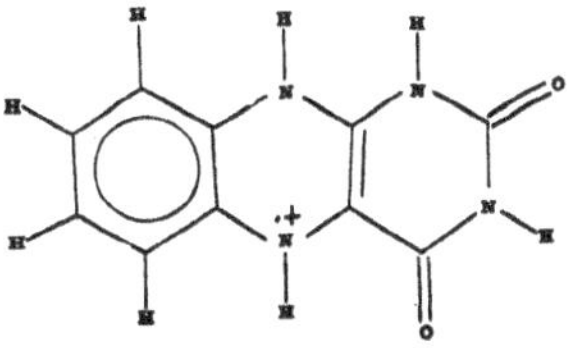

(5) CATION RADICAL
-748.39942

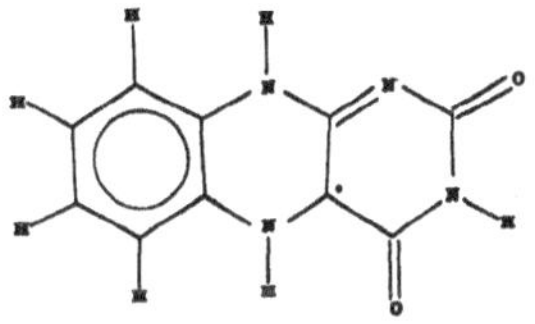

(4a) NEUTRAL RADICAL
-747.97434

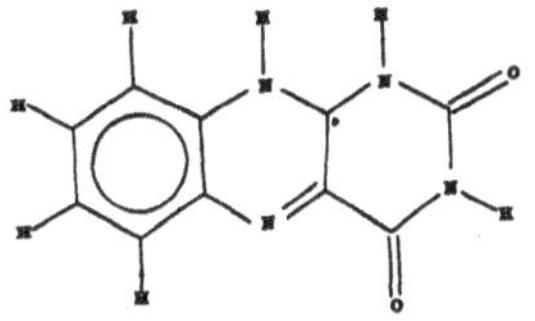

(4b) NEUTRAL RADICAL
-747.97279

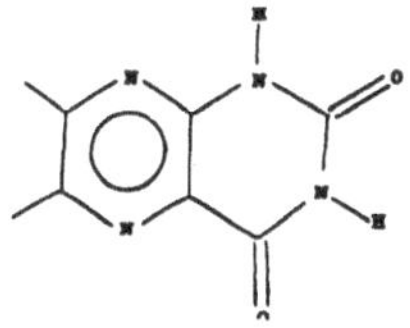

(7) LUMAZINE
-595.16374

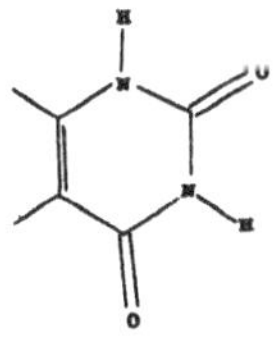

URACIL
MB
-411.07831
(6)

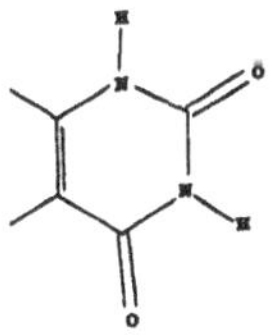

URACIL
DZ
-412.23909
(6)

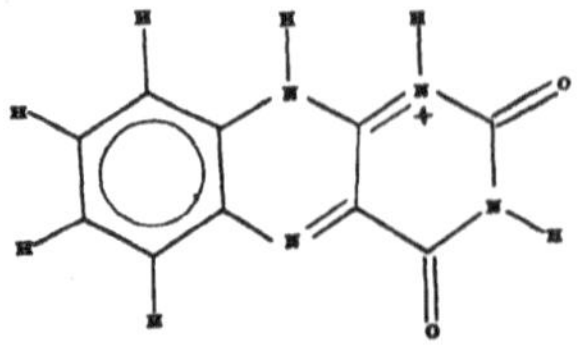

(9) 1H-PROTONATED
-747.75892

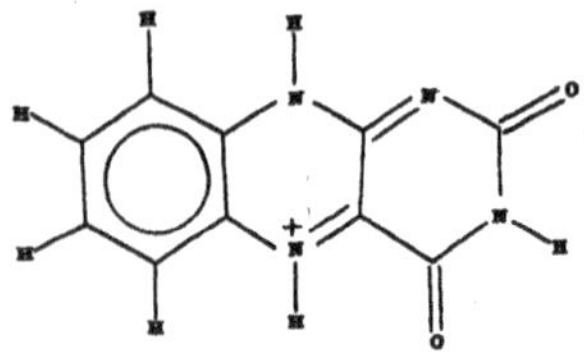

(10) 5H-PROTONATED
-747.69439

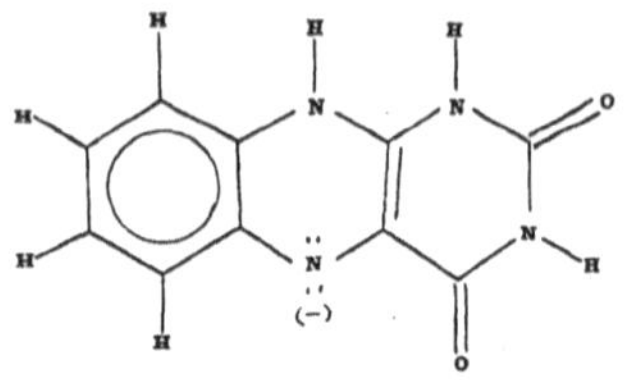

(11) 1,10H_2-ANION
-747.93571

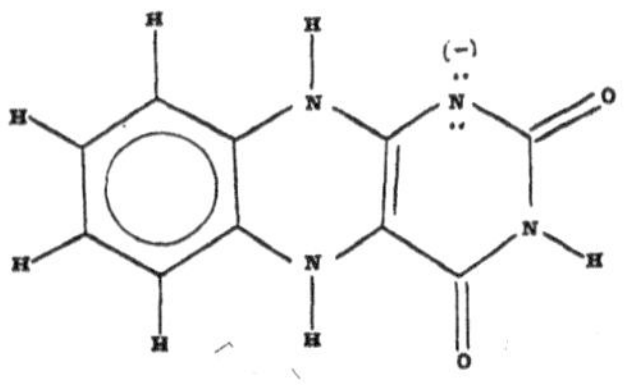

(12) 5,10H_2-ANION
-747.99679

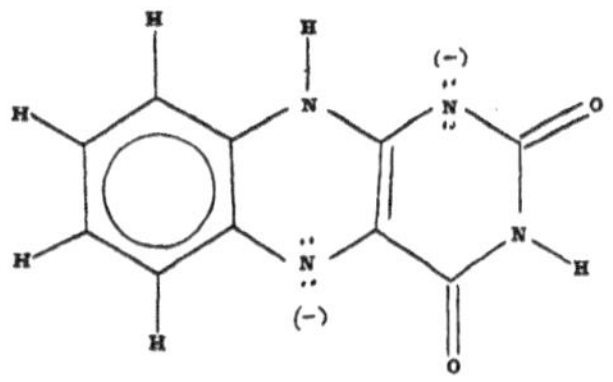

(13) DIANION
-747.26494

Equilibrium Geometry of the Radicals (3-5) and Reduced Flavin (2b).

The experimental crystal structures [16, 17] are usually heavily substituted and thus subject to steric intractions, for example between the 9-bromo- and 10-methyl groups in the reduced compound studied [16, 17]. Also there will be solid state lattice forces. We thus set out to optimise the folding ("butterfly") angle (Φ) across the centre of ring B in the V-shaped molecules 2b (Fig. 1) etc. The minimum for 2b (Fig. 2), the radicals 3, 4a, 5, 11 and 13 was near 164° in all cases. The barrier to inversion in the reduced neutral compound (2b) was

TABLE 1. Molecular Total Energies (a.u.)[a]

Molecule	E_T	E_T
10H-Isoalloxazine	(1c)	-747.40595
10Me-Isoalloxazine	(1b)	-786.30000
10H-Isoalloxazine Dianion	(13)	
Interplanar angle 148°		-747.23351
" 164°		-747.26494
" 180°		-747.23441
Uracil (6) (minimal basis)		-411.07831
Uracil (6) (double zeta)		-412.23909
Lumazine (7)		-595.16374
1H-Protonated-isoalloxazine	(9)	-747.75892
5H " "	(10)	-747.69439
1,10H_2-Alloxazine anion	(11)	-747.93571
5,10H_2-isoalloxazine anion	(12)	-747.99679
1H-Alloxazine	(8)	-747.49551
1,5,10H-Dihydroalloxazine	(2b)	
Interplanar angle 148°		-748.59844
" 164°		-748.62732
" 180°		-748.57929

The Radicals		E_T RHF	E_T UHF
1,10H_2 Neutral radical (4b) (180°)		-747.97279	
1,5H_2-Neutral radical (4a)			
	148°	-747.94946	
	164°	-747.98737	
	180°	-747.97434	-748.00039
10H-Anionic radical (3)			
	148°	-747.42143	
	164°	-747.45343	
	180°	-747.43101	-747.45957
1,5,10H_3-Cationic radical (5)			
	148°	-748.33941	-748.35652
	164°	-748.38153	-748.39942
	180°	-748.33951	-748.35558

[a] 1 a.u. = 2626 kJ mol^{-1}

Table 2 Total and π-Populations for Uracil, Lumazine, 1H-Alloxazine and 10H-Isoalloxazine.

(a) URACIL

	0-2	0-4	N-1	N-3
π	1.4076	1.3516	1.7368	1.7293
Total	8.5148	8.4899	7.4658	7.4609
	C-2	C-4	C-5	C-6
π	0.8626	0.8610	1.1371	0.9123
Total	5.3804	5.5388	6.2322	5.9702
	H-1	H-3	H-5	H-6
Total	0.6725	0.6661	0.8079	0.8004

(b) LUMAZINE

	0-2	0-4	N-1	N-3	N-5	N-8
π	1.4116	1.3106	1.7407	1.7279	0.9405	1.0669
Total	8.5143	8 4622	7.4486	7.4444	7.1961	7.2613
	C-2	C-4	C-4a	C-6	C-7	C-8a
π	0.8654	0.8683	1.1114	1.0466	0.9576	0.9524
Total	5.4008	5.5224	5.9499	6.0661	6.0350	5.7463
	H-1	H-3	H-6	H-7		
Total	0.6609	0.6604	0.8127	0.8185		

Table 2 Total and π-Populations (cont)

(c) 1H-ALLOXAZINE

	0-2	0-4	N-1	N-3	N-5	N-10
π	1.4155	1.3125	1.7450	1.7306	0.9394	1.0894
Total	8.5189	8.4656	7.4509	7.4460	7.1945	7.2719

	C-2	C-4	C-4a	C-5a	C-6	C-7
π	0.8642	0.8661	1.1130	1.0642	0.9613	0.9929
Total	5.4028	5.5240	5.9543	5.9441	6.1280	6.1627

	C-8	C-9	C-9a	C-10a
π	0.9492	1.0002	0.9845	0.9720
Total	6.1484	6.1426	5.9140	5.7603

	H-1	H-3	H-6	H-7	H-8	H-9
Total	0.6640	0.6626	0.8028	0.8185	0.8136	0.8076

(d) 10H-ISOALLOXAZINE

	0-2	0-4	N-1	N-3	N-5	N-10
π	1.332	1.325	1.258	1.731	0.898	1.680
Total	8.490	8.473	7.347	7.454	7.193	7.404

	C-2	C-4	C-4a	C-5a	C-6	C-7
π	0.894	0.870	1.090	1.118	0.934	1.038
Total	5.448	5.531	5.939	5.942	6.123	6.182

	C-8	C-9	C-9a	C-10a
π	0.926	1.063	0.942	0.900
Total	6.123	6.178	5.857	5.727

	H-3	H-6	H-7	H-8	H-9	H-10
Total	0.670	0.809	0.826	0.819	0.818	0.666

Table 2 (cont) Total and π-Populations

(e) 10 METHYL-ISOALLOXAZINE (partial results)

	0-2	0-4	N-1	N-3	N-5	N-10
π	1.339	1.327	1.276	1.731	0.901	1.641
Total	8.496	8.477	7.357	7.454	7.175	7.280

	C(Me)	H(Me)
Total	6.353	0.801

(f) REDUCED ALLOXAZINE (2b)

	0-2	0-4	N-1	N-3	N-5	N-10
π	1.436	1.406	1.695	1.693	1.716	1.693
Total	8.535	8.512	7.439	7.443	7.447	7.441

	C-2	C-4	C-4a	C-5a	C-6	C-7
π	0.866	0.876	1.188	0.981	1.044	1.006
Total	5.405	5.549	5.977	5.875	6.183	6.166

	C-8	C-9	C-9a	C-10a
π	1.032	1.013	1.038	0.931
Total	6.175	6.162	5.902	5.715

	H-1	H-3	H-5	H-6	H-7	H-8
Total	0.665	0.675	0.702	0.833	0.835	0.833

	H-9	H-10
Total	0.831	0.700

calculated as 30kcal mol^{-1}, to be compared with experimental NMR results of about 20 kcal mol^{-1} [19]. Of course no additional geometric relaxation was allowed on flattening, so our value is bound to be high. The calculations clearly show that the equilibrium structures for 3, 4a, 5, 11 and 13 are all non-planar, but that the inversion barrier will be low for the mono- and di-anions 11 and 13. This has considerable implications on the ESR results for the radicals, and their interpretation.

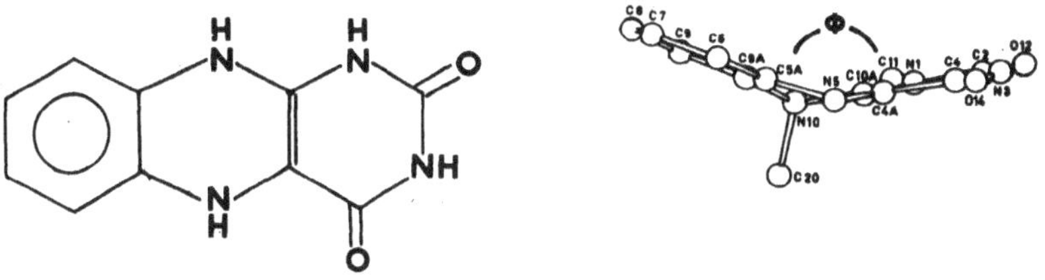

Fig. 1. Reduced alloxazine and the 'butterfly angle'

The Photoelectron Spectrum of 10-Methylalloxazine (1b).

The He(I)/He(II) PES of 10-methylisoalloxazine (1b) shows (Fig.3) only 10 discrete IP's in the 8-23eV range. The valence shell contains 42 doubly occupied M. O. 's, so that there is no question of an orbital by orbital assignment. However, the orbitals of primary interest at those corresponding to the first few IP's, and especially the LP_N and LP_O levels. At the high binding energy end the IP's near 21 and 23eV which must correspond to $2s_C$ levels we have attempted to arrange the M. O. 's in groups to be consistent with the overall envelope (Table 3). Thus the first two IP's are assigned to 1 and 2 π-levels respectively. The present calculations suggest that the LP_N levels are split by $\sim$ 2eV, whereas the LP_O levels differ by $\sim$ 1eV [24]. These lead to assignments of 10.37/12.6eV for LP_N and 10.37/10.73 for LP_O. Although the LP_O levels occur in the calculations as linear combinations $(LP_{0-4} \pm LP_{0-2})$, the LP_N separation is sufficiently large for the 2 IP's to refer to effectively separate levels LP_{N-1} (10.37) and LP_{N-5} (12.6eV). The lower binding energy of LP_{N-1} relative to LP_{N-5} is consistent with the more energetically favourable protonation of N-1 above; the higher value for LP_{N-5} is also consistent with the more positive character of N-5.

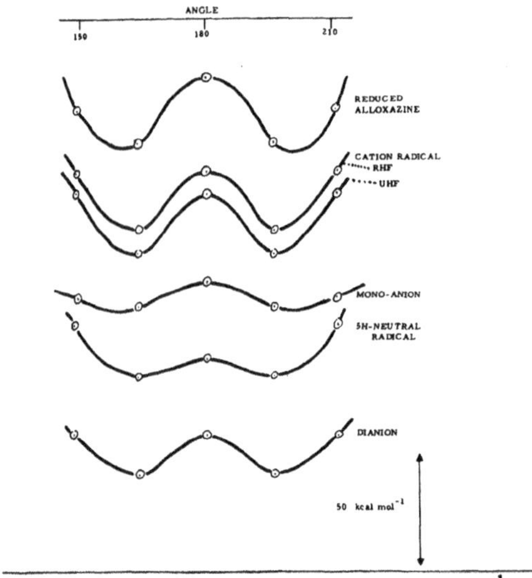

Fig. 2. Inversion Barriers (kcal mol^{-1})

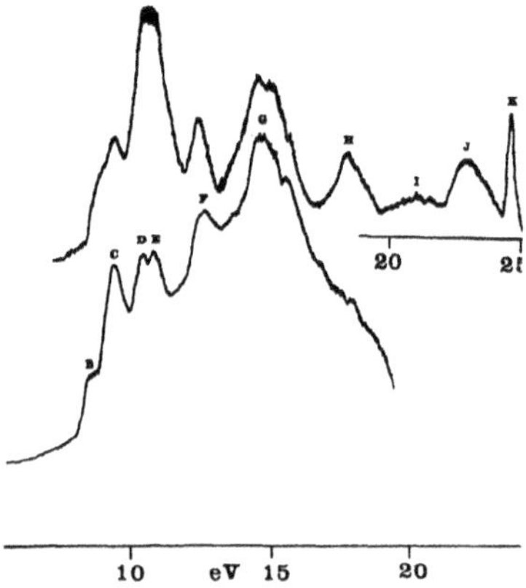

Fig. 3. Photoelectron Spectrum of 10-Methyl-isoalloxazine.
(Upper trace - He(II); Lower trace - He(I))

Electron Distribution in Uracil, Lumazine and 1H-Alloxazine

The only previous all electron calculation for these molecules was the FSGO calculation on uracil [21]. Few results were given, but the dipole moment was calculated as 2.27D, to be compared with the solution value of 4.16D [22]. The present calculated figure is 4.54D, and clearly arises from a delicate balance of the bond internal dipoles. There seems to be general agreement between these results and the semi-empirical studies [23] on the dipole direction (Figure 4), which we find to make an angle of 13° to the N-3/H-3 bond a towards C-4/O-4. Of course a gas-phase dipole moment would be better for comparison with the calculated data.

Direct comparison of the dipole moments for the series uracil (6), lumazine (7), 1H-alloxazine (8) and 10H-alloxazine (1c) is shown in Figure 4. The principal difference is in the magnitudes (shown to scale) which are much higher for the 10H- and 10Me-compounds (calculated values 7.8 and 8.0D).

In general the atomic populations change by comparatively small amounts (Fig. 5) on going from uracil (6) to lumazine (7) etc. This is especially true of

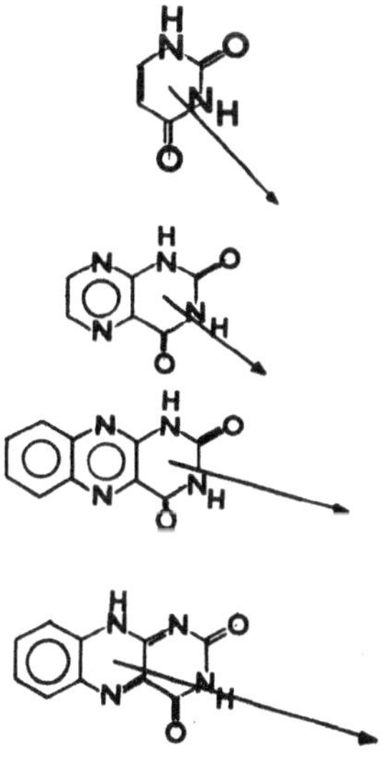

Fig. 4. Dipole Moments, Relative Magnitudes, and Directions

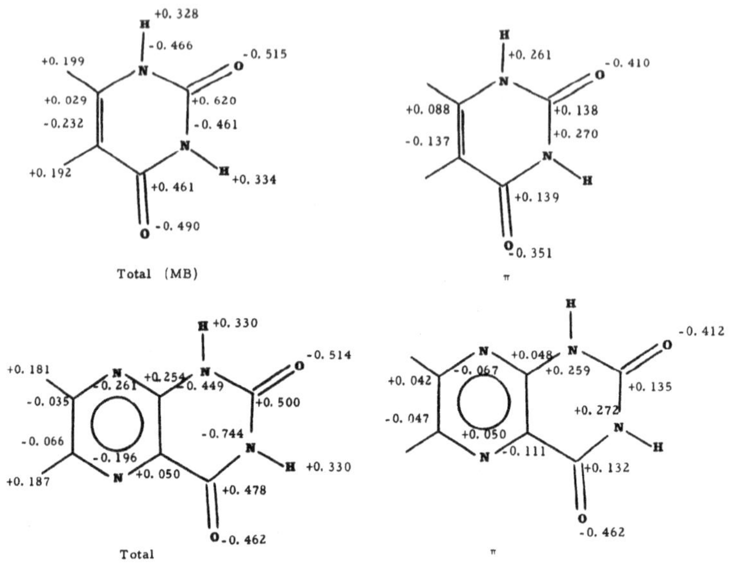

Fig. 5a. Atomic populations for uracil and lumazine

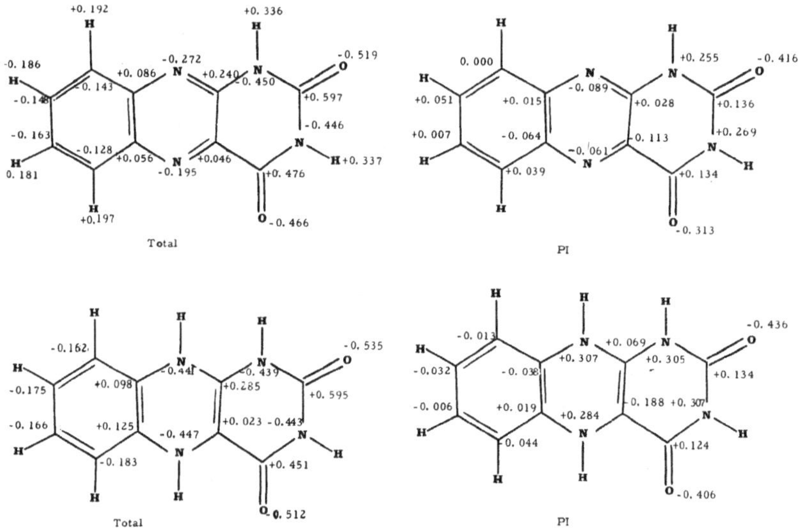

Fig. 5b. Charge distribution in 1H-Alloxazine (8) and the reduced compound (2b)

the π-density (Table 2). In the σ-system replacement of the $\overset{\delta-}{C}$ - $\overset{\delta+}{H}$ dipole (δ = 0. 2e) by C-C lowers the C-4a/C-8a σ-populations by about 0. 25e. The change from lumazine (7) to 1H-alloxazine (8) is similar except that the low polarity of the benzenoid ring brings small changes in C-6/C-7 of 7 on fusion to 8 (about 0. 12e). The energy difference b ween the 1H-and 10H-forms, alloxazine (8) and the isoalloxazine (1c) is suffiently large (56 kcal mol^{-1}) (Figure 6), that it seems unlikely that stable 10H-species will exist withour a substituent at N-10 in the neutral state.

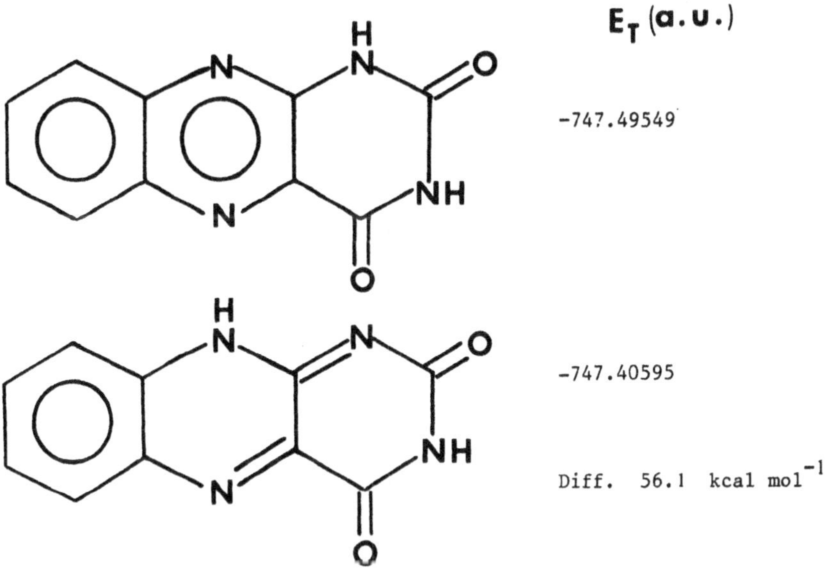

Fig. 6. Comparison of alloxazine and isoalloxazine in energy

The reduced alloxazine (2b) shows very similar populations (Table 2). Each of the -NH- groups is like those in aniline or pyrrole, σ-acceptor and π-donor. The principal beneficiary of the π-donation is C-4a (-0. 19e). This could well be relevant to the reactivity of the reduced alloxazine (2b) with oxygen in the natural reoxidation step.

10H- and 10-Methyl isoAlloxazines (1b, 1c).

With the obvious exceptions of N-1/N-10, the populations (Table 2) for the 10H-compounds (1b, 1c) are very similar to that for the 1H-isomer (8). Although both N-1 and N-5 are nominally pyridine-like in nature there is a difference in total populations of 0.15e, with N-1 being the more electron rich (Table 2, Figure 7a). The π-electron population difference (Figure 7b) is even higher (0.36e with N-1 being more rich). Clearly N-5 is unusual; these observations are not an artefact of the 10H, since the 10-Me compound (1b) shows the same phenomenon, although the 10-Me is a weaker donor (0.24e) to

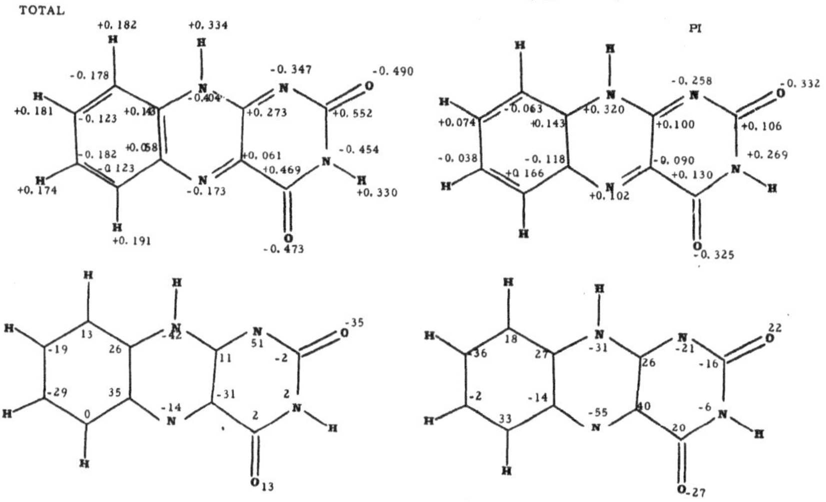

55 HOMO Fig. 7. 10H-Alloxazine 56 LUMO

the σ-system than the 10H (0.34e). Thus although N-5 is overall negative, the level is much less than usual and the π-component is not positive. The LP_N IP's also indicate that this is the case (Table 3). Overall the 10H-series behave as a series of semi-independent groups composed of -CO-NH-CO-, -N=C-C=N-, and $-C_6H_4$-NH- groups. The great reactivity is of course associated with the diaza-1,3-butadiene group.

TABLE 3

10-Methylisoalloxozine

IP	PE Band	Correlation* Orbital Energies
8.72	B	9.47 (π)
9.51	C	11.14 (π), 11.57 (π)
10.37	D	11.84 (LP_{N-1}), 12.15 (LP_O)$^-$
10.73	E	12.75 (π), 13.07 (LP_O)$^+$
12.6	F	13.98 (LP_{N-5}), 14.48 (π), 15.59 (π)
14.0 to 15.3	G	16.26 (σ) to 22.67 (σ) (including several π-levels, 17 IPs in total).
18.4	H	23.79 (σ), 24.90 (σ)
(18.9)	(H)	26.04 (σ)
21.1	I	27.2 (σ)
23.0	J	29.3 (σ) to 29.89 (σ)

*A,K = He^+ (24.58 eV binding energy).

Reactivity of the Oxidised Flavin (1c).

The HOMO and LUMO of the oxidised form have a large amount of buta-1,3-diene character as mentioned elsewhere. The distortion of the -N=C-C=N- portion towards N-1 is shown in Figure 7c for the HOMO, 7d for LUMO, where eigenvectors X100 are given (without sign). This is consistent with ready protonation at N-1, initially via π-coordination followed by rotation into the plane. Protonation of N-5, already less probable on total density and binding energy grounds looks very improbable on HOMO distribution grounds; this is of relevance to the formation of the neutral radical from the oxidised flavin (1c).

In fact protonation of 10H-isoalloxazine (1c) is thermodynamically much more favourable at N-1 yielding 9 than N-5 yielding 10 on total energy grounds. It is easy to explain the marked preference for N-1 on classical resonance terms involving N-10, but the energy difference calculated (40 kcal mol^{-1}) is large, and about equal to the resonance energy of a benzene ring, using the present basis set. A further interesting feature of the H-1 protonated form (9) is that its π^* LUMO, which is presumably the most accessible to an incoming nucleophile, has a very high amplitude at N-5, with rather less at C-10a. This is again all restricted to the -N=C-C=N- system, as indeed is the next higher LUMO which is again π^* and localised largely at C-10a, with alternating sign along the -N=C-C=N- system [c.f. butadiene]. The latter, C-10a, is the position attacked by methanol in the quaternary salt of type 9 [25]. The LUMO itself then suggests a possible route for reduction of oxidised flavins (1c, 1d etc):- (i) protonation at N-1 yielding 9, (ii) nucleophilic addition at N-5, either of the NADH or less likely of $H^{\ominus}$. Further alternatives are electron capture yielding the neutral radical 4b, followed by hydrogen atom transfer yielding 2b.

Of course the biological activity of riboflavin and its derivatives does not have to occur via a protonated state. The LUMO of the oxidised compounds (1b, 1c) have a very high amplitude at N-5 with rather less at C-4a. These are the sites for nucleophilic attack. If we place an electron in the LUMO, the radical anion (3) has a very high concentration of spin density at N-5 (Fig. 8c); in fact the UHF calculation shows that 41% of the single electron is placed at N-5. This is still true of the dianion. However, although some discussion of a possible mechanism involving 2-electron transfer followed by an internal $H^{\oplus}$ shift direct from NADH has been given, the first step seems unlikely even though deuterium exchange reactions favour direct H transfer. Transfer of one electron from NADH requires the IP (8eV), the second would be prohibitive in energy, even though solvation effects would help. Furthermore, there appears to be no driving force as far as NADH is concerned, for the aromatic energy gain only comes in a subsequent step.

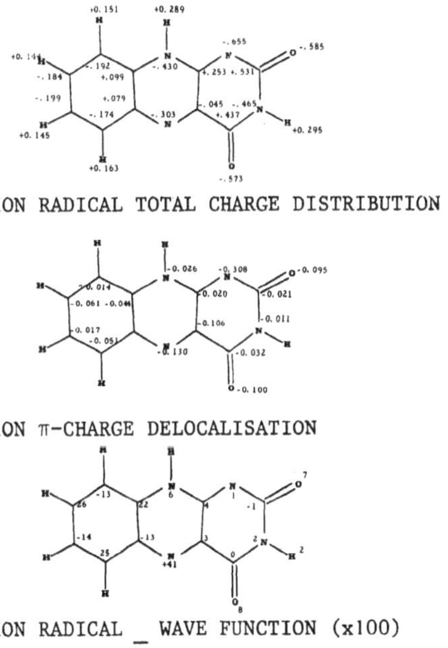

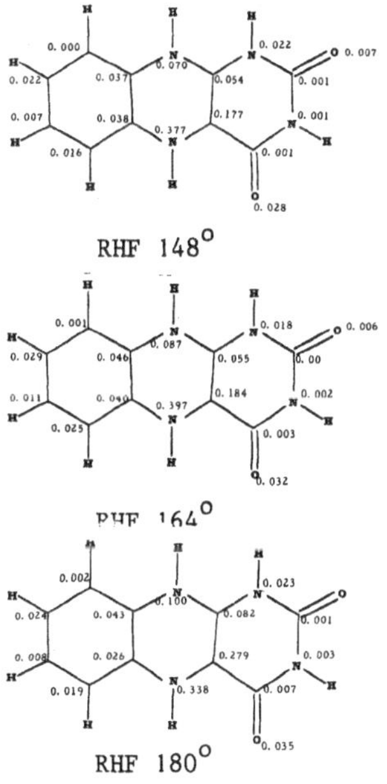

Fig. 8. The radical anion

Fig. 9. Spin density versus angle

The Radicals and their Representation

These compounds (3-5) have generally been assumed planar, and this is now thought not to be the case. Various workers [26, 27] have recorded the ESR hyperfine splitting of anionic (3), neutral radical (4a) and cationic radical (5). The compounds are usually heavily substituted, but a summary is given in Table 4. The principal coupling is with N-5, N-10 and C-4a. The PPP-π UHF calculations [11] do not give reasonable values for the N-5/N-10 coupling ratio. The corresponding values from the present UHF calculations are given; they are much improved, particularly in their internal ratios 2.7:2.5:1.5 for anion:neutral:cation radical N-5/N-10 ratio. These are again based upon the planar forms. It is interesting to note that the ratios for the non-planar forms are all higher; for instance in the cation radical (Table 5) the ratios at angles of 148°, 164° and 180° are respectively 5.4, 4.6 and 3.4 in the RHF calculations (Fig. 9), which are reduced in the UHF cases to 4.0, 3.5 and 2.8. Clearly optimisation of the coupling with basis set is important as well as with geometry. The most obvious feature of the present limited results is the loss of spin density at C-4a as the molecules go non-planar.

Fortunately the total populations at the centres are much less subject to change with geometry. The neutral radical conventionally drawn as 14 is much better represented by 4a, since 35% of the spin density is found on C-4a, 26% on N-1 and only 11% on N-5 (Fig. 10). In contrast, the cation radical is well represented by 5, since 40% of the spin density is localised there, as well as 40% of the positive charge (Fig. 11). The mono-anionic radical is more difficult to represent by a single structure, but the high spin density at N-5 (41%) (Fig. 8) makes 3 the choice.

TABLE 4 ESR Coupling Constants (Gauss) for Isoalloxazine Derivatives

10-Methyl-Radicals

Coupling Constants (a_x)	Anion (+e) $LH^{\cdot -}$	Neutral (+H·) $LH_2^{\cdot}$	Cation (-e) $LH_3^{\cdot +}$
X			
N-1	-	-	-
N-3	-	-	-
N-5	7.3	8.0	8.5,9.0,7.39
N-10	3.2	3.6	4.3, ,4.85
C-4a	-(or +)3	-(or +)5.5	-(or +)1.8
H-6	3.5	1.7	-
H-8	4.0	2.4	3.4
H-9	0.9	-	-
H-5	-	7.6	7.75
H-10(Me)	3.0	3.9	4.9
Ratio N-5/N-10	2.3	2.2	2.0 , ,1.52
PPP-UHF	5.6	18.5	1.3
Present-UHF (planar)	7.2	6.6	4.0
Relative-ratios	2.7	2.5	1.5

TABLE 5. Reduced isoalloxazine cation radical spin densities versus angle, Methods and coupling constants.

METHOD		ANGLE 148°	164°	180°
RHF	N-5	0.377	0.397	0.338
	N 10	0.070	0.087	0.100
	C-4a	0.133	0.184	0.279
	C-6	0.016	0.025	0.019
	C-8	0.022	0.029	0.024
	N-5 / N-10	5.39	4.56	3.38
UHF	N-5	0.301	0.339	0.277
	N-10	0.076	0.105	0.098
	C-4a	0.133	0.160	0.217
	C-6	0.196	0.158	0.209
	C-8	0.206	0.172	0.228
	N-5 / N-10	3.96	3.23	2.83
	$\langle S^2 \rangle$	1.268	0.929	1.197

NEUTRAL RADICAL TOTAL CHARGES

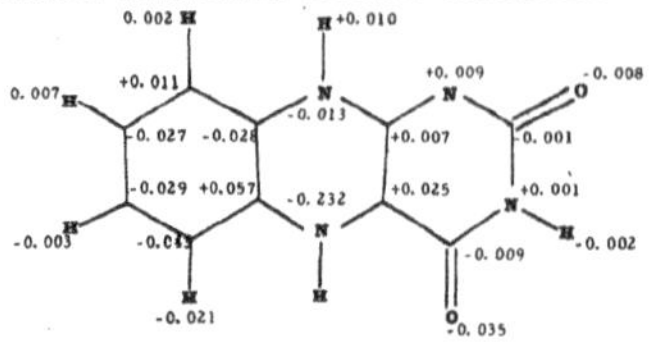

NEUTRAL RADICAL π-CHARGES

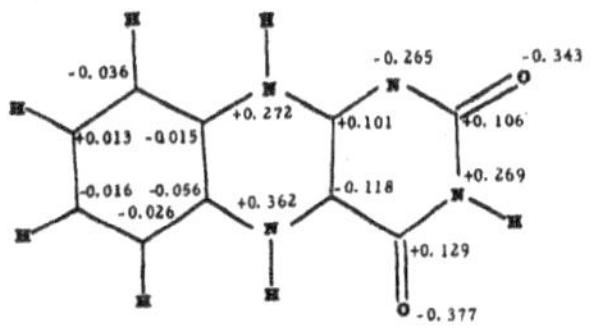

NEUTRAL RADICAL SPIN DENSITY

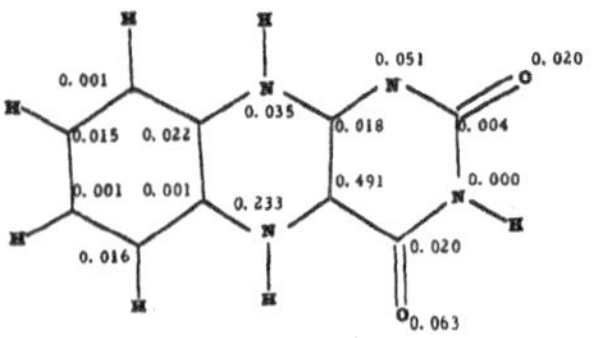

Fig. 10. The neutral radical

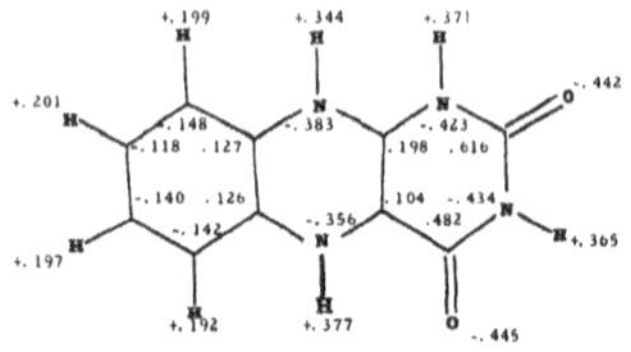

CATION RADICAL TOTAL CHARGES

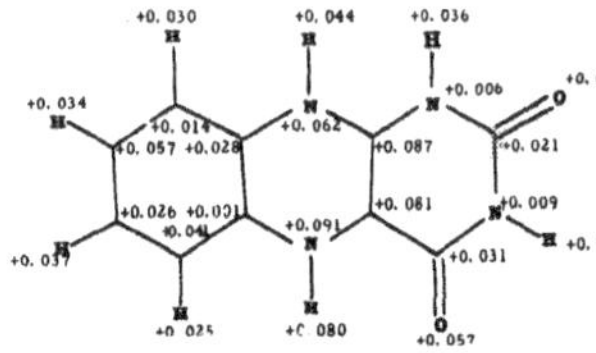

CATION RADICAL π-CHARGE DELOCALISATION

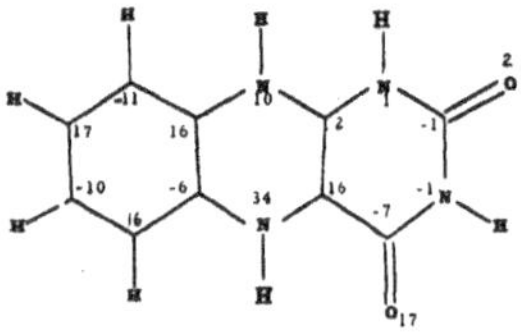

CATION RADICAL WAVE FUNCTION (x100)

Fig. 11. The cation radical

REFERENCES AND FOOTNOTES

* A preliminary account of this work was given at the SON Conference, Discussion Group in Biomolecular Chemistry, 23rd October, 1978, Lunteren, The Netherlands.

1. Ed. H. Kamin "Flavins and Flavoproteins," Butterworths, London, 1971.

2. Ed. T. P. Singer, "Flavins and Flavoproteins," Elsevier, Amsterdam, 1976.

3. C. Walsh, Ann. Rev. Biochem., 1978, 47, 881.

4. J. L. Fox and G. Tollin, Biochemistry, 1966, 5, 3865, 3874.

5. C. Kemel, T. W. Chan and T. C. Bruice, J. Amer. Chem. Soc., 1977, 99, 7272.

6. B. Pullman and A. Pullman, Proc. Nat. Acad. Sci. USA, 1959, 45, 136.

7. G. Karreman, Bull. Math. Biophys., 1961, 23, 55.

8. Pill-Soon Song and M. Sun, Jerusalem Symposium on Quantum Chem. Biochem., 1974, 6, 407; Biochem. React. Proc. Int. Symp., 1973).

9. Pill-Soon Song, J. Phys. Chem., 1968, 72, 536.

10. R. Norrestam, P. Kierkegaard, B. Stensland and L. Torbjornsson, J. C. S. Chem. Comm., 1969, 1250.

11. B. Graebe, Acta Chem. Scand., 1972, 26, 4084; ibid., 1974, 28A, 363; Biopolymers, Symposia, No. 1, 1964, 283.

12. M. Sun and Pill-Soon Song, Biochemistry, 1973, 12, 4663.

13. R. F. Stewart and L. H. Jensen, Acta Cryst., 1967, 23, 1102.

14. R. Norrestam, B. Stensland and E. Soderberg, Acta Cryst., 1972, B28, 659.

15. C. J. Fritchie, Acta Cryst., 1975, B31, 802.

16. M. Leijonmarck, Chem. Communications (Stockholm), 1977, No. 8, 1; R. Norrestam and B. Stensland, Acta Cryst., 1972, B28, 440.

17. R. Norrestam and M. V. Glehn, Acta Cryst., 1972, B28, 434.

18. M. H. Palmer, R. H. Findlay, S. M. F. Kennedy, W. Moyes and J. D. Nisbet, "Quantum Chemistry - The State of the Art," Ed. V. R. Saunders and J. Brown, Publ. Science Research Council, 1975, London, p. 229; see also references therein.

M. H. Palmer, R. H. Findlay and A. J. Gaskell, J. Chem. Soc. Perkin II, 1974, 420.

19. F. Müller, personal communication. It has also been observed that the oxidised form (1) ^{1}H nmr spectra give shifts on polymethyl substitution which are best understood in terms of non-planarity; see H. J. Grande, C. G. van Schagen, T. Jarbandhan and F. Müller, Helv. Chim. Acta, 1977, 60, 348.

20. F. P. Colonna, G. Distefano, V. Galasso, K. J. Irgolic, C. E. King and G. C. Pappalardo, J. Organomet. Chem., 1978, 146, 235.

21. T. J. O'Donnell, P. R. LeBreton and L. L. Simpson, J. Phys. Chem., 1978, 82, 343.

22. I. Kulakowska, M. Geller, B. Lesyng and K. L. Wierzchowski, Biochim. Biophys. Acta, 1974, 361, 119.

23. A. Pullman in "Quantum Aspects of Heterocyclic Compounds in Chemistry and Biochemistry." The Jerusalem Symposia on Quantum Chemistry and Biochemistry, Volume 2, March 1969; published Academic Press, 1970.

24. The photoelectron spectra of a series of flavin molecules including 1d and 5b have been obtained independently by workers in Amsterdam. There is substantial agreement for the IP's of 1b. Their tentative assignment of IP_1 and IP_2 in the 10-methyl compound (1b) to linear combinations $LP_N \pm LP_O$ on the basis of CNDO-2 calculations are

not borne out by the present work. We are grateful to them for access to unpublished results.

25. F. Müller in "Flavins and Flavoproteins," ed. H. Kamin, Butterworths, London, p. 363.

26. K. H. Dudley, A. Ehrenberg, P. Hemmerich and F. Müller, Helv. Chim. Acta, 1964, 47, 1354.

27. F. Müller, L. E. G. Eriksson and A. Ehrenberg, Eur. J. Biochem., 1970, 12, 93; F. Müller, P. Hemmerich, A. Ehrenberg, G. Palmer and V. Massey, ibid., 1970, 14, 185.

ACKNOWLEDGEMENT

This work originates from the Department of Biochemistry, Agricultural University of Wageningen, The Netherlands, where one of the authors (RJP) has worked and has been made possible via a grant of NATO (no. 912 (D 1935)) to Prof.Dr. C. Veeger.

FLAVIN-CATALYSIS, THE EXPERIMENTAL APPROACH

H.J. Grande, F. Müller, C. van Schagen, W.H. Scouten, C. Veeger*, A.J.W.G. Visser.
Department of Biochemistry, Agricultural University, De Dreijen 11, 6703 BC Wageningen, The Netherlands.

It is impossible to review fully the important contributions of many co-operating research teams, which has led to our present insight in redox reactions catalyzed by flavin containing proteins. The pioneering research and the development of insight in the molecular events is adequately summarized in review articles and symposium proceedings, to which the interested reader is referred [1-7].

The major goal of this contribution is to discuss the problem how and why flavoproteins are able to catalyse such a variety of different reactions. No doubt that the answer, "it is the protein which dictates the type of reaction", is correct in its simplicity. However the major question how the protein influences the flavin nucleus and at which sites cannot be answered, because our present views have developed from model studies. The model studies with substituted flavins, as well as enzymological studies with natural and modified flavins show that the nucleus positions of major importance (cf. Fig 1) are N(1), C(4a), N(5), C(8) and C(10a). There are a few high resolution X-ray structures of flavoproteins, preliminary structures of p-hydrobenzoate hydroxylase and glutathione reductase have been presented recently [6]. Thus one can only derive from these model studies proposals how and to which extent the interaction of the flavin with the specific protein at either one of these positions directs the flow and number of reducing equivalents from donor to acceptor and thus explain the physiological function.

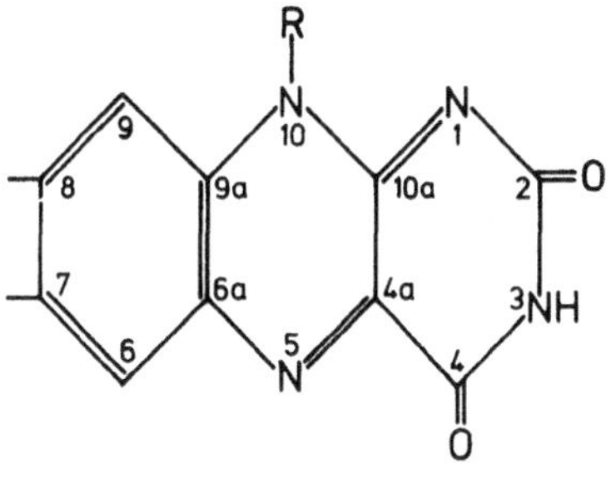

Fig. 1. The flavin nucleus

* Who presented this contribution.

B. Pullman (Ed.): Catalysis in Chemistry and Biochemistry. Theory and Experiment. 171–190.

The early days of flavin chemistry can be characterized by the general believe that flavoprotein catalysis proceeds via 1e-transfer. Thus flavin radicals were postulated as intermediates [8], a postulate which lacks any evidence in the majority of flavin-catalyzed reactions. In fact in a number of flavoproteins the red flavin-anion radical $Fl^{\overline{\cdot}}$, generated by photo-reduction, shows no catalytic activity. In certain flavoproteins specifically the blue neutral flavin radical (5-HFl·) can be generated, which is involved in biological activity. Actually it is proposed [7], that regio-specific hydrogen bridges between flavin and apo-enzyme determine the actual mechanism; e.g. blocking of the N(5) lone-pair determines one-electron transfer, while blocking in the N(1)-O(2) region implies two electron transfers.

Although this might be true to some extent the role of the protein is more complex. For instance reduced free flavins easily exchange H^+. But certain flavoproteins do not exchange upon reduction H^+ derived from the electrondonor with the solvent [9]; actually, depending on the enzyme, rates of H^+-exchange from fast to zero have been observed. For instance the enzyme transhydrogenase transfers 2 electrons and the B-side H^+ from one pyridine nucleotide to the same B-side of the other pyridine nucleotide with little H^+-exchange with the solvent.

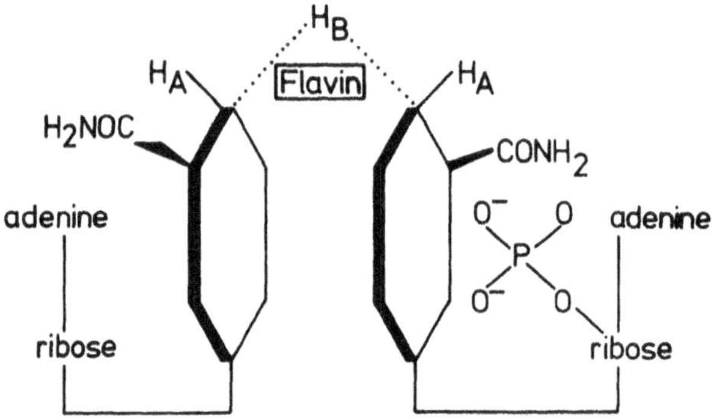

Fig. 2. Schematic proton-transfer by transhydrogenase

The mechanism of action of this 2e-input, 2e-output enzyme shows that the two pyridine nucleotides have each at least one binding site on the protein. H^--ion transfer proceeds reversibly through the flavin from one pyridine nucleotide to the other in a ternary complex on the enzyme [10].

Van den Broek and Veeger explained the lack of H^+-exchange with the medium and the direct H^+-transfer from donor to acceptor by specific adduct formation between NADH and flavin in the N(5)- C(4a)-region. From spectral considerations it was derived that the positive charge of the other nucleotide ($NADP^+$) is in the vicinity of N(1). Although Knappe [11] has recently provided some evidence that C(4a)-adduct formation between reduced nicotinamide and flavin models might be possible at low temperature, our proposal means that the attack on the flavin N(5) occurs through the A-side of the pyridine nucleotide, which is not the H^+ to be transferred. Alternatively the attack on the flavin could occur through the B-side, which makes the presence of a strong basic group to bind H_B^+ in the flavin vicinity essential.

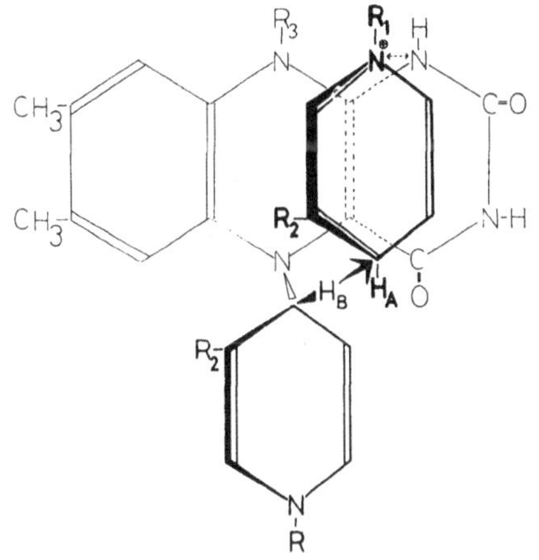

Fig. 3. Proposed model for a specific 4B-4B hydrogen transfer [cf. ref. 10].

The mechanism of action of the group of flavoproteins catalyzing the reaction PNH + H^+ + R_1-S-S-R_2 ⇌ PN^+ + R_1-SH + R_2SH in which PNH is a reduced pyridine nucleotide, and R_1-S-S-R_2 can be a disulphide bridge of a protein, such as the enzyme lipoamide dehydrogenase (discussed below), has been critically reviewed recently [12]. Without going into details, the experimental evidence indicates that the electron transfer

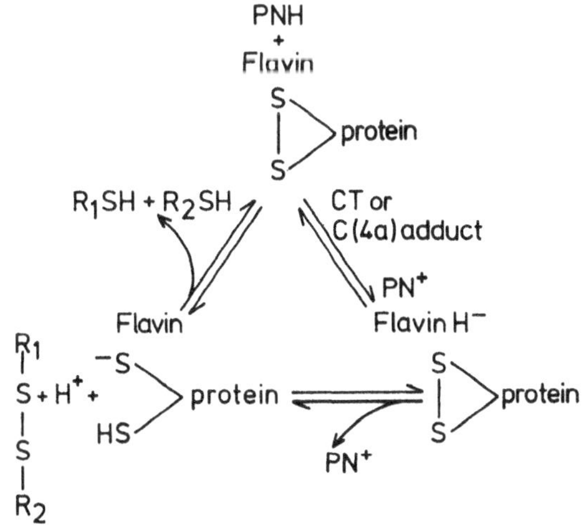

Fig. 4. Reaction mechanism of lipoamide dehydrogenase

(2e-input-2e output) occurs through a sequence of intermediates. It is essential that direct 2e-transfer to the acceptor-disulphide takes place via a S-S bridge of the enzyme. The latter in turn is reduced by the reduced flavin, which is expected to be near the enzyme S-S bridge, under these conditions.

Recent insights obtained from studies with lipoamide dehydrogenase will be discussed in more detail here. Our studies show that in this enzyme rather mobile protein structures are present, in agreement with Weber's concept [13] that the protein is "a kicking and screaming stochastic molecule". Support for this view comes from time resolved fluorescence experiments.

Lipoamide dehydrogenase is a flavoprotein containing a highly fluorescent FAD, which is firmly, but non-covalently attached to the protein. The FAD functions as an internal probe for the structure of the catalytic centre. We have studied the temperature dependence of the fluorescence lifetimes of the bound FAD in detail employing sophisticated laser pulse techniques. Two lifetimes are present in the temperature range 0°-40°C (Fig. 5, Table I).

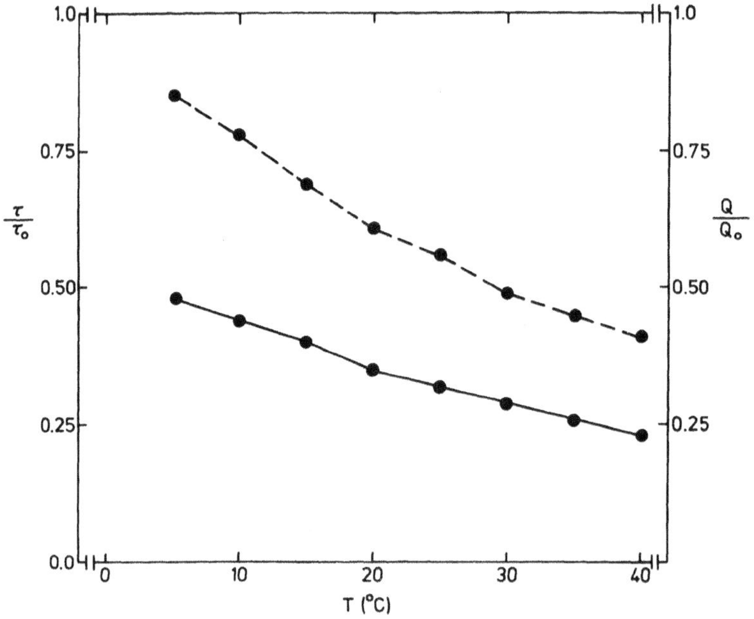

Fig. 5. Temperature dependence of lifetimes and quantum yields of the flavin fluorescence in lipoamide dehydrogenase (LPD) from pig heart. The ratio of quantum yields and lifetimes of lipoamide dehydrogenase and FMN is plotted. The lifetimes were determined with a phase fluorometer operating at 15 MHz and represent the average values of the heterogeneous fluorescence of lipoamide dehydrogenase. The quantum yields were determined according to standard procedures. The solid line represents Q_{LPD}/Q_{FMN}; the dotted line τ_{LPD}/τ_{FMN}.

Table I. Parameters calculated [cf. ref. 14] from the total fluorescence decay and from the time dependent anisotropy of the flavin in lipoamide dehydrogenase

Source	τ_1 (ns)	τ_2 (ns)	α_1	α_2	Γ^2	τ (ns)	r^2	τ_c (ns)
Pig heart [a] free	0.43	3.42	0.46	0.54	0.0028	2.65	0.043	43
A.vinelandii [b] 3 comp.	0.3	2.8	0.30	0.70	0.0042	2.58	0.037	36
A.vinelandii [c] 4 comp.	0.2	2.8	0.44	0.56	0.0040	2.44	0.035	36
E.coli [d] 3 comp.	0.4	3.8	0.47	0.53	0.0047	2.93	0.045	45

a) concentration 0.9 mg/ml, temperature 21°C; values are comparable with phase and modulation lifetimes of the free lipoamide dehydrogenase from E.coli and A.vinelandii [cf. ref. 15].
b) Pyruvate dehydrogenase complex consisting of 3 components; concentration 0.24 mg/ml, temperature 11°C [cf. ref. 16].
c) Pyruvate dehydrogenase complex consisting of 4 components; concentration 0.19 mg/ml, temperature 11°C [cf. ref. 16].
d) Pyruvate dehydrogenase complex consisting of 3 components, conentration 0.24 mg/ml, temperature 11°C.

The relative decrease in lifetime with increasing temperature parallels the relative decrease in quantum yield of the fluorescence. An energy of activation for the quenching of about 5 kcal/mole is found, which is typical for a diffusion-limited rate of quenching. The quenching is induced by collisional contact of neighbouring molecules during the short fluorescence lifetime of the flavin. The rapid collisions, that are required, are caused by local structural fluctuations of the protein. The diffusion limitation of quenching can be shown by changes in lifetime if water is exchanged for deuterium oxide. The averaged lifetime measured at room temperature is in D_2O about 1.2 times longer than the lifetime in aqueous solution, a result to be expected since the viscosity of D_2O is 1.2 times that of H_2O.

The heterogeneity in lifetimes was initially explained by the assumption that this dimeric protein contains two catalytic FAD sites in different environments [17], an explanation justified by the equal contributions to the fluorescence decay. From the present results it is possible to consider an alternative model. The quenching groups move towards the fluorophor, somewhat hindered because of the constraints imposed by the protein framework and a reversible chemical reaction in the excited state occurs. In the excited state the flavin is very electrophilic and rapid reversible electron transfer occurring between flavin and protein residues, e.g. a sulphhydryl, tryptophanyl, or tyrosyl moiety cause deviation from mono-exponential decay. Explanations of this type for non-exponential decay have been proposed by Gafni and Brand [18] and Robinson et al [19] for compounds other than flavins.

Rapid mobility of the flavin itself can be determined from the time dependence of the emission anisotropy:

$$r(t) = \frac{I_{\parallel}(t) + I_{\perp}(t)}{I_{\parallel}(t) + 2I_{\perp}(t)}$$

A semilogarithmic plot of the experimental anisotropy (as measured by using psec laser pulses for excitation) versus time gives a straight line (Fig. 6) the slope of which leads to a correlation time of 37 ns

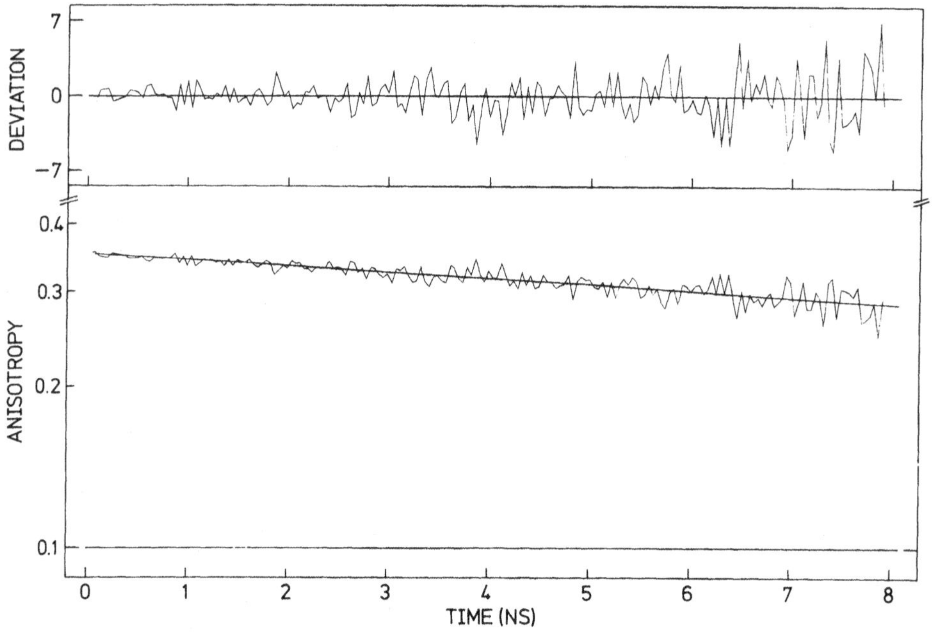

Fig. 6. Decay of emission anisotropy of FAD in lipoamide dehydrogenase. The noisy curve is the experimental anisotropy and the straight line reflects the calculated time dependence of the anisotropy with a correlation time of 37 ns. The deviation function (DV(i)) of the weighted residuals is displayed on top of the decay curves.
DV(i) is defined as: $DV(i) = (A(i) - A_c(i))/\sqrt{A(i)}$, A(i) is the experimental anisotropy in channel i, $A_c(i)$ is the computed one. The temperature was 30.0^{o}C and the concentration of enzyme was 1 mg/ml in 30 mM Na-phosphate + 0.3 mM EDTA, pH 7.2. The mode locked Ar-ion laser line at 476 nm was used for excitation, polarized emission components were transmitted through a Schott KV 520 cut-off filter and a Balzers 554 nm interference filter.

(extrapolation to 11°C gives a value of 49 ns). This correlation time is larger than can be expected for the Brownian rotation of a globular protein with molecular weight of 100,000 daltons [20] and indicates a large shell of hydration. It simultaneously indicates that under these conditions the internal rotational motion of the FAD can not be resolved from the motion of the whole protein.

The time resolved fluorescence measurements were also conducted with several pyruvate dehydrogenase multi-enzyme complexes in which lipoamide dehydrogenase is a constituent. Independent of the source of isolation, the lipoamide dehydrogenase is usually bound in the bacterial pyruvate dehydrogenase complexes together with two other functional enzymes (Fig. 7). However the complex from A.vinelandii contains an extra acetyl-accepting protein, the physiological function of which has not been yet elucidated and which can only be removed with considerable difficulty [16,21,22].

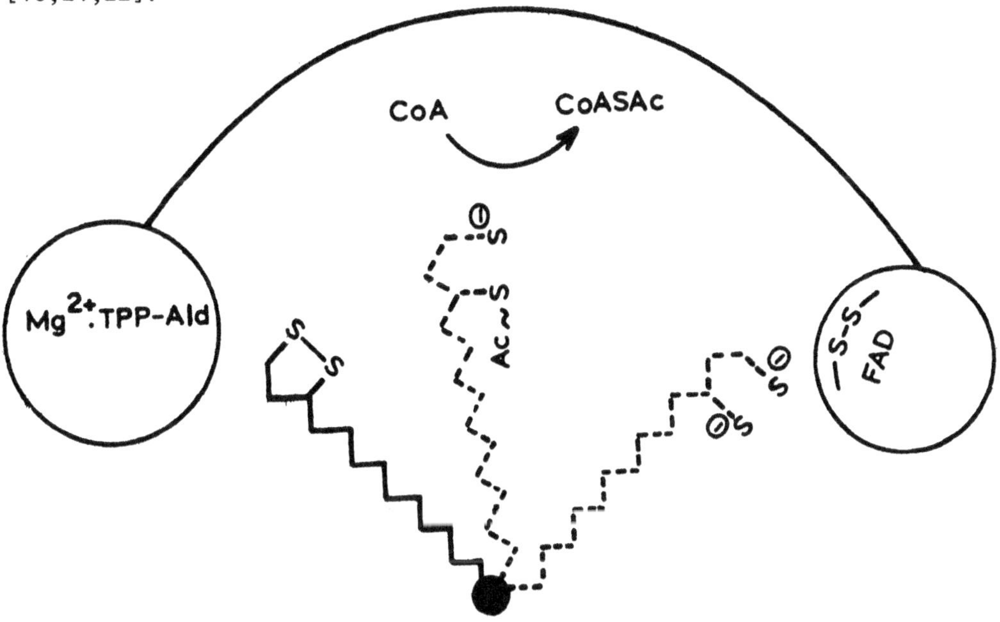

Fig. 7. Proposed mechanism of acetyl- and 2e-transfer by the 3-component pyruvate dehydrogenase complex [cf. ref. 21]. It is proposed in this model that pyruvate is decarboxylated by the first enzyme (pyruvate dehydrogenase, PDH) in the presence of Mg^{2+}, thiamin pyrophosphate (TPP). By the aid of a swinging lipoylarm of 1.4 nm the second enzyme (dihydrolipoyl-transacetylase, LTA) reacts with the enzyme-bound hydroxyethyl, in which the S-S bridge oxidizes this compound, transfers the acetyl group to coenzyme A (CoA), forming acetyl coenzyme A (CoAS-ac), and simultaneously transfers 2 reducing equivalents to the S-S bridge of the third enzyme (lipoamide dehydrogenase, LPD), which leads finally to the reduction of NAD^+ to NADH by the mechanism given in Fig. 4.

The molecular weights of the complexes vary between 6 x 10^5-3.5 x 10^6 daltons [16,22]; the chain stoichiometries of the enzymes of the 3-component complexes are the same [16,21].

Also in these experiments the fluorescence decay of the bound FAD is heterogeneous, and the average lifetimes are even shorter than the ones of isolated lipoamide dehydrogenase (Fig. 8). This could indicate a more efficient dynamic quenching process in the multi-enzyme complexes resulting from higher mobility of the protein structures responsible for the quenching. The anisotropy experiments repeated with the multi-enzyme

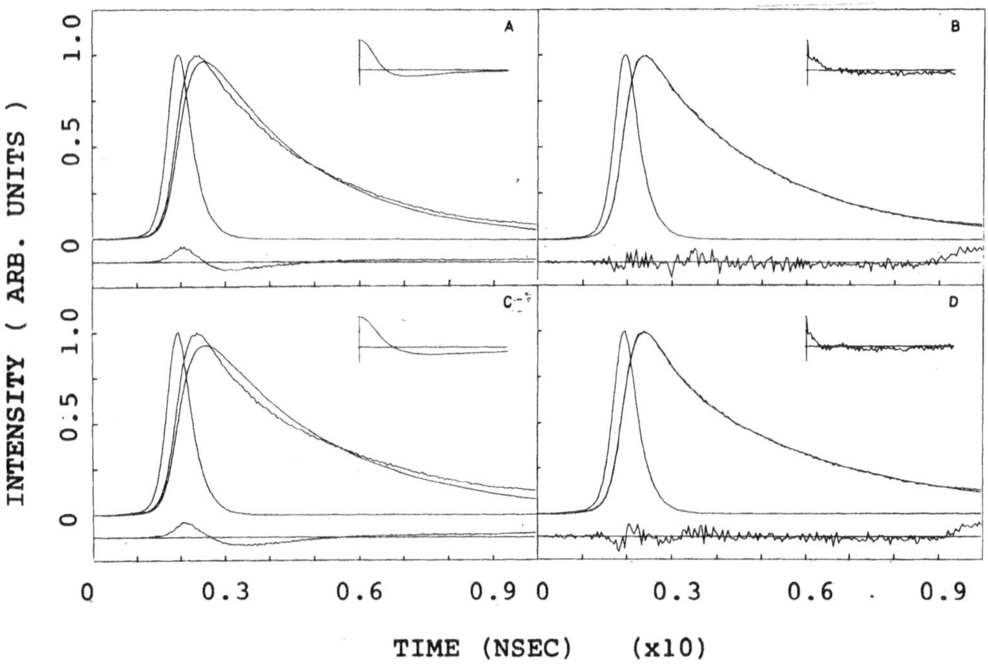

Fig. 8. Fluorescence decay curves of E.coli and A.vinelandii pyruvate dehydrogenase complexes. A, single exponential fit to 4 component A.vinelandii complex (0.19 mg/ml); B, double exponential fit to 4 component A.vinelandii complex; C, single exponential fit to 3 component E.coli complex (0.24 mg/ml); D, double exponential fit to 3 component E.coli complex. The temperature was 10°C. The excitation and emission wavelengths were the same as in Fig. 6. The calculated parameters are given in Table I.

show very strikingly that the rotational correlation time is about 40 ns at 11°C, in sharp contrast to the expected correlation time in the μsec time-region for motion of these large multi-enzyme complexes. This indicates a very large flexibility of a protein fragment (including the cofactor FAD) inside the complex.

The results have some impact on the catalytic mechanism within this multi-enzyme complex, since it can be shown from energy transfer studies (see below) with fluorescent probes, that the catalytic centres of the three participating enzymes involved, are remote from each other. The probe used was eosin maleimide, which was specifically incorporated into the complex (bound to the lipoyl group of transacetylase). It serves two functions. In the first place the probe can be selectively excited with light (without excitation of the flavin) and the anisotropy can be monitored without interference of flavin emission. The time-dependent anisotropy measured was constant and gives rise to an apparently infinite rotational correlation time (Fig. 9) independent of the presence or absence of substrates (CoA, TPP, Mg^{2+}, pyruvate and NAD^+).

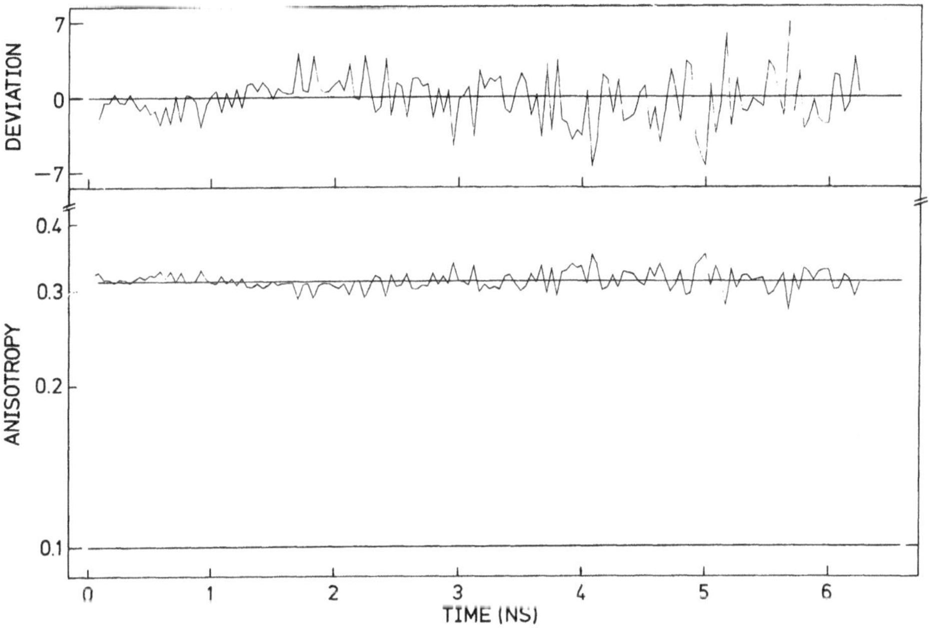

Fig. 9. Decay of emission anisotropy of eosin-pyruvate dehydrogenase complex conjugate. Eosin maleimide had been reacted specifically on the lipoyl residues of transacetylase of pyruvate dehydrogenase of E.coli. The eosin was selectively excited using the 514 nm line of the mode locked Ar ion laser. The best fit to the experimental curve corresponds to a straight horizontal line. An infinitely long correlation time results from the slope of the line, illustrating that the probe undergoes the relatively slow rotational diffusion. Details and conditions are described in the legend of Fig. 6.

This observation can be considered to be an important control experiment, since it shows that instrumental artefacts causing anisotropic flavin fluorescence decay are improbable. The eosin maleimide is bound very rigidly to the protein and follows the slow rotational diffusion of the complex. Another reason to use this probe is that it can act as an acceptor of excitation energy of the flavin, since there is a good overlap between the fluorescence spectrum of the donor (flavin) and the absorption spectrum of the acceptor (eosin). The critical transfer distance (R_o) was calculated from the overlap and quantum yields to be 4.0 nm. The occurrence of energy transfer can be demonstrated by measuring the fluorescence lifetimes of the donor molecule in the presence and absence of the acceptor. Only a small difference (∿5%) in the lifetime of the flavin was found, confirming that the distance between the lipoyl group and the flavin is at least 4.5 nm.

From these experiments we can conclude the following.

1. The fluorescence of the flavin is quenched dynamically. This implies that rapid displacements (in the order of 1 nm during the fluorescence lifetime) take place. Channels in the protein structure might be created allowing solvent and other small molecules to penetrate towards interior groups, since FAD is completely accessible to D_2O. It has been shown by Cooper [23] that this type of stochastic processes is leading to sizeable thermodynamic fluctuations in single protein molecules. In a population of many protein molecules these fluctuations cancel to give an average value for the thermodynamic parameters. Consequently, the interior must be considered as a fluid medium and when the fluctuations are concentrated into one area, for example around FAD, accessibility of substrates is very possible.
2. From energy transfer studies it has been shown that the distance between the reaction centers of the LTA- and LPD-components of the multi-enzyme complex is larger than 4.5 nm which is also found for the distances between the other catalytic centers involved [24,25]. The theory of the flexible lipoyl moiety (swinging sidearm of 1.4 nm) on LTA functioning as carrier of reaction intermediates between PDH and LPD must be modified. Not only since the maximum distance reached is 3.0 nm, but also in view of the rigidly bound eosin. In contrast to the latter observation is the relative mobility shown by lipoate-bound spinlabels [15].
3. The fact that in the multi-enzyme complex lipoamide dehydrogenase exhibits a rotational diffusion similar to that found in the isolated lipoamide dehydrogenase indicates that the multi-enzyme complex cannot be considered as a closed packing of different protein spheres, but that there exist many degrees of motional freedom. As judged from the decay of emission anisotropy of eosin, the lipoyl residues to which the probe is attached, are rigidly bound to the surface of the core of the complex A simple model can be deduced from the internal motion of lipoamide dehydrogenase. FAD is located at the surface of a sphere with radius 4.2 nm [26]. Considering full rotational freedom for lipoamide dehydrogenase the root mean-square displacement of FAD in time t is given by $(t/\tau_c)^{\frac{1}{2}}$, with τ_c = 1/6D the correlation time for rotational diffusion, D the diffusion coefficient for a sphere. It takes about 120 ns to cover by rotation the maximum distance of 8.4 nm between flavin and lipoate-bound eosin.

The absence of energy transfer between the catalytic centers can be explained on the basis of this model. The utilization of energy transfer as spectroscopic ruler is more or less confined to cases where donor and acceptor keep a constant average distance. If the probe molecules are located on different proteins, which can rotate independently within the constraints of the complex the situation is different.

Based on this mobility in lipoamide dehydrogenase an alternative model can be put forward [Fig. 10]. Instead of a rapid random movement

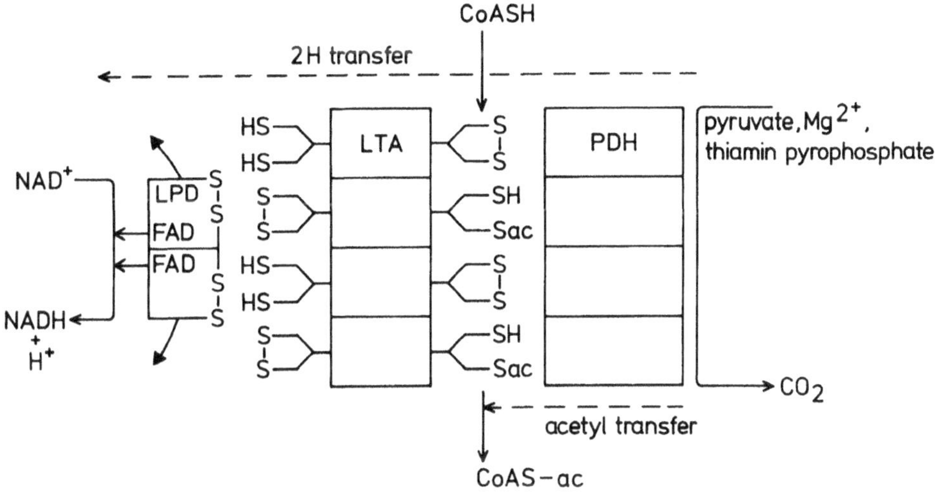

Fig. 10. Alternative model for acetyl- and H-transfer in the 3-component pyruvate dehydrogenase complexes of A.vinelandii and E.coli. In view of the uncertain physiological role of the fourth protein in the A.vinelandii 4-component complex any proposal concerning this type of complex is omitted. The different component enzymes are given in their respective chain ratio's [22], while the sizes represent the respective molecular weights.

of lipoyl moieties, the catalytic center S-S bridge of LPD approaches the fixed lipoyl residues on the transacetylase core. In this way the disulphide bridge of LPD can be brought into contact with the reduced lipoyl groups of the LTA. For this model a restricted movement of lipoamide dehydrogenase (or a segment) is necessary. This wobbling is much more probable than large rotations of the enzyme within the complex, since more or less specific interactions keep the complex together. Movements of the order of 2.5 nm together with the experimental correlation time of 40 ns can give rise to a collisional frequency of about $5 \times 10^7 s^{-1}$. The lack of energy transfer or the low efficiency can be due to a long average distance between donor and acceptor molecules or an unfavourable orientation factor (κ^2). Dale and Eisinger have analyzed in detail several models allowing for orientational freedom of donor-acceptor couples [27]. They showed for example that for more or less perpendicularly oriented transition moments relatively low values of κ^2 are obtained. In general very specific combinations of orientation angles can also give low values of κ^2 [27]. Since the essential disulphide bridge and

FAD do not need to be in direct contact in the oxidized flavoprotein, - only the reduced disulphide bridge needs to be in close contact with the flavin - the movement of the disulphide bridge is not necessarily identical with that of the FAD, but coupled. Therefore the mobility of FAD provides the encounters between the disulphide of the active center of LPD and the lipoylmoieties at a relatively high frequency. In this way efficient catalysis at the output site of the complex of 4100-4400 for the 4- and 3-component complexes of A.vinelandii and 6500 s^{-1} for th complex of E.coli [cf. ref. 16] can be explained.

Proteins like transhydrogenase and lipoamide dehydrogenase, whose strong bound flavin is highly fluorescent, upon reduction lose all flavin fluorescence. Although reduced flavins can be also fluorescent [28], under reducing conditions after reaction with the donor these flavoproteins lack any reduced-flavin fluorescence emission.

In sharp contrast are in this respect flavoproteins like lactate oxidase, D-amino acid oxidase and L-amino acid oxidase all of which cata lyse the reaction: $SH_2 + O_2 \rightarrow S + H_2O_2$. These enzymes do not show any fluorescence of enzyme-bound flavin, but are fluorescent in their reduced forms. Similarly only the reduced form of the electron transport flavoprotein flavodoxin is fluorescent.

The class of flavoproteins called oxidases has one important kinetic feature. Preceeding the reaction with the flavin, deprotonation of the substrate takes place. The carbanion then attacks the flavin probabl by forming a N(5)-adduct [29], which is attacked by O_2 (either π-complex or dioxethane).

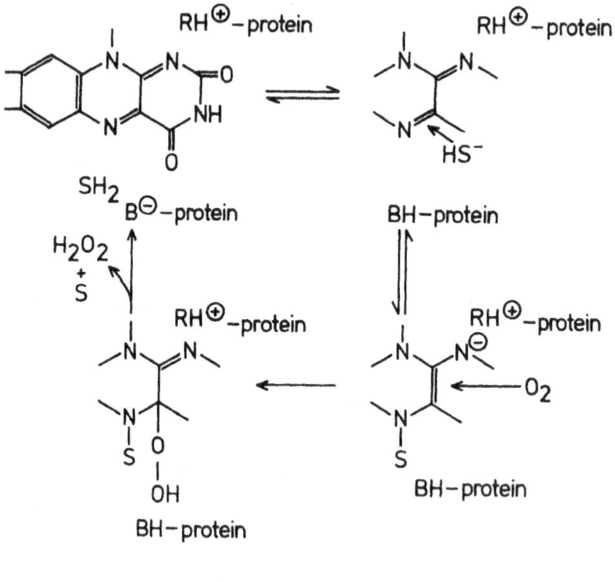

Figure 11. Schematic reaction mechanism of flavoprotein oxidases; S stands for substrate.

Enzymes which resemble the oxidases in their reactivity with O_2, but in fact react in a fully different way, are the pyridine nucleotide dependent oxygenases, catalyzing $PNH + H^+ + S + O_2 \rightarrow PN^+ + SOH + H_2O$. Many aspects of the mechanism are unclear, but a striking feature is the inability of the pyridine nucleotide to reduce the enzyme in the absence of the compound to be hydroxylated, even in the absence of O_2. The enzyme p-hydroxybenzoate hydroxylase in the oxidized form shows flavin fluorescence, which is fully quenched under reducing conditions, as outlined above.

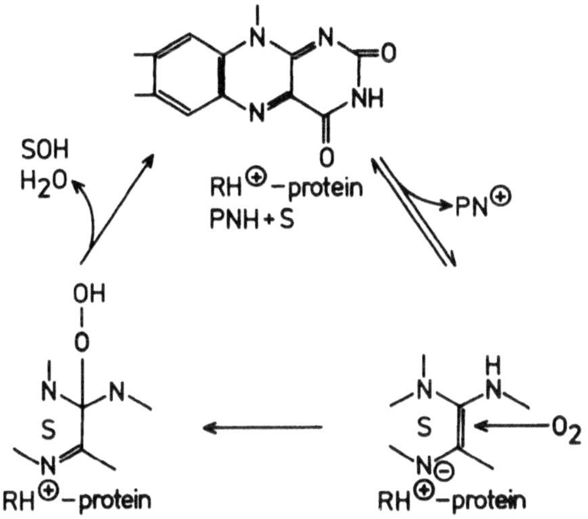

Fig 12. Schematic reaction mechanism of flavoprotein oxygenases; S stands for substrate

In contrast to the oxidases, oxygenases need splitting of the O_2 molecule in order to incorporate one O-atom as hydroxyl in the substrate. Since this process proceeds without the formation of $O_2^{\cdot-}$, the proposal has been made that O_2 forms a C(10a)-peroxide, which leads to oxene splitting by opening of the N(10)-C(10a) bond [cf.ref. 7]. Striking in this respect is that although traces of $O_2^{\cdot-}$ are not formed either in the oxygenase or in the oxidase reactions, the class of 2e-input - 2e-output flavoproteins involved in S-S bridge reduction by pyridine nucleotides (and thus supposed to react without O_2), actively form $O_2^{\cdot-}$ in the presence of O_2. Similarly $O_2^{\cdot-}$ is very rapidly formed from flavodoxin hydroquinone. The resulting semiquinone is rather resistant to reaction with O_2. This makes it impossible to draw firm conclusions as to the position at which O_2 reacts with the flavin in either the hydroquinone form or in the substrate-flavin adduct form.

Fluorescent N(5)- or C(4a)-adducts of flavin can be produced either by reaction with substrate or by chemical or by photo-chemical reaction. Table II shows that the decay curves of the fluorescence are in all but one case not mono-exponential. Analysis in terms of mono-exponential and

Table II. Fluorescence decay characteristics of reduced flavoproteins[a]

Enzyme	Coenzyme	Reduction method	α_1	τ_1 (ns)	α_2	τ_2 (ns)	$<\tau>$[b] (ns)	RMS[c]	p[d] (excitation wavelength (nm))
L-Amino acid oxidase	FAD	Substrate, $Na_2S_2O_4$	1.00	2.4				0.026	0.37(430)
			0.85	1.3	0.15	4.7	2.6	0.004	0.12(340)
D-Amino acid oxidase N(5)-acyl adduct	FAD	$CH_3COCOOH$-hν	1.0	3.0				0.048	0.33(420)
			0.83	0.94	0.17	5.6	3.5	0.010	0.15(330)
D-Amino acid oxidase C(4a)-benzyl adduct	FAD	$C_6H_5CH_2COOH$-hν	1.00	3.2				0.026	0.42(420)
			0.86	2.1	0.14	7.6	4.1	0.005	0.28(320)
D-Amino acid oxidase	FAD	$NaBH_4$	1.00	3.1				0.030	0.41(420)
			0.74	1.4	0.26	5.4	3.7	0.006	0.31(340)
Flavodoxin	FMN	EDTA-deazaflavin-hν	1.00	2.1				0.007	0.35(450)
			0.72	2.1	0.28	2.1	2.1	0.007	0.08(380)
Lactate oxidase	FMN	Substrate ($CH_3CH(OH)COOH$)	1.00	3.7				0.014	0.41(420)
			0.40	0.96	0.60	4.7	3.8	0.007	0.32(350)
Lactate oxidase N(5)-carboxymethyl adduct	FMN	1) $CH_3CH(OH)COOH$	1.00	9.7				0.011	0.40(400)
		2) $BrCH_2COOH$-hν	0.20	3.0	0.80	10.8	10.1	0.007	0.27(320)

a) For general references concerning the used buffer, concentrations and properties of the listed enzymes see [28]. Temperature was 22°C. For methods see refs. 14 and 30.
b) Averaged lifetime.
c) RMS is root mean square of the deviation.
d) Degree of polarization observed at different excitation wavelengths. The first value corresponds to the maximum observed p, the second value refers to another region in the polarization excitation spectrum where p is more or less constant; the latter value is defined less precisely.

bi-exponential decay curves shows, (as judged from the RMS values), in all cases that bi-exponential fitting leads to better results. Only in the case of flavodoxin, the analysis with both fitting procedures leads in both cases to the unambiguous conclusion: one lifetime.

The contributions of the lifetimes to the total decay are not equal which is different from what is observed in lipoamide dehydrogenase (cf. Table I). This excludes previous proposals like non-identical flavin binding sites [17]. However it is in agreement with the proposal given above, that reversible electron transfer in the excited state is responsible for this phenomenon. However in contrast to the oxidized flavin which is electrophilic and can act as acceptor in electron transfer reactions with amino acid residues [cf.ref. 31], the reduced flavin and its adducts are electron donating probably to the rings of the tryptophanyl, histidyl or tyrosinyl residues.

Table II includes the maximum degree of fluorescence polarization. Due to its low molecular weight (15,000 daltons) the low value for reduced flavodoxin is not surprising. Surprisingly low are the polarization values of the substrate-reduced form of L-amino acid oxidase and the N(5)-acyl adducts of D-amino acid oxidase. Also the values found for adducts of lactate oxidase are low in view of its high molecular weight (330,000). This suggests a certain mobility of the reduced flavins in these flavoproteins, especially in view of our observation [30], that the polarization values obtained with model compounds in glycerol are higher than the ones in flavoproteins.

The flavoprotein flavodoxin belongs to the class of electron transfer proteins. It serves in important bacterial processes like $N_2 \rightarrow NH_3$ (A.vinelandii), $SO_4^{2-} \rightarrow S^{2-}$ (D.vulgaris) and $2H^+ \rightarrow H_2$ (M.elsdenii) as low-potential one electron donor [32,33], shuttling between the 2-equivalent-reduced (hydroquinone) and 1-equivalent-reduced (blue neutral semiquinone, 5-HFl·) forms. In anaerobic systems such a shuttling is not much of a problem, since it is very probably in interaction with one-electron donating proteins like ferredoxin. In aerobic organisms the major problem is the low E_o' (pH 7) of this redox couple (-500 mV), much too low to allow reduction by pyridine nucleotides (-330 mV). In addition pyridine nucleotides are 2-electron donors, 1-electron donation is impossible [34]. Nevertheless 2-electron reduction of oxidized flavodoxin by pyridine nucleotide is possible, since for this reduction E_o' (pH 7) is about -300 mV. In other words flavodoxin shows in fact electron-pair-splitting properties: hydroquinone/semiquinone about -500 mV, semiquinone/quinone about -100 mV. Nevertheless one is faced with the problem how to oxidize flavodoxin semiquinone to its quinone; direct oxidation by O_2 is very slow. Hemmerich [7] has proposed that 1-electron transfer proceeds via the C(7) -C(8)- region of the isoalloxazine nucleus, since according to X-ray crystallography this part is in contact with the solvent. The evidence is not very convincing, since the arguments were derived from experiments in which flavodoxin models served as electron acceptor; more convincing experiments in which these models act as 1-electron donor are still lacking.

From the preceeding fluorescence studies, it is clear that mobility and interaction of the flavin in either form maybe highly important for the mechanism of catalysis. In order to understand in more detail the diversity of reactions catalyzed by flavoproteins it is necessary to obtain a better insight into the interactions between the flavin prosthetic group and the apo-enzyme. These interactions are primarily responsible for the particular physical and chemical properties exhibited by a certain flavoprotein. One of the most powerful tools to achieve this goal is the NMR technique, yielding information on the molecular and submolecular structure of the molecule investigated. The overwhelming number of protons in a protein makes it impossible to obtain informatio on the bound flavin. Thus we set up a program where use is made of selectively ^{13}C-enriched flavin in interaction with various apo-enzymes In addition ^{13}C chemical shifts represent, in general, the π-electron densities in a molecule much better than other nuclei.

Table III. ^{13}C-chemical shifts in ppm of free and protein-bound flavins in the oxidized and two-electron reduced state[a].

Compound	Oxidation State	pH	C-atom[b] 4	10a	2	4a
TARF[c]	ox.		161.4	149.2	155.9	136.0
	red.		157.0	137.1	150.6	105.2
FMN	ox.	6.5	162.6	151.2	158.8	135.2
	red.	5.0[d]	157.3	144.0	151.1	102.4
	red.	9.0[e]	157.3	155.0	157.6	101.1
M.elsdenii flavodoxin	ox.[f]	6.5	162.6	153.1	159.4	135.4
	red.[f]	6.5	154.9	154.2	156.5	103.1
A.vinelandii flavodoxin	ox.	8.0	161.7	154.8	159.3	135.3
	red.	8.0	154.9	154.9	158.1	102.2
Old yellow enzyme	ox.	7.5	173.6	153.0	160.4	136.8
	ox.[g]	7.5	163.2	152.9	160.4	135.0
	red.	7.5	174.0	157.7	159.3	94.9

a) Chemical shifts relative to TMS
b) The carbon atoms were enriched at least by 90 atom %
c) Tetra-acetylriboflavin in $CDCl_3$
d) Neutral form
e) Anionic form
f) Independent of pH in the range of 6.0-8.0
g) In the presence of 4-hydroxy-N-n-butylbenzamide yielding a green coloured "charge transfer" complex [35].

The ^{13}C-chemical shifts of some compounds investigated are shown ir Table III. The difference between TARF and FMN in the oxidized state is caused mainly by the larger polarisation of the carbonyl function of the flavin molecule in aqueous than in apolar solvents and is to a less extent due to stacking interaction in aqueous solution. Upon reduction the difference in chemical shifts between the two carbonyl functions of TARF and FMN disappears almost completely, but the chemical shifts of C(4a) and C(10a) in TARF and in FMN at pH 5.0 become different. A similar effect is observed for the ^{13}C-^{13}C coupling constant (Table IV) where the C(4a)-C(10a) coupling increases drastically as compared to reduced TARF. The

Table IV. Direct ^{13}C-^{13}C coupling constants (in Hz) and the calculated S-character (s%) of the corresponding bonds in free and protein-bound flavin in the oxidized and two -electron reduced state.

Redox State	Compound	Coupling constant $^1J_{C(4)-C(4a)}$	(% s-character) $^1J_{C(4a)-C(10a)}$
Oxidized	TARF[a]	76.5 (35.8)	53.3 (31.0)
	FMN[b]	75.4 (35.6)	55.9 (31.6)
	M.elsdenii flavodoxin[b]	76.5 (35.8)	58.8 (32.2)
Reduced	TARF[a]	79.2 (36.3)	84.5 (37.3)
	FMN (pH 5.0)[c]	88.0 (37.9)	93.0 (38.8)
	FMN (pH 8.0)[d]	74.2 (35.4)	74.2 (35.4)
	M.elsdenii flavodoxin[b]	70.6 (34.6)	79.4 (36.4)

a) Tetra-acetylriboflavin in $CDCl_3$
b) in phosphate buffer, pH 7.5
c) neutral form
d) anionic form

position of the resonances of C(4a) and C(10a) and the calculated percentage s-character of this bond indicate a fairly localized double bond. The observation that in neutral reduced FMN the C(4) - C(4a) bond also possesses a high degree of s-character suggest that the molecule must be quite planar with respect to the C(4)-C(4a)-C(10a) bonds. Deprotonation of N(1)H of reduced FMN influences the chemical shifts of C(10a) and C(2), whereas those of C(4a) and C(4) are affected to a minor degree (Table III). From this it can be concluded that the negative charge of the ionized reduced FMN is localized mainly on N(1). Ionization of the molecule leads to a decrease in the direct coupling constants and percentage s-character of the considered bonds in reduced FMN. Two effects contribute probably to the latter observation: a) the negative charge of the molecule and b) a change in the conformation of the molecule; i.e. less coplanar.

Binding of FMN to M.elsdenii flavodoxin affects the chemical shifts of C(10a) and C(2) as compared to those of free FMN, indicating interaction of the C(2) carbonyl function with the apo-enzyme (weak hydrogen bond). A comparison of the chemical shifts of reduced FMN and reduced M.elsdenii flavoprotein reveals that the chemical shifts of the latter resemble those of reduced FMN at pH 9.0, indicating that the protein-bound FMN is in the anionic form (Table III). However the corresponding coupling constants (Table IV) differ from those of ionized reduced FMN. From crystallographic studies it is known that reduced FMN bound to apo-flavodoxin of M.elsdenii possesses a planar structure almost identical to that of the oxidized form. The calculated % s-character of the C(4a)-

C(10a) centers in the oxidized state is very similar to that of the C(4 C(4a) centers of the reduced form. A reversed relationship with respect to percentage s-character exists for the C(4)-C(4a) centers in the oxidized and the C(4a)-C(10a) centers in the reduced state. These observations indicate that the NMR data are in agreement with the crystallographic data. The chemical shifts obtained with A.vinelandii flavodoxin closely resemble those obtained with M.elsdenii flavodoxin, indicating that in the former the reduced flavin is also in the anionic form.

Binding of FMN to the apoenzyme of old yellow enzyme has a dramati effect on the chemical shift of C(4) which is shifted downfield by abou 12 ppm as compared to the other two flavoproteins (Table III). The dras tic downfield shift indicates a strong interaction of the CO(4) functio with the apo-enzyme. Reduction of the enzyme does not disrupt this inte action but leads to a higher shielding of C(4a). An interesting observation is that addition of a strongly complexing molecule to oxidized old yellow enzyme results in an upfield shift of the resonance due to C(4). The resulting chemical shift is then of the same magnitude as the corresponding chemical shifts of the other flavoproteins. If our interpretation is correct, the addition of this complexing molecule will disrupt the hydrogen bond. It is also possible that the amino acid residue involved in the interaction with the flavin becomes involved in the bindi of the complexing molecule which is bound only in the ionized form. One can interpret this finding in two ways:

1) binding of the complexing molecule around CO(4) disrupts the specifi flavin-protein interaction which leads to inhibition, by making reactio with the substrate impossible;

2) the binding of the complexing molecule reflects the binding of the reducing substrate (NADPH) and mimics the approach of the reduced pyridine nucleotide in its disruption of the CO(4)-protein interaction.

It can be stated with certainty that the factor contributing significantly to the specific catalytic action of flavoproteins is regio-specific binding of the flavin to the protein. This binding determines the conformation of the flavin and the accompanying electron-distribution in any of the catalytic forms and/or intermediates, their specific short range interaction with the protein by a process of rapid displacements of (part) of the interacting structures and in addition freedom for mobility of either the tightly bound flavin or a segment of the flavin-bindi region of the protein. The factors seem to be dependent on the class of proteins according to present information.

It is clear from these data that specific conclusions can be drawn for the two types of flavodoxin studied; conclusions in agreement with X-ray crystallographic data. For other flavoproteins where such structural information is not available the proper interpretation of physical data will meet many more difficulties, not to speak of the translation of these specific flavin-binding data into a specific enzyme-mechanism.

Acknowledgements

We wish to thank Mr. B.J. Sachteleben for drawing the figures and Mrs. J.C. Toppenberg-Fang for the preparation of the manuscript. These studies were supported in part by the Netherlands Organization for Chemical Research (S.O.N.) with financial aid from the Netherlands Organization for the Advancement of Pure Research (Z.W.O.).

References

1. Beinert, H., in The Enzymes (P.D. Boyer, H. Lardy, K. Myrbäck, eds.), (1960) Academic Press, New York, vol. 2, 2nd ed. p. 339-416.
2. Flavins and Flavoproteins, Proceedings 1st Symposium, (E.C. Slater, ed.), (1966) Elsevier Amsterdam.
3. Flavins and Flavoproteins, Proceedings 2nd Symposium (K. Yagi ed.), (1968) University of Tokyo Press, Tokyo.
4. Flavins and Flavoproteins, Proceedings 3rd Symposium (H. Kamin ed.), (1971) University Park Press, Baltimore.
5. Flavins and Flavoproteins, Proceedings 5th Symposium (T.P. Singer ed.), (1976) Elsevier, Amsterdam.
6. Flavins and Flavoproteins, Proceedings 6th Symposium (K. Yagi and T. Yamano, eds.) Japan Scientific Societies Press, to be published.
7. Hemmerich, P., in Transport by proteins (G. Blauer, H. Sund eds.), (1978) Walter de Gruyter & Co., Berlin 123-149.
8. Michaelis, L., Schubert, M.P. and Smythe, C.V. (1936) J.Biol.Chem. 116, 587-607.
9. Drysdale, G. in Flavins and Flavoproteins, Proceedings 1st Symposium (E.C. Slater ed.), (1966) Elsevier Amsterdam, p. 159-170.
10. Van der Broek, H.W.J. and Veeger, C. (1971) Eur.J.Biochem. 24, 63-71.
11. Knappe, W.R. in Pyridine nucleotide-dependent Dehydrogenases (H. Sund ed.) (1977), Walter de Gruyter & Co, Berlin, p. 199-202.
12. Williams, C.H. in The Enzymes (P.D. Boyer, ed.) (1976) vol. XIII, 3rd ed., p. 89-173.
13. Weber, G. (1975) Adv. Protein Chem. 29, 1-83.
14. Grinvald, A. and Steinberg, I.Z. (1974) Anal.Biochem. 59, 583-598.
15. Grande, H.J., van Telgen, H.J. and Veeger, C. (1976) Eur.J.Biochem. 71, 509-518.
16. De Abreu, R.A., De Kok, A. and Veeger, C. (1977) F.E.B.S. Lett. 82, 89-92.
17. Wahl, Ph., Auchet, J.C., Visser, A.J.W.G. and Veeger, C. (1975) Eur.J.Biochem. 50, 413-418.
18. Gafni, A. and Brand, L. (1976) Biochemistry 15, 3165-3171.
19. Robinson, G.W., Morris, J.M., Robbins, R.J. and Fleming, G.R. (1977) SPIE Annu.Tech.Symp.Proc., 22nd 113, 13-24.
20. Chen, R.F. in Fluorescence: Theory, Instrumentation and Practice (G. Guilbault, ed.) (1969) Marcel Dekker, New York, p. 469.
21. Koike, M., Reed, L.J. and Carroll, W.R. (1963) J.Biol.Chem. 238, 30-39.
22. De Abreu, R.A., De Kok, A., De Graaf-Hess, A.C. and Veeger, C. (1977) Eur.J.Biochem. 81, 357-364.

23. Cooper, A. (1976) Proc.Natl.Acad.Sci. 73, 2740-2741.
24. Scouten, W.H., De Graaf-Hess, A.C., De Kok, A., Grande, H.J., Visser, A.J.W.G. and Veeger, C. (1978) Eur.J.Biochem. 84, 17-25.
25. Moe, O.A.Jr, Lerner, D.A. and Hammes, G.G. (1974) Biochemistry 13, 2552-2557.
26. Visser, J. and Veeger, C. (1968) Biochim.Biophys.Acta 159, 265-275.
27. Dale, R.E. and Eisinger, J. in "Concepts in Biochemical Fluorescence" (R.F. Chen and H. Edelhoch eds.) (1975) Marcel Dekker, New York, p. 115-284.
28. Ghisla, S., Massey, V., Lhoste, J.M. and Mayhew, S.G. (1974) Biochemistry 13, 589-597.
29. De Kok, A., Veeger, C. and Hemmerich, P. in Flavins and Flavoproteins Proceedings 3rd Symposium (H. Kamin ed.), University Park Press, Baltimore, p. 63-80.
30. Visser, A.J.W.G., Ghisla, S., Massey, V., Müller, F. and Veeger, C., Eur.J.Biochem. submitted for publication.
31. Kaptein, R., Dijkstra, K., Müller, F., van Schagen, C.G. and Visser, A.J.W.G. (1978), J.Magn.Reson. 31, 171-176.
32. Scherings, G., Haaker, H. and Veeger, C. (1977) Eur.J.Biochem. 77, 621-630.
33. Mayhew, S.G., Van Dijk, C. and Van der Westen, H.M. in Hydrogenases: Their catalytic activity, structure and function (H.G. Schlegel, ed.) E. Goltze, K.G., Göttingen, in press.
34. Blankenhorn, G. in Pyridine nucleotide-dependent Dehydrogenases (H.Sund, ed.), (1977) Walter de Gruyter & Co., Berlin, p. 185-198.
35. Abramovitz, A. and Massey, V. (1976) J.Biol.Chem. 251, 5321-5336.

MOLECULAR ORBITAL CALCULATIONS OF METAL MICROCLUSTER PROPERTIES

R. C. Baetzold
Research Laboratories, Eastman Kodak Company, Rochester, N.Y. 14650

Abstract

Molecular orbital calculations for clusters of Ag, Pd, Cu and Ni are discussed. The lifetime broadened density of states (DOS) for these clusters is shown to agree closely with experimental DOS and to display a similar dependence with size. The surfaces of these clusters show a negative charge relative to the interior and this effect is analyzed in terms of d orbital occupancy. As the size of the cluster increases its equilibrium bond length increases in accord with experiment. The origin of this effect is explained in the electronic interactions involved in band formation. These electronic effects are all related to factors possibly important in heterogeneous catalysis.

INTRODUCTION

Small metal particles containing a few atoms have been actively investigated by several groups over the past few years.[1-4] These theoretical efforts have been matched by active experimental programs which have measured various metal properties. Some of the properties measured recently include Fe_N binding energy,[5] Na_N ionization potential,[6] Pd_N and Pt_N density of states,[7] various metal ion intensities produced by ion sputtering,[8] and spectroscopic properties of Ag_N and Cu_N clusters.[9] Generally these efforts have focused on the electronic properties of clusters in the absence of interaction with support. Thus, these efforts serve as a first approximation to modeling the catalytic properties of metal clusters. As such, these studies are primarily aimed at elucidating those electronic properties which may be significant in catalysis. Properties such as binding energy, ionization potential, valence density of states and electron configuration have been studied.

Catalytic experiments have shown that small particles of Ag, Au or Pd containing only a few atoms have interesting properties. For example,

B. Pullman (Ed.): Catalysis in Chemistry and Biochemistry. Theory and Experiment. 191–206.

Hamilton and Logel[10] have shown that Ag_4 and Au_2 are critically sized clusters which catalyze silver metal deposition from solutions of silver ion and reducing agent. Clusters smaller than the above sizes are inactive, whereas larger ones are active catalysts. Thus, even very small particles are important catalytically.

This report is concerned with electronic properties of clusters calculated by CNDO and extended Hückel methods. We shall investigate the limitations and strengths of these methods and make comparison with other calculations in the literature. In addition, frequent comparisons with experiment will be offered. Clusters of silver and palladium will be considered here.

A primary limitation of semiempirical methods involves the need for parameter input. Parameters must be chosen based upon experimental data for similar systems. For transition metal clusters, experimental data have been relatively scarce and have only recently become available. Since binding energies, ionization potentials, energy gaps and bond lengths of the homonuclear diatomic metal clusters such as Ag_2 have been available[11] our efforts have involved a study of these properties in larger clusters. However, a parameter selection based upon these properties need not reproduce other properties in the larger clusters. This is particularly true for the density of states. Use of single zeta d orbitals with the exponents of Clementi et al.[12] produced reasonable binding energy, ionization potential, energy gap, and bond length changes with cluster size in EH and CNDO calculations. However, the d bands calculated were too narrow and subsequent work has shown that the d orbitals must be represented by more diffuse orbitals of the double zeta type. We extend the prior work here and make comparison with experimental quantities.

In this report we consider EH and CNDO calculations separately. We examine Ag_N valence density of states (DOS) using a lifetime broadening mechanism to show the changes as the cluster grows from two atoms to bulk. We also examine such effects for Pd_N clusters. For Pd we have examined some irregularly shaped clusters which show strong variations in properties that are dependent upon coordination number. These effects are analyzed in terms of band effects and the surface potentials which arise. Finally, the CNDO and EH methods are compared. Let us begin with the methods.

METHODS

The EH procedure as originally developed by Hoffmann[13] is used here. The secular equation is solved using valence orbital ionization potential parameters for the diagonal Hamiltonian terms. These are determined by atomic spectra. Off-diagonal Hamiltonian terms are defined by the Wolfsberg-Helmholz[14] formula

$$H_{ij} = \tfrac{1}{2}KS_{ij}(H_{ii} + H_{jj}) \qquad (1)$$

with K = 1.75. The overlap terms (S_{ij}) are calculated using double-zeta d Slater orbitals. The parameters used here have been defined before.[4]

We extend the EH calculation to an infinite plane or bulk using Bloch functions in the standard manner. The molecular orbitals

$$\Psi_i = \sum_{\ell} e^{ik\cdot R_\ell} \sum_j C_{ij}\chi_j \qquad (2)$$

involve terms in the wave vector k, position vector R_ℓ, coefficient C_{ij} and atomic orbital χ_i. Solving for the matrix elements yields complex Hamiltonian elements of the form

$$H_{ij}(k) = \sum_{\ell} H_{ij}^{o,\ell} e^{ik\cdot R_\ell} \qquad (3)$$

where the superscripts denote the unit cells on which the respective orbitals are centered. The Wolfsberg-Helmholz equation is used to evaluate the terms involving different orbitals or similar orbitals in different cells. The overlap matrix can be calculated in similar fashion. Then the complex secular equation is solved throughout the Brillouin zone to achieve a uniform sampling of k space. An equivalent of 2048 points is considered in the bulk calculation.

The CNDO method of Pople[15] is extended to treat atoms with d orbitals having various degrees of occupation. The secular equation employs a diagonal overlap matrix and Hamiltonian elements as follows:

$$\begin{aligned} F_{\mu\mu} &= -\tfrac{1}{2}(I+A)_\mu + \gamma_{ss}^{AA}(\tfrac{1}{2} - P_{\mu\mu}) + \qquad (4) \\ &\quad \sum_B (P_{ss}^B - N_B)\gamma_{ss}^{AB} + (P_{dd}^B - M_B)\gamma_{sd}^{AB} \\ F_\mu &= \beta_{AB}^o S_{\mu\upsilon} - P_{\mu\upsilon}\gamma_{\mu\upsilon}^{AB} \quad . \end{aligned}$$

A separate equation is solved for electrons of both spins. This diagonal $F_{\mu\mu}$ term refers to s orbitals but it can be converted to the form appropriate to d orbitals by interchanging s with d and N_B with M_B. The terms in equation (4) are as follows:

I - ionization potential of the atom, orbital μ
A - electron affinity of the atom, orbital μ

γ_{ss}^{AB} - electron repulsion between s orbital on atom A and s orbital on atom B.
$P_{\mu\mu}$ - bond density element

N_B - number of s + p valence electrons on atom B.
M_B - number of d valence electrons on atom B.

β_{AB}^{o} - empirical resonance parameters.

RESULTS

Extended Hückel

The bulk experimental structure of silver is face-centered-cubic (fcc) which we employ for these calculations. This structure comprises shells of atoms as follows: 1 atom, 12 atoms, 6 atoms, 24 atoms, etc. Thus, in the bulk each atom has 12 nearest neighbors. We calculate potential energy curves for each cluster size within this structure. Most of the bonding arises from the 5s orbitals of the silver atom in the $4d^{10}5s^1$ configuration. We find a lengthening of the equilibrium bond length as size increases. We calculate 2.0_0A for Ag_2, 2.2 A for triangular Ag_3, 2.4 A for bipyramidal Ag_5 and 2.5 A for Ag_8 cubes and larger clusters. The ionization potential as calculated by the highest occupied molecular orbital decreases from 8.60 eV at Ag_2 to 6.62 eV at Ag_{39} and continues to drop to 6.17 eV at the bulk. There is a gap between the highest occupied molecular orbital and the next unoccupied molecular orbital. Values decrease with increasing size from 2.26 eV for Ag_2, 0.36 eV for Ag_{12} to 0.14 eV for Ag_{39}. Even at 39 atoms this gap is larger than the spacing between energy levels in the bulk of ~0.02 eV at room temperature (i.e., kT).

Let us consider the density of states determined from the EH calculation. These data are shown in Figure 1 where we have compared clusters of varying size and the bulk. Only the occupied energy levels are shown. To obtain smooth DOS curves, Gaussian functions of 0.1 eV full-width at half-maximum (FWHM) are placed upon each molecular orbital and their resulting intensities are then summed. The lifetime broadened DOS curves are obtained following a procedure used for silver bulk calculations.[16] Here a skewed Gaussian function of the form

$$e^{-(a\Delta E/(1 + a\beta\Delta E))^2} \quad a = 5\ eV^{-1},\ \beta = -.1 \qquad (5)$$

is convoluted with the molecular orbitals and the lifetime function

$$I\alpha \frac{\cos(\pi\alpha/2 + (1-\alpha)\ \arctan\ (\Delta E/\gamma))}{(\Delta E^2 + \gamma^2)^{(1-\alpha)/2}} \qquad (6)$$

$$\alpha = 0.07 \qquad \gamma = 0.10$$

Here ΔE is measured from the threshold of the unbroadened line, γ is the lifetime width, and α is an asymmetry parameter. These values and equations are taken directly from reference 16. These curves are normalized to the peak amplitude. The intense portion in these curves is due to d molecular orbitals which have also partially hybridized with s molecular orbitals. We note that the d bands spread from ~0.8 eV width at Ag_2, to 1.8 eV at Ag_{39} to 2.3 eV at the bulk considering the FWHM. In addition, the separation between d band threshold and s band threshold increases from 2.6 eV for Ag_2 to 4.1 eV for the bulk. There is also an apparent splitting of the d band showing up at 13 atoms and larger sizes. This effect is due to the pattern of orbital mixing and is similar to a crystal field effect. This type of splitting has been noted in experimental studies of silver clusters on carbon.[17] Although carbon orbitals may mix with the silver orbitals to accentuate such an effect, the effect also appears to occur in these isolated clusters.

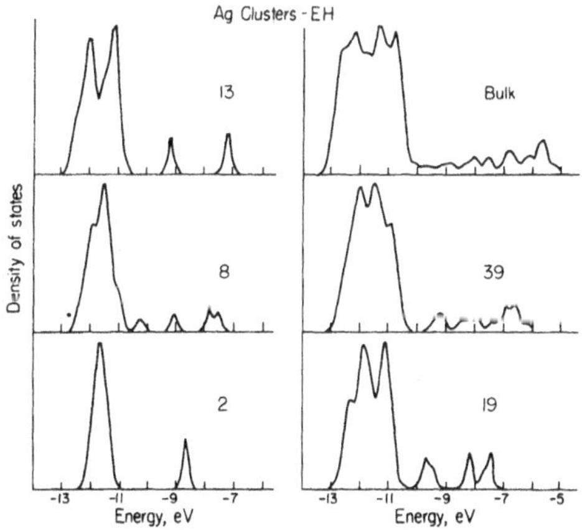

Figure 1: Lifetime broadened DOS for occupied energy levels of silver clusters including cubic Ag_8 and fcc structures for larger clusters with 2.5 A bond length.

These DOS curves compare favorably with other calculations. Previous work[4] has shown good agreement with relativistic augmented plane wave (RAPW) calculations in terms of shape, although the d bandwidth is 20% more narrow in this calculation. Comparison can also be made with experimental ESCA spectra as shown in Figure 2. Here there is good agreement in terms of curve shape, but the d band is too narrow in our calculation. Of course, instrumental resolution effects could be partially involved in this discrepancy. Despite this difference, the overall agreement is good and when the bond length is shortened only 0.2 A the bandwidth becomes the same as experimental.

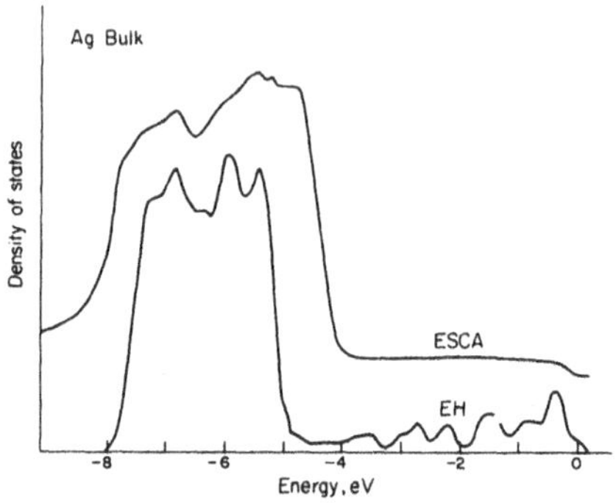

Figure 2: Lifetime broadened DOS calculated for bulk silver compared with ESCA spectrum for silver from reference 16.

The atomic configuration used for Pd is $4d^{10}$, but we also include 5s and 5p orbitals in the calculation. The bonding in Pd clusters is largely due to the 5s orbitals, which are more diffuse than the 4d orbitals. Thus, even in Pd_2 the 5s orbitals overlap strongly and form a molecular orbital more strongly bonding than the 4d molecular orbitals and leading to a $4d^95s^1$ configuration. This electron configuration approaches $4d^{9.4}5s^{0.6}$ as the cluster size increases. This configuration is attained at about 8-10 atoms. Thus, even the small clusters have unoccupied d molecular orbitals and their presence may be important in catalysis.

The calculated DOS for various Pd clusters is shown in Figure 3. The Gaussian function and lifetime broadening function discussed earlier were employed to prepare these plots. Each cluster was restrained to have a 2.5 A nearest-neighbor bond length. We observe a broadening of the d band as cluster size increases and the appearance of distinct s molecular orbitals below the d band for Pd_{19}. The d band increases with increasing size and the single planar layer resembles the Pd_{19} cluster. The DOS is considerably broader for the bulk than for the single planar or other cluster sizes, as shown by the relative ratios 4.6, 2.0, 2.8, 2.7, 2.8, 1.0 for FWHM for bulk, planar to smaller clusters considered in Figure 3.

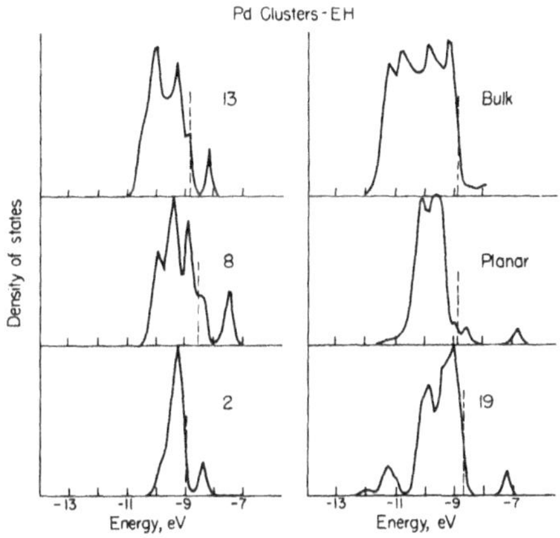

Figure 3: Lifetime broadened DOS for palladium clusters including cubic Pd_8, fcc planar Pd and fcc structure clusters and bulk with 2.5 A bond length. The dotted line denotes the highest filled orbital.

We examine the density of states for Pd_{16} clusters in Figure 4. Here the local density of states is obtained from the projection onto atoms with varying numbers of nearest neighbors. These curves arise from summing Gaussian functions of 0.1 eV. FWHM positioned on each molecular orbital, but are not normalized. We observe strong variations in DOS depending upon the number of nearest neighbors. The density of states at the Fermi energy $D(E_F)$ is at a maximum at the atom with 7 co-ordination. This effect has also been found in Pd_{18} and Pd_{20} clusters, where $D(E_F)$ is largest at atoms with 7-9 co-ordination.

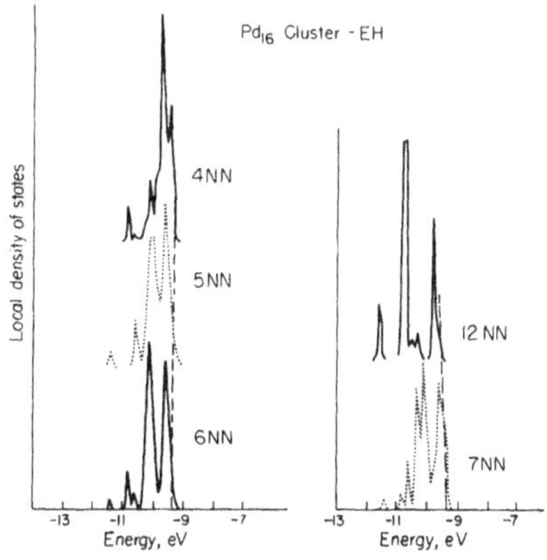

Figure 4: The DOS of the Pd_{16} fcc clusters is projected onto atoms with various numbers of nearest neighbors (NN). The dotted line denotes the highest filled orbital.

The causes of high density of states at intermediate co-ordination can be traced in our calculation. We consider the local density of states for Pd_{19} clusters in Figure 5. The first cycle results (non-self-consistent) and final converged results (self-consistent) are shown. We see that the d band is broader at the center atom which has the greatest number of nearest neighbors in the first cycle of the calculation. Thus, more of these orbitals will lie above the Fermi energy and will be unoccupied. This leads to a positive charge on the center atom in Pd_{19} of +2.1 and a charge of -0.28 on the outer atoms after the first cycle. In succeeding cycles of the calculation the orbitals shift so as to make the charge distribution more uniform. Orbitals on outer atoms are shifted towards the vacuum relative to the center atoms and this leads to a surface potential. As a result of this, there are two opposing forces which operate to move orbitals relative to the Fermi energy. Apparently, for Pd the balance of these two forces positions orbitals on atoms with intermediate co-ordination near the Fermi energy. These effects can be observed in Figure 5 where the orbitals on the center atom appear at much greater binding energy than those on the middle or outer atom.

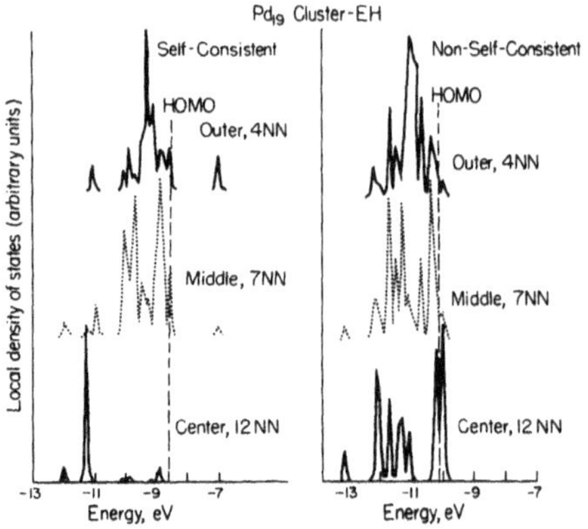

Figure 5: The DOS of the Pd_{19} fcc cluster is projected onto atoms with various numbers of nearest neighbors (NN) for the self-consistent and non-self-consistent EH calculation.

The surface potential and pattern of charge distribution for Pd clusters can be seen in Table I. The orbitals on atoms with lower co-ordination number are shifted towards the vacuum. For the Pd_{19} cluster, the s and d orbitals are shifted nearly 1 eV towards the vacuum relative to center atoms. The shifts are smaller for the less symmetric Pd_{16} cluster. In addition the center atom in these clusters retains a more positive charge than the outer atom. There are some exceptions to these generalizations with Pd_{13}.

Table I

Self-Consistent $-H_{ii}$ Values
Pd Clusters

Model	Atom	$-H_{ii}$ Value (eV) S	P	D	Charge
Pd_{13}	Center (12NN)	6.42	0.91	9.46	-.15
	Outer (4NN)	6.90	0.87	9.63	+.02
Pd_{16}	Center (12NN)	6.76	1.23	10.33	+.29
	Middle (7NN)	6.50	0.99	9.99	+.02
	Middle (6NN)	6.49	1.02	9.86	-.05
	Middle (5NN)	6.54	0.90	9.84	+.00
	Outer (4NN)	6.58	1.22	9.75	-.06
Pd_{19}	Center (12NN)	7.69	1.84	10.99	+.07
	Middle (7NN)	6.73	0.73	9.40	-.02
	Outer (4NN)	6.80	0.57	9.24	+.00

CNDO

We have examined Ag_N clusters using various types of parameterization in which the bond length and resonance parameters (β) were varied. The equilibrium bond length is 3.4 A using a parameterization in which $\beta = -1$ and bond energies are calculated properly. This bond length is ~0.5 A larger than the bulk, although shorter equilibrium bond lengths may be achieved by using more negative β values. However, under such a parameterization the binding energy is much too large. Thus, we have retained the 3.4 A bond length in our calculations. Calculated binding energies and HOMO values are shown in Table II for this parameterization. Binding energies increase with size and generally with co-ordination number, but HOMO values show the reverse trend. Each of these trends is reasonable and would be expected on the basis of experimental values known for Ag_2 and bulk silver.

A parameterization with different β values for s and d orbitals has been discussed by Blyholder[18] and employed by Head and Mitchell.[19] Using this approach with $\beta_s = -1$ and $\beta_d = -3$, we calculate the values shown in Table II for Ag clusters. We do not find significant changes in the binding energy or HOMO using this approach as opposed to keeping $\beta = -1$ for s and d orbitals. However, this approach has a dramatic effect on the density of states, as shown in Figure 6. For Ag_{19} clusters the d bandwidth is roughly doubled using $\beta = -3$ as opposed to $\beta = -1$ for the d orbitals. The s band, however, is not significantly changed by this parameter.

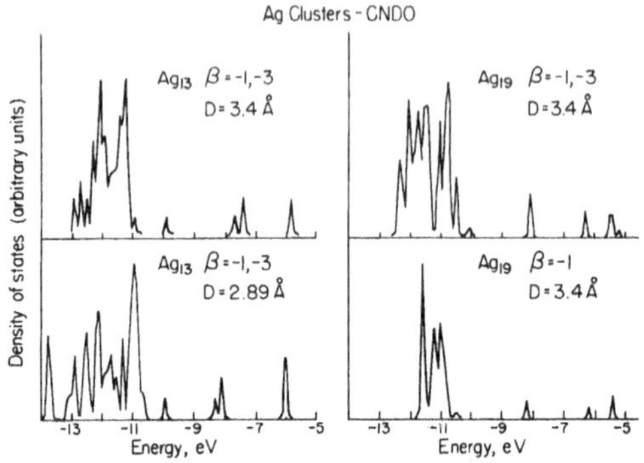

Figure 6: The DOS for the occupied energy levels of fcc silver clusters containing 13 and 19 atoms is shown for various parameters in the CNDO calculation. The bond length is denoted by R.

Table II

Properties of Ag_N Clusters - CNDO

Cluster	$\beta_s = \beta_d = -1$		$\beta_s = -1, \beta_d = -3$	
	BE/N(eV)	HOMO(eV)	BE/N(eV)	HOMO(eV)
Ag_2	1.51	7.01	1.50	6.85
Ag_3	1.25	6.46		
Ag_4	1.50	5.80		
Ag_5	1.18	5.98		
Ag_6	1.88	6.25		
Ag_7	1.66	6.36		
Ag_8	1.68	6.24	1.76	6.27
Ag_9	1.55	5.43		
Ag_{10}	1.79	5.86		
Ag_{11}	1.79	5.76		
Ag_{12}	1.87	5.49		
Ag_{13}	1.91	5.82	1.99	5.68
Ag_{16}	1.80	5.70	1.76	5.66
Ag_{19}	1.86	5.41	2.06	5.14

The effect of working at nonequilibrium bond length is also investigated in Figure 6 for Ag_{13} clusters. The bulk experimental bond length at Ag_{13} gives a broader bandwidth with relatively minor changes in shape of the band. However, the binding energy per atom is significantly reduced at 2.89 Å to 0.54 eV as opposed to the value 1.99 eV shown for R = 3.4 A in Table II. In addition, there is a reversal in charge distribution upon changing the bond length from 3.4 to 2.89 A. The center atom in Ag_{13} has the calculated charge of +0.14 for R = 3.4 A and -0.53 for R = 2.89 A. This discrepancy may be due to the equal partitioning of charge in a Mullikan analysis, but it points out the importance of using equilibrium bond lengths in calculations as opposed to the procedures of reference 19, which ignore this effect as well as the binding energy.

The density of states for Ag_N clusters calculated by CNDO is shown in Figure 7 for several clusters. Only occupied states are considered and the lifetime broadening function is employed. The features in these plots are very similar to the features calculated by EH in Figure 1. These features include band broadening with size, splitting of the d band and shift of the HOMO towards vacuum with increasing size.

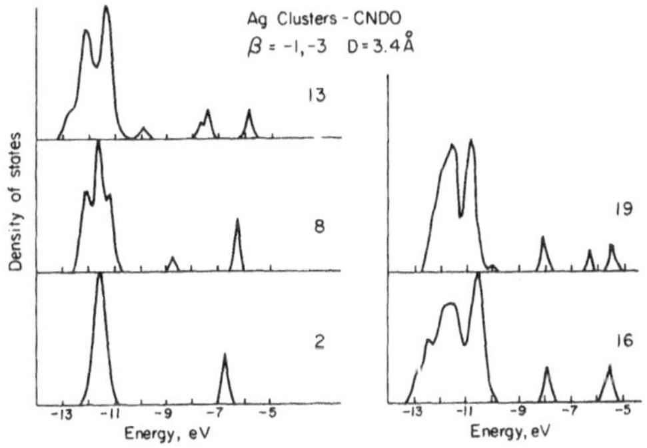

Figure 7: The lifetime broadened DOS for the occupied energy levels of silver clusters is shown for the CNDO calculation.

Palladium clusters were examined using β = -3, -5 for s and d orbitals, respectively. The lifetime broadened DOS for the occupied states are shown in Figure 8. The centroid of the d band shifts to lower energy with increasing size. There is a complex structure in the d band showing several peaks and the s orbitals significantly overlap the d band. A major deficiency appears to be the lowering of DOS intensity at the top of the d band. It would be expected, on basis of bulk experimental ESCA spectra, that a higher intensity would exist in this region.

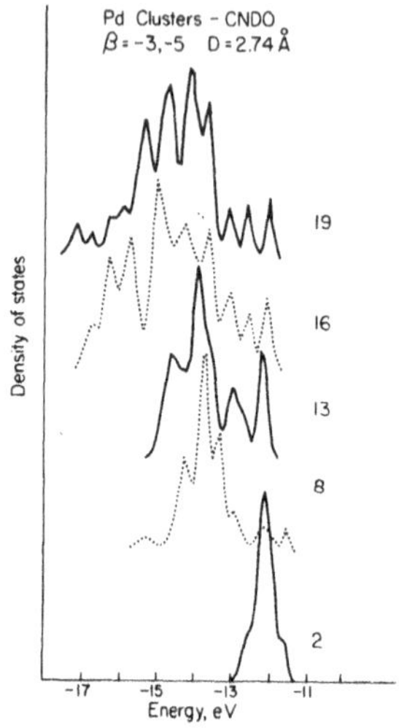

Figure 8: The lifetime broadened DOS for occupied levels of palladium clusters shown for cubic Pd_8 and fcc structure larger clusters is calculated by CNDO.

The binding energy of Pd_N clusters calculated by CNDO shows the expected increase with size from 0.73 eV/atom at Pd_8 to 1.92 eV/atom at Pd_{19}. The d configuration calculated for these clusters changes from $d^{9.0}$ for Pd_2 to $d^{9.5}$ at Pd_{19}. This trend is similar to the trend observed in EH calculations for these clusters. The β_s parameter is most critical in determining d occupation since we find d^{10} configurations in Pd_2 if $\beta_s > -3$. From binding energy considerations, however, we believe that $\beta_s^s = -5$ is the best choice for this parameter. The d orbitals show a pattern of occupancy which varies within a given cluster as shown in Table III. Apparently surface atoms have a larger d occupation than interior atoms.

DISCUSSION

The properties of metal clusters show significant variations as size increases up to the largest sizes we have considered. Significant changes in binding energy per atom, ionization potential and DOS are observed in silver and palladium clusters with EH and CNDO calculations.

Table III

d Orbital Occupation in Pd Clusters

Cluster	Atom	Occupation
Pd_8	-	$d^{9.69}$
Pd_{13}	center - 12NN	$d^{9.67}$
	outer - 4NN	$d^{9.60}$
Pd_{16}	center - 12NN	$d^{9.28}$
	middle - 7NN	$d^{9.26}$
	6NN	$d^{9.47}$
	5NN	$d^{9.31}$
	outer - 4NN	$d^{9.49}$
Pd_{19}	center - 12NN	$d^{9.08}$
	middle - 7NN	$d^{9.39}$
	outer - 4NN	$d^{9.62}$

Significant splitting of the d band is observed in both calculations for Ag clusters, and Pd clusters show variations in the d occupancy with size and within a given cluster.

Calculations using EH and CNDO place a small positive charge on the center atom in 19-atom clusters with the exception of the CNDO calculation on Pd_{19} as shown in Table IV. A positive charge on the center atom is in accord with physical expectations. The wave functions on outer atoms overlap with fewer orbitals and thus do not share electron density. This result is in accord with ab initio calculations for Li clusters, where surface atoms of body-centered cubic Li_9 were found to have a slight negative charge.[20] X_α calculations[3] for 13 atom clusters of Cu, Ni, Pd, and Pt predict a positive surface charge in opposition to these results.

Table IV

Charges of M_{19} Clusters

Cluster	Calculation	Reference	Charges Center	Middle	Outer
Cu_{19}	EH	4	+.09	-.01	+.001
Pd_{19}	EH	4	+.07	-.02	+.002
Ag_{19}	EH	4	+.35	-.04	+.03
Pd_{19}	CNDO	This work	-.33	-.002	+.06
Ag_{19}	CNDO R=3.4	This work	+.22	-.11	+.18
Ag_{19}	CNDO	19	+.21	-.10	+.16
Ag_{19}	X_α	19	-.18	+.40	+.78

The equilibrium bond length calculated using EH methods shows an elongation with increasing size, as shown for several clusters in Table V. The diatomic bond length is generally quite short, followed by a change in bond length that becomes constant after a few atoms. Absolute values of these bond lengths are a strong function of the orbital exponents of the atomic Slater orbitals. The reason for a minimum in potential energy curves calculated by EH is the asymmetrical spreading of bands leading to a greater antibonding component. Thus, the dimers in Table V have relatively short bonds because all the repulsions come from d bands since no antibonding s orbitals are occupied (i.e., $d^{10}s^1$ configurations are found for Ag_2 and Cu_2, whereas d^9s^1 configurations are found for Pd_2 and Ni_2). At larger cluster sizes the antibonding s molecular orbitals become occupied, providing the repulsive force leading to longer bonds. This argument ignores nuclear repulsions, which, if considered, would markedly elongate the very short dimer bond lengths. The trend of bond elongation with increasing cluster size has been observed by Apai and Hamilton[21] in recent EXAFS experiments on Cu and Ni clusters.

Table V

Calculated Equilibrium Bond Lengths - Extended Hückel, K = 1.75, Three-Dimensional Geometries

Size	Silver R_{eq} (Å)	Palladium R_{eq} (Å)	Copper R_{eq} (Å)	Nickel R_{eq} (Å)
2	2.0	1.4	1.75	1.7
3	2.2	2.0	2.2	2.0
4	2.2	2.2	2.3	2.3
5	2.4	2.2	2.4	2.4
6	2.4	2.3	2.5	2.5
7	2.5	2.3	2.5	2.5
8	2.5	2.3	2.5	2.5
9	2.5	2.3	2.5	2.6
10	2.5	2.3	2.5	2.6
13	2.5	2.3	-	2.6
19	2.5	2.3	-	2.6

These results show that both EH and CNDO are useful methods for calculating electronic properties of metal clusters. CNDO is somewhat more complicated because of the greater number of parameters involved. The use of different β values for s and d orbitals in CNDO is useful in obtaining broad DOS data. It appears that a parameterization based upon one set of properties need not yield accurate data when calculations on other properties are considered.

REFERENCES

1. Desjonqueres, M.C., and Cyrot-Lackman, F.: 1975, J. Phys. F. 5, p. 1368. 1976, Solid State Commun. 18, p. 1127. 1976, J. Chem. Phys. 64, p. 3707.

2. Anderson, A.B., and Hoffmann, R.: 1974, J. Chem. Phys. 61, p. 4545

3. Messmer, R.P., Knudson, S.K., Johnson, K.H., Diamond, J.B., and Yang, C.Y.: 1976, Phys. Rev. B: p. 1396.

4. Baetzold, R.C.: 1971, J. Chem. Phys. 55, p. 4363. Baetzold, R.C., and Mack, R.E.: 1976, ibid., 62, p. 1513. Baetzold, R.C.: 1978, ibid., 68, p. 555. Baetzold, R.C.: 1978, J. Phys. Chem. 82, p. 738.

5. Freund, H.J., and Bauer, S.H.: 1977, J. Phys. Chem. 81, p. 994.

6. Herrmann, A., Schumacher, E., and Woste, L: 1978, J. Chem. Phys. 68, p. 2327.

7. Mason, M.G., Gerenser, L.J., and Lee, S.-T.: Phys. Rev. Lett. 39, p. 288.

8. Joyes, P.: 1974, J. Phys. B. 7, p. 516.

9. Ozin, G.A., and Huber, H.: 1978, Inorg. Chem. 17, p. 155. Moskovits, M. and Hulse, J.E.: 1977, J. Chem. Phys. 67, p. 4271.

10. Hamilton, J.F., and Logel, P.C.: 1974, Photogr. Sci. Eng. 18, p. 507. 1973. J. Catal. 29, p. 253.

11. Verhaegen, G., Stafford, F.E., Goldfinger, P., and Ackerman, M.: 1962, Trans. Faraday Soc. 58, p. 1926.

12. Clementi, E., Raimondi, D.L., and Reinhardt, W.P.: 1967, J. Chem. Phys. 47, p. 1300.

13. Hoffmann, R.: 1963, J. Chem. Phys. 39, p. 1397.

14. Wolfsberg, M., and Helmholtz, L.: 1953, J. Chem. Phys. 20, p. 837.

15. Pople, J.A., Santry, D.P., and Segal, G.A.: 1965, J. Chem. Phys. 43s, p. 129.

16. Christensen, N.E.: 1972, Phys. Status Solidi B. 54, p. 551. Barrie, A., and Christensen, N.E.: 1976, Phys. Rev. B. 14, p. 2442.

17. Mason, M.G., and Baetzold, R.C.: 1976, J. Chem. Phys. 64, p. 271.

18. Blyholder, G.: 1974, Surf. Sci. 42, p. 249.

19. Head, J.D., and Mitchell, K.A.R.: 1978, Mol. Phys. 35, p. 1681.

20. Hermann, K., and Bagus, P.S.: 1978, Phys. Rev. B. 17, p. 4082.

21. Apai, G.A., and Hamilton, J.F.: unpublished work.

DISCUSSION

Peter POLITZER : We have carried out extended-Hückel calculations for a model system representing a single CO molecule chemisorbed on the (100) surface of nickel. One aim was to determine how many nickel atoms must be included to describe reasonably well the properties of the adsorbed CO. We found that inclusion of just the nearest-neighbors of the interaction site, and in only the surface plane, was sufficient. This is of course a different question from that of how many metal atoms are needed to reproduce bulk properties. From your work in general, do you have any comments regarding the description of surface interactions ?

R. BAETZOLD : I am familiar with your work which indicated that the CO vibration frequency attains a limiting value upon binding to only a small number of nickel atoms. You may know that M. Moskovitz (U. of Toronto) has measured CO vibration frequency when the molecule is adsorbed to Cu clusters in a matrix. The experiments agree with your calculation. Yet how do we explain the fact that small clusters have properties different from the bulk ? I am not sure this result would be found for all properties and it may depend upon the strength of the interaction. Thus, in cases of large charge transfer, I would expect a larger effect.

STRUCTURE SENSITIVITY AND SUPPORTED PLATINUM CATALYSTS

Robert L. Burwell, Jr.
Ipatieff Laboratory, Department of Chemistry
Northwestern University, Evanston, IL 60201 USA

Abstract: The hydrogenolysis of cyclopropane and methylcyclopropane, the hydrogenation of propylene, and the isotopic exchange between deuterium and cyclopentane and 3,3-dimethylbutane have been investigated on sets of Pt/SiO_2 and Pt/Al_2O_3 catalysts of a wide range of percentage exposed. The catalysts have been characterized by hydrogen chemisorption and in the case of Pt/SiO_2 by X-ray line profile analysis and transmission electron microscopy. The temperature of pretreatment by hydrogen has a larger effect upon turnover frequencies per surface atom of platinum than does variation in percentage exposed. The reverse tends to be true for selectivities.

The importance of supported noble metal catalysts has led to their receiving extensive study. Such catalysts consist of particles of metals like platinum, palladium, and rhodium with diameters of 1 to 10 nm dispersed on the internal surface of such materials as silica gel and γ-alumina. The supports are usually in the form of high area (100-500 m^2 g^{-1}) granules or pellets whose texture is much like that of a loose, cemented gravel bed. Pore diameters are in the range of 2-16 nm. Typically about 50% of the volume of a granule is void and only of the order of 0.1% is metal.

Inevitably one would like to know how the adsorptive and catalytic characteristics of the catalyst depend upon the support and the surface morphology of the metal particles. One could classify variation in surface morphology in terms of the crystal planes exposed, the nature and concentration of surface defects, and the proportions of atoms at vertices, at edges, and within surface planes. There are no current methods for studying the first two items, but, since we can make some estimate of the last by determining the particle diameters, there have been many studies of the last item and often in conjunction with measurements of catalytic activity. The techniques used for the determination of particle diameters have been primarily transmission electron microscopy (TEM), X-ray linewidth analysis and small angle scattering, and chemisorption measurments in systems in which it is

B. Pullman (Ed.): Catalysis in Chemistry and Biochemistry. Theory and Experiment. 207–216.

believed that the ratio of atoms or molecules adsorbed to surface atoms of metal is about unity. The last method ideally measures the percentage exposed of the metallic constituent (the percentage of the atoms of metal which are on the surface), it gives only average particle diameters and even those only after one has made some assumption about particle shape. Line profile analysis can provide information about particle shape and the distribution of particle sizes but only for particles whose average diameter is greater than about 2.5 nm. TEM can miss very small particles and it provides little information about shape for particles in the most interesting range of sizes. In the special case of nickel, magnetic measurements can also provide information about particle size and one of the earliest (but probably premature) attempts to correlate particle size and catalytic activity involved Ni/SiO_2 (Burwell and Tuxworth, 1956).

If the rate of some reaction varies little with particle size, the reaction is said to be structure insensitive; if it varies substantially, it is said to be structure sensitive. It must be emphasized that sensitivity or insensitivity applies only to a particular reaction on a particular type of catalyst under the particular conditions and pretreatment procedures studied. In addition a reaction on a set of catalysts may be structure sensitive with regard to product selectivity or to competitive selectivity even though rates may vary little (Kung et al., 1976).

THE CATALYSTS AND THEIR CHARACTERIZATION

Most of the work on Pt/SiO_2 and Pt/Al_2O_3 catalysts reported in this paper represents collaborative research involving J. B. Butt (Department of Chemical Engineering), J. B. Cohen (Department of Materials Science), R. L. Burwell, Jr. (Department of Chemistry), and several workers in other institutions. A number of papers on this general subject had appeared during the last decade, but, in general, they involved the study of a rather restricted group of catalysts pretreated under only one condition and only one or two catalytic reactions. We thought that it would be useful to make a much more extensive investigation. A set of ten Pt/SiO_2 catalysts containing 0.5-2% by weight was prepared of a wide range of percentage exposed and characterized as carefully as possible by hydrogen chemisorption (Uchijima et al., 1977), X-ray line profile analysis (Sashital et al., 1977), TEM (Aika et al., 1976), and X-ray photoelectron spectroscopy (XPS) (Angevine et al., 1976). The agreement among the determinations of particle size by the first three methods listed above was good. At this stage, the Pt/Al_2O_3 catalysts have been characterized only by chemisorption techniques.

The Pt/SiO_2 catalysts on Davison silica gel Grade 62 (nominal pore diameter 14 nm) were prepared by two methods, ion exchange with $Pt(NH_3)_4^{2+}$ and impregnation with H_2PtCl_6. Percentages exposed as measured by hydrogen chemisorption at 25°C assuming that H/Pt_s = 1 (to be denoted D_h) varied from 6.2 to 81%. 81-SiO_2-IonX indicates the catalyst with D_h = 81% prepared by ion exchange and 6.3-SiO_2-PtCl indicates the catalyst with D_h = 6.3% prepared by impregnation. No differences in the behavior of the IonX and the PtCl catalysts have been observed. After initial preparation, the catalysts were reduced in hydrogen, cooled in nitrogen, and stored in air. To prepare a clean surface of platinum free of hydrogen for hydrogen chemisorption, catalysts were pretreated in situ; O_2,300°;H_2,300°;He,450° where He,450° signifies that the catalyst was heated to 450°C and cooled in helium following treatment with hydrogen at 300°C. This will be called the standard pretreatment, but a number of other pretreatments were also investigated in catalytic experiments. The highest temperature reached in the initial preparation of any of the catalysts was 435°C. Thus, after the standard pretreatment the condition of the silica surface should be the same for all catalysts.

A catalyst which had received the standard pretreatment rapidly adsorbed hydrogen and oxygen at 25°C. The oxygen adsorbed in 15 min was removed by a pulse of hydrogen. However, on more protracted exposure to oxygen, more oxygen adsorbed and reaction with hydrogen became much slower. The coverage (O/Pt_s) after 15 min was about 0.8; after storage in air for a year it was about unity; and after O_2,300° it exceeded unity. The reaction with oxygen at 300°C was structure sensitive. Reaction of 81-SiO_2-IonX with oxygen at 300°C gave O/Pt_s = 1.33, that is, the particles of metallic platinum were completely destroyed. Nevertheless, reduction by hydrogen at 300°C generated particles of platinum in their original states. At D_h = 40%, O/Pt_s became only about 0.49 and the metallic core of the particles survived. Treatment with hydrogen at 100°C appeared completely to remove surface oxide. At 25°C reaction with hydrogen was much slower for the particles of D_h = 81% than for those of D_h = 40% and those of D_h = 40% reacted more slowly than those of D_h = 6.3%. This was true both for stored catalyst and for catalysts treated O_2,300°. Percentage exposed did not appear to depend upon the temperature of pretreatment by hydrogen between 25 and 450°C except for the catalysts with the largest particles. For these, pretreatment at 25-100°C gave percentages exposed about 50% larger than those which resulted from the standard pretreatment. However, the pretreatment, H_2,25-100°;He,450°, led to normal values of D_h.

As determined by X-ray line profile analysis, the particles in the catalysts with D_h = 21.5, 27, and 40% are perfect FCC single crystals which have nearly the same diameter in four crystallographic directions, i.e., the particles can be approximately inscribed in a sphere. Figure 1 shows the distribution of particle sizes in the (100) direction for 21.5-SiO_2-IonX (Sashital et al., 1977). Other results were similar in appearance. However, particles of 7.1-SiO_2-PtCl were somewhat elongated in the (100) direction and they exhibited microstrains. The particles of this catalyst were about as big as the pores of the silica. The data of Figure 1 were determined on catalyst as stored in air. Subsequent examination in a controlled atmosphere cell showed that the catalyst reduced and cooled in hydrogen gave the same results. Thus,

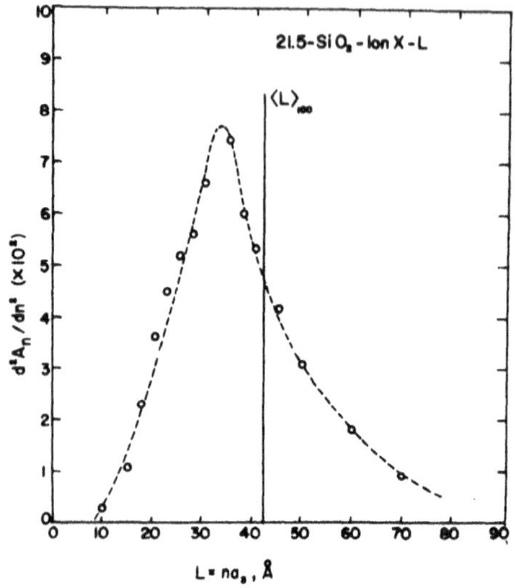

Figure 1. Distribution of particle diameters in the (100) direction for 21.5-SiO_2-IonX. Relative abundance on y-axis, diameter on x-axis. The vertical line is the area weighted average diameter.

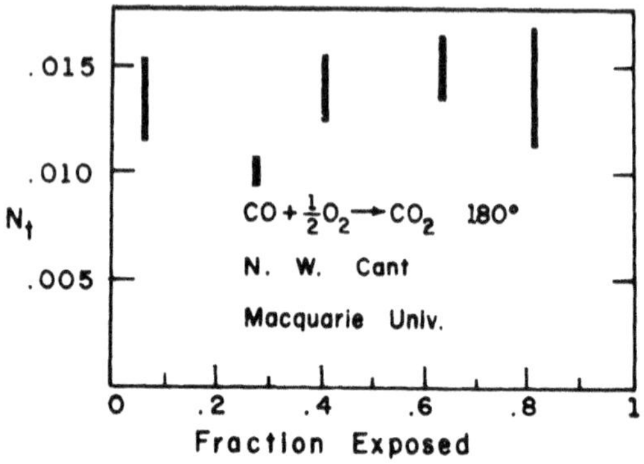

Figure 2. N_t on y-axis vs. fraction exposed on x-axis for the oxidation of carbon monoxide by oxygen at 180°C. From N. W. Cant.

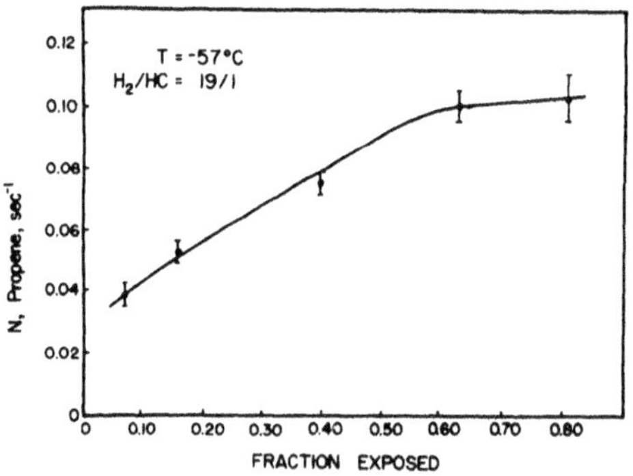

Figure 3. N_t for the hydrogenation of propylene at -57°C (y-axis) vs. fraction exposed for Pt/SiO_2 catalysts (x-axis). The mole ratio, hydrogen/propylene, was 19.

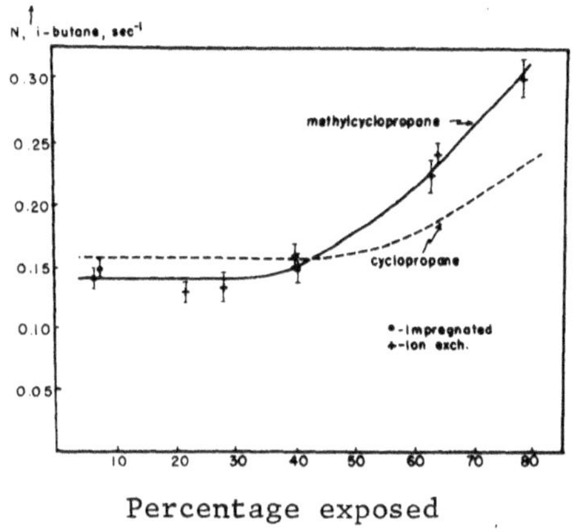

Figure 4. N_t for formation of isobutane in the hydrogenolysis of methylcyclopropane on Pt/SiO_2 catalysts at 0°C (y-axis) vs. percentage exposed (x-axis). The mole ratio, hydrogen/methylcyclopropane was 14. N_t for the hydrogenolysis of cyclopropane is also shown.

the surface "oxide" layer on such rather large (5 nm) particles does not affect the structure of the metal core.

The preparation of Pt/Al_2O_3 catalysts started with impregnation by an aqueous solution of $Pt(NH_3)_2(NO_2)_2$ or $Pt(NH_3)_3NO_2^+$. The presence of chloride ion was avoided. D_h for the catalysts varied between 27 and 107%. In addition some catalysts with D_h of 4% were prepared. Since the particles of these catalysts are bigger than the pores of the alumina, migration of platinum to the external surface of the alumina granules must have occurred. The alumina employed, Aerocat 1000 of the American Cyanamid Company, was a γ-alumina with a reported purity of about 99.99% and a nominal pore diameter of about 6 nm. Platinum loadings were 0.3-0.4 wt.%.

CATALYTIC REACTIONS ON Pt/SiO_2

Figure 2 exhibits the rates of oxidation on our catalysts of carbon monoxide by oxygen at 180°C measured by N. W. Cant of Macquarie University in Australia whom we thank for permission to use the figure. A helium carrier containing a few per cent of each gas was used in these experiments and the pretreatment of the catalysts was the standard one. The data are plotted as turnover frequencies, N_t (the number of molecules of carbon monoxide oxidized per sec per surface platinum atom). Presumably everyone would classify the reaction under these conditions as structure insensitive. In the oxidation of hydrogen on a different set of Pt/SiO_2 catalysts, the reaction was found to be structure insensitive when oxygen was in excess but structure sensitive when hydrogen was in excess (Hanson and Boudart, 1978).

Figure 3 shows Nt for the hydrogenation of propylene at -57°C on our Pt/SiO_2 catalysts (Otero-Schipper et al., 1977). N_t increases by a factor of 2.5 as D_h varies from 6.3 to 81%. Workers might disagree in their classification of this reaction as sensitive or insensitive to structure.

Figure 4 shows N_t for the hydrogenolysis of cyclopropane and methylcyclopropane (Otero-Schipper et al., 1977). Again there might be some disagreement as to classification. The hydrogenolysis of methylcyclopropane gives two products, isobutane and butane according to whether the ring is opened opposite or adjoining the methyl group. Figure 5 shows how the selectivity, expressed as the ratio, isobutane/butane, varies with D_h. The extreme variation with D_h is a factor of 2.

In Figures 3-5, the catalysts had been pretreated in hydrogen to 470°C for the experiments with propylene and to 350°C for those with cyclopropanes. Figure 6 shows N_t for the formation of isobutane from methylcyclopropane as a function of D_h and T in the pretreatment, O_2,300°;H_2,T (Otero-Schipper et al., 1978). As may be seen, the conditions of pretreatment have more effect upon rates than does percentage exposed. In particular, the effect of increased temperature of

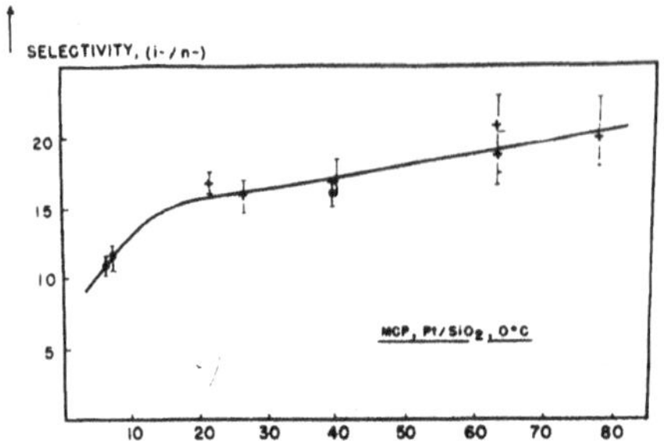

Percentage exposed

Figure 5. Selectivity in the hydrogenolysis of methylcyclopropane vs. percentage exposed.

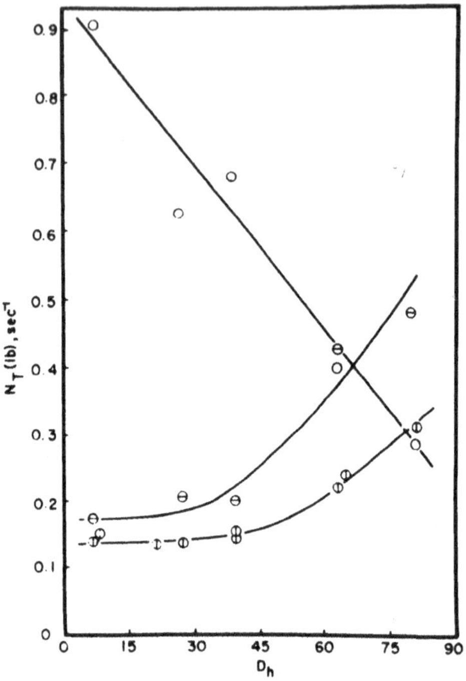

Figure 6. At right: N_t for isobutane in the hydrogenolysis of MCP on Pt/SiO_2 at 0^0C vs. D_h for three pretreatments. The line which descends with increasing D_h is for H_2,25^0; the middle curve is for H_2,480^0; and the lower curve is for H_2,370^0.

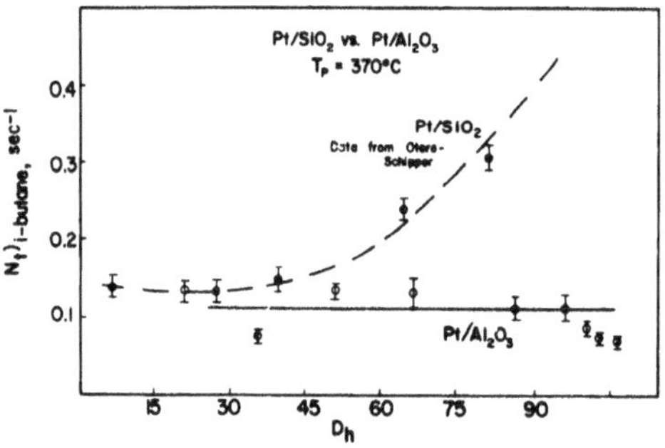

Figure 7. N_t for isobutane vs. D_h on Pt/SiO_2 (upper line) and on Pt/Al_2O_3 (lower line), both pretreated H_2,370^0.

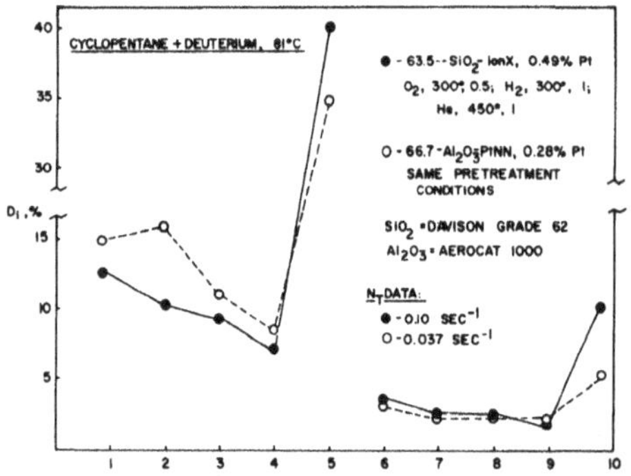

$\underline{i}$ in cyclopentane-d_i

Figure 8. The initial isotopic distribution of cyclopentane exchanged with D_2 at 81^0C on 63.5-SiO_2-IonX (full line) and 66.7-Al_2O_3-PtNN (dashed line). The fraction of cyclopentane-d_i in total exchanged cyclopentane is plotted on the y-axis. The catalysts received the standard pretreatment.

pretreatment by hydrogen is inverted in going from large to small particles of platinum. The very high activity of large particles pretreated at low temperature is particularly striking.

We have also investigated the initial isotopic distribution in exchange of cyclopentane with deuterium at 81°C (Inoue et al., 1978) and of 3,3-dimethylbutane and deuterium (V. Eskinazi, unpublished). The effect of pretreatment conditions of Pt/SiO_2 upon the rate of formation of exchanged species resembles that of Figure 6. In addition, there is some structure sensitivity in selectivity, i.e., in the proportions of cyclopentane-d_1, -d_2 .... -d_{10} formed from cyclopentane and in those of 3,3-dimethylbutane-d_1, etc. Although variation of the isotopic distribution patterns with D_h was not large overall, there was a substantial variation in the ratio d_6/d_{10} with cyclopentane. The ratio is about unity for 6.3-SiO_2-PtCl and 3 for 81-SiO_2IonX. The effect of pretreatment upon the ratios is minor. The results accord with the species d_6-d_{10} being largely formed via 1,1-diadsorbed cyclopentane on large particles of platinum and via rollover of 1,2-diadsorbed cyclopentane on small particles.

CATALYTIC REACTIONS ON Pt/Al_2O_3

These studies have been repeated on the Pt/Al_2O_3 catalysts. Figure 7 exhibits a comparison of N_t for formation of isobutane from methylcyclopropane on Pt/SiO_2 catalysts (the data of Figure 4) and on Pt/Al_2O_3 catalysts (S. Wong, unpublished). The selectivity, isobutane/butane, is also less structure sensitive on Pt/Al_2O_3; it is about 14 with no clear variation with D_h. The Pt/Al_2O_3 catalysts are affected by T in O_2,300°;H_2,T in much the same was as were the Pt/SiO_2 catalysts. However they showed an effect not seen with Pt/SiO_2. At higher temperatures of pretreatment in hydrogen (370 and 450°C), catalysts which were originally reduced at higher temperatures (400 or 500°C) gave larger values of N_t than those reduced at lower temperatures (about 300°C). On the other hand, the temperature of reduction seemed to be without effect when the pretreatment temperature was 25°C. In general, then, the phenomena noted for Pt/SiO_2 are also observed with Pt/Al_2O_3 and, although there are differences between the two sets of catalysts, the differences are small and they appear mostly with catalysts of large D_h.

Figure 8 presents a comparison between the initial isotopic exchange patterns of cyclopentane on Pt/SiO_2 and on Pt/Al_2O_3 with D_h's of about 65%. The patterns are of the same general type. However, variation of d_6/d_{10} with D_h on Pt/Al_2O_3 is the opposite of that found with Pt/SiO_2. N_t for exchange on catalysts given the standard pretreatment is relatively structure insensitive, but, at D_h of about 100%, N_t is 10 times as large for pretreatment in hydrogen at 450°C as at 100°C. On the contrary, at D_h of about 4%, pretreatment at 100°C gives rates about twice as fast as pretreatment at 450°C (M. Kobayashi and Y. Inoue, unpublished). This behavior parallels that in hydrogenolysis of methylcyclopropane.

DISCUSSION OF THE EFFECT OF SUPPORT, D_h, AND PRETREATMENT

Overall, Pt/SiO_2 and Pt/Al_2O_3 behave very similarly in the reactions which we have investigated. There is no sign of any large difference between the electronic effects of the supports on the platinum particles. There does appear to be some influence of the support upon the morphology of the platinum particles particularly at large D_h. It has been suggested in the past that the platinum particles on alumina consist of single layers when D_h is near 100%.

In general agreement with previous work (Boudart, 1976), variation in N_t and D_h is small for the reactions which we have studied when the catalysts have been given the standard pretreatment. There are, of course, reactions which are structure sensitive. For example, Brunelle et al., 1976, examined the simultaneous hydrogenolysis and isomerization of pentane on a set of Pt/SiO_2 catalysts which were prepared and characterized rather like those of the present paper. When pentane in hydrogen was passed over catalysts which had been reduced in situ at 410°C, isomerization behaved as a structure insensitive reaction. However, the ratio, hydrogenolysis/isomerization, increased by a factor of 10 as the percentage exposed was increased from small to large values. Thus, hydrogenolysis was structure sensitive. Few structure sensitive reactions have been found so far and most of these are higher temperature reactions.

Almost surely there is a large change in the ratio of the number of platinum atoms at edges and in faces over the large range of D_h which we have studied. How then can so many reactions be nearly structure insensitive? It has been suggested that structure insensitive reactions involve just one surface atom of the metal, whereas structure sensitive reactions involve clusters of surface atoms. However, the coordination number to platinum is substantially different for an edge atom and a face atom. In catalysis by molecular complexes, changes in ligands have a pronounced effect. It has also been suggested that, during catalytic reactions of hydrocarbons, the surface of the metal becomes covered with carbonaceous residues and that the actual catalytic sites are on the carbonaceous residues and not on the metal. However, this does not appear to be the case for the hydrogenolysis of cyclopropane or for the hydrogenation of ethylene at room temperatures or below (Hattori and Burwell, 1979). Corrosive chemisorption (Hanson and Boudart, 1978) or extractive chemisorption (Kung et al., 1976) might in some way minimize differences between edge and face atoms. Overall, however, we do not appear to have a satisfactory explanation of why there are so many structure insensitive reactions.

Previous studies of structure sensitivity have not involved investigation of various pretreatments. At least for platinum catalysts, varying pretreatment conditions can affect rates more than varying percentage exposed. As judged by hydrogen-oxygen titration, the percentage exposed of the Pt/SiO_2 catalysts of the largest particle size is substantially larger for pretreatment by hydrogen at 25 to 100°C than for pretreatment by hydrogen at 300-450°C. This effect is probably related to the exalted activity of low D_h catalysts pretreated with hydrogen at low temperatures, and it may be just one aspect of differing surface morphology occasioned by differing conditions of pretreatment. The low activity of 81-SiO_2-PtCl pretreated at low temperatures may be connected with incomplete reduction of surface oxide although our measurements of hydrogen consumption during pretreatment (Uchijima et al., 1977) indicate that the degree of incompleteness cannot have been very large. Also, catalysts cooled in hydrogen from 450°C contain hydrogen in excess of that chemisorbed by clean platinum at 25°C. Another possible effect of pretreatment conditions and one which is currently under investigation is some effect on the degree of segregation of inorganic impurities in the supports onto the surface of the platinum particles (Fuentes and Figueras, 1978). The impurities of particular interest in commercial silica gel are iron and titanium. The level of impurity in our alumina is at least an order of magnitude lower than that in silica gel and there is more doubt that impurity effects can explain the different catalytic activities of Pt/Al_2O_3 catalysts which result from different conditions of pretreatment.

STRUCTURE SENSITIVITY AND ADSORBATE-SURFACE HINDRANCE

Related information should be provided by investigation of reactions between hydrogen and hydrocarbons of a structure such that adsorbate-surface hindrance would develop upon adsorption. The hydrogenations of di-*t*-butylacetylene (dba) and trans-di-*t*-butylethylene (tdbe) are of this type (Kung, et al., 1976).

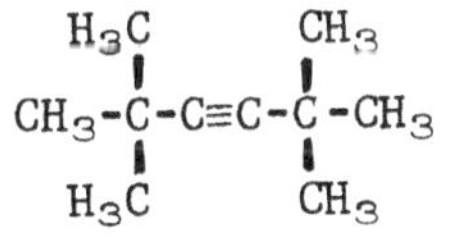

di-*t*-butylacetylene

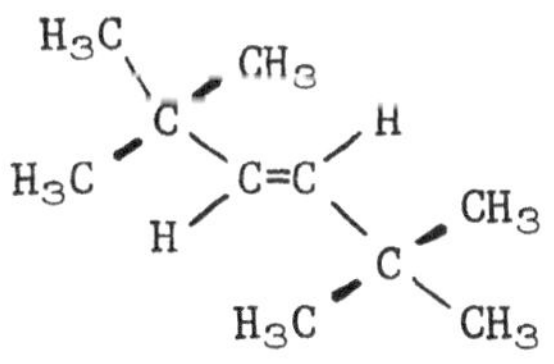

trans-di-*t*-butylethylene

In the approach of either species to more closely-packed planes of platinum, substantial steric hindrance should develop between the *t*-butyl groups and the surface atoms of platinum. The tdbe is hydrogenated very slowly and its hydrogenation is essentially completely inhibited by the presence of cyclopentene (a relatively unhindered olefin). The hydrogenation of dba is strongly inhibited by cyclopentene, whereas ordinarily an acetylene is hydrogenated competitively much more rapidly than an olefin. The degree of inhibition is much larger on Pt/SiO_2 and Pt/Al_2O_3 catalysts of small D_h than on those of large D_h. This result accords with the view that adsorbate-surface hindrance would be smaller at edge atoms than at atoms in (111) or (100) faces. However, the rate of hydrogenation of dba alone is relatively structure insensitive and N_t is as large as N_t for the hydrogenation of the relatively unhindered diethylacetylene alone. It was conjectured that extractive chemisorption may reduce the adsorbate-surface hindrance i.e., bonding of the -C≡C- unit to the surface pulls one or two surface platinum atoms somewhat above the plane of the surface.

ACKNOWLEDGEMENT. This work was supported by the Office of Basic Energy Sciences of the U.S. Department of Energy, the Petroleum Research Fund of the American Chemical Society and the Ipatieff Fund of Northwestern University.

REFERENCES

Aika, K., Ban, L. L., Okura, I., Namba, S., and Turkevich, J.: 1976, J. Res. Inst. Catal. Hokkaido Univ. 24, pp. 54-64.

Angevine, P. J., Vartuli, J. C., and Delgass, W. N.: 1976, Proceedings of Sixth International Congress on Catalysis, pp. 611-618.

Boudart, M.: 1976, Proceedings of Sixth International Congress on Catalysis, pp. 1-9.

Brunelle, J.-P., Sugier, A., and LePage, J.-F.: 1976, J. Catal. 43, pp. 273-291.

Burwell, R. L., Jr., and Tuxworth, R. H.: 1956, J. Phys. Chem. 60, pp. 1043-1049.

Fuentes, S., and Figueras, F.: 1978, J.C.S. Faraday Trans. 74, pp. 174-181.

Hanson, F. V., and Boudart, M.: 1978, J. Catal. 53, pp. 56-67.

Hattori, T., and Burwell, R. L., Jr.: 1979, J. Phys. Chem. 83, pp. 241-249.

Inoue, Y., Herrmann, J. M., Schmidt, H., Burwell, R. L., Jr., Butt, J. B., and Cohen, J. B.: 1978, J. Catal. 53, pp. 401-413.

Kung, H. H., Pellet, R. J., and Burwell, R. L., Jr.: 1976, J. Am. Chem. Soc. 98, pp. 5603-5611; Proceedings of Sixth International Congress on Catalysis, pp. 108-116.

Otero-Schipper, P. H., Wachter, W. A., Butt, J. B., Burwell, R. L., Jr., Cohen, J. B.: 1977, J. Catal. 50, pp. 494-507.

Otero-Schipper, P. H., Wachter, W. A., Butt, J. B., Burwell, R. L., Jr., and Cohen, J. B.: 1978, J. Catal. 53, pp. 414-422.

Sashital, S. R., Cohen, J. B., Burwell, R. L., Jr., and Butt, J. B.: 1977, J. Catal. 50, pp. 479-493.

Uchijima, T., Herrmann, J. M., Inoue, Y., Burwell, R. L., Jr., Butt, J. B., and Cohen, J. B.: 1977, J. Catal. 50, pp. 464-478.

RELATION BETWEEN CATALYTIC AND ELECTRONIC PROPERTIES OF TRANSITION METALS

F. Cyrot-Lackmann

Groupe des Transitions de Phases, C.N.R.S., B.P. 166, 38042 Grenoble Cedex, France.

Catalytic properties of transition metals are related to their electronic properties through their chemisorptive bond energies. The electronic properties of surface metal atoms and of clusters of various sizes and geometries are shown to have different characteristics from those of the bulk. In particular, there exists in the local densities of states of the active sites for adsorption having a low coordination number, extra resonant surface states. These electronic features of the surface catalyst have a strong influence on the chemisorption process. Two examples are stressed. The first one concerns the anisotropy of the binding energy of 4d atoms on a (100) and (110) Mo surface ; the second one concerns the change of magnetic properties of nickel clusters atoms when hydrogen is chemisorbed as a function of the geometry of the cluster.

1. INTRODUCTION

It is well established that the catalytic function of surfaces is to reduce the activation energy of a reaction by temporarily forming chemical bonds with the adsorbing molecules (Balandin, 1969). For instance, let us compare the bond energy Q_{AB} between two atoms A and B in the molecule AB to the energy of the same bond when the molecule is chemically adsorbed on the catalyst K. If the bond energy between the atom A and the catalyst K is denot Q_{AK}, to break the bond AB into atoms adsorbed on K, one will need the energy

$$Q_{AB}^{K} = Q_{AB} - Q_{AK} - Q_{BK}$$

Thus the bond will be weakened on the catalyst if $Q_{AB} > Q_{AK} - Q_{BK}$. If $Q_{AB} < Q_{AK} + Q_{BK}$, the A-B bond will be broken during adsorption. In principle, this scheme permits to calculate the activation energy of a reaction through the energies Q_{AK} which will depend of course on the other species adsorbed near A during the reaction. A zero order theory of the catalytic reaction will be to replace these quantities by the bond Q_{AK}^{o} of one atom A on the catalyst without any other species on its surface. The energetics of heterogeneous catalysis is thus simplified to the ener-

B. Pullman (Ed.): Catalysis in Chemistry and Biochemistry. Theory and Experiment. 217–226.

getics of chemisorption.

The ability of the surface to break some of the strong chemical bonds o the reactant molecules, such as H-H, C-H,...) depends on the surface considered for a particular catalyst. For example, it has been recently shown that there exist various active sites where chemical bond scissio is easier (Somorjai, 1977). The structure and the chemical composition the working catalyst is non investigated on the atomic scale, and kinks steps, terraces,... have been shown to have a different catalytic activ in many reactions. This shows that a Pauling type calculation of the bo energies is not sufficient. Indeed, the energy depends strongly on the electronic structure of the site on which the adatom sits.

To answer to the main questions arising on catalysis concerning the specificity of a metal, its selectivity and its structure sensibility as a catalyst, we are led to first consider the local electronic structure o its surface. One way of doing this is to study the local densities of states on the different active sites prior to adsorption and their influence on the strength of the bond with the adsorbate.

2. ELECTRONIC STRUCTURE OF METAL CLUSTERS

Through the study of the electronic properties of transition metal clus of various size and geometry, some questions of fundamental importance the comprehension of catalysis emerge. We are particularly interested i the following concepts :

- at what size does a cluster of metal atoms behave as a bulk meta
- what are the relations between electronic properties of a metal cluster and its atomic structure ?
- are we able to predict the stability of clusters when their size and atomic structures vary ?

2.1 Calculational procedure

The electronic structure of a transition metals is well described by a narrow d band with a high density of states within a broad sp band. Man of the peculiar properties of transition metals have been shown to be related to the d character of their valence states. One can thus reasona bly neglect the sd hybridization, as a first approximation, when descri bing the electronic properties of surface and cluster transition metals The small overlap between the d atomic orbitals leads naturally to stud the d band in the tight binding approximation or LCAO. It involves matr elements between atomic orbitals and the lattice potential, either compu ted directly or obtained from interpolation schemes on band structure calculations.

As any molecular orbital theory, this approach is approximate and, there fore, its application is limited to a semiquantitative explanation of phenomena. But the use of physical concepts permits an usuful insight understanding of several electronic properties (Friedel, 1969).

The physical properties of a cluster of bonded atoms are determined from equilibrium conditions. The potential energy curve is constructed and the bond length minimizing this energy is taken as the equilibrium internuclear distance. The attractive part due to the electronic bond contribution is obtained through the electronic density of states $n_T(E)$

$$E_B = -10 \int^{E_F} (E-E_d) n_T(E) dE$$

where E_F is the Fermi level. $n_T(E)$ is the sum of all the local densities of states (LDS), $n_i(E)$, which give information on the spatial distribution of electrons in the various orbitals. These LDS are obtained using the moment method and the continued fraction expansion described in previous articles (Cyrot-Lackmann, 1970, and Gaspard and Cyrot-Lackmann, 1973).

2.2 Results for semi-infinite transition metals

One can resume the detailed results obtained for various cleavage planes and various crystallographic structures in the following way (Desjonquères and Cyrot-Lackmann, 1975).

- For a dense cleavage plane (i.e. (111) for an FCC lattice, (110) for a BCC one), the LDS on the surface behaves like the bulk DOS (Fig. 1a).

- For a non dense cleavage plane such as the (110) for a FCC lattice or the (100) for a BCC lattice, it exists a virtual bound state peak in the middle of the band. This peak is due to the resonance of the surface atoms interacting weakly with the bulk ones (Friedel, 1976) (Fig. 1b).

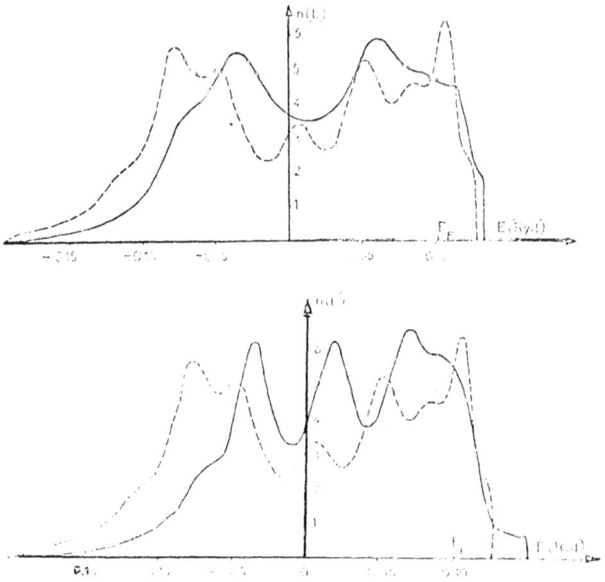

Fig. 1. LDS on the surface of nickel compared with the bulk one (broken curve)

a) (111) plane

b) (100) plane.

These features of the LDS on surfaces exist for all the transition metal and have been observed by spectroscopy. They lead to a slight increase in the asphericity of the charge density of the d electrons when going from a bulk to a surface site giving indications on the possible adsorption sites.

In the case of stepped surfaces, there exist also extra vertical bound states on the LDS of the edge atoms. Spin orbit coupling may play an important role in that case (Desjonquères and Cyrot-Lackmann, 1976b).

2.3 Results for Ni clusters

We have studied Ni clusters between 13 atoms and 2089 atoms, of differen geometries, icosahedra, cubooctahedra, truncated octahedra (Cyrot-Lackma et al, 1977 and Gordon et al, 1978). The comparison between the propertie of icosahedron and cubooctahedron clusters are particularly interesting, as they have the same total number of atoms but a different symmetry. Indeed, the cubooctahedron is formed by shells of close packed FCC atoms and the icosahedron is obtained through a distortion leading to a five fold symmetry. Their respective compactness is 0.74 and 0.688, the BCC being 0.680.

Experimentally one has observed transition metal clusters of five fold symmetry for small size with usually a contracted lattice parameter, the clusters making a phase transition towards an FCC lattice when the size is increased (Renou and Gillet, 1977 ; Moraweck el al, 1979 ; Solliard and Buffat, 1977). These clusters, widely used in catalysis, have surely chemisorptive properties related to the peculiar role played by those surface atoms which are not located within low index planes with a high coordination number. Indeed these atoms might possibly act as active centers in adsorption (Somorjai, 1977). It is a situation similar to stepped surfaces, but enhanced in clusters by a high rate of surfaces to bulk atoms, which varies for example from 0.92 for a 13 atom clusters to 0.31 for a 2057 atom cluster.

Some of our results are presented in figures 2 and 3. They show the following trends for the LDS :

- a width for the icosahedron band smaller than the cubooctahedron one,

- an opening of a through in the center of the icosahedron band, related to its weak compactness. It is a characteristic which reminds some of the band structure of BCC lattice,

- the band features of clusters approach those of an infinite surfac solid only for large clusters containing at least of the order of 1000 atoms, i.e. clusters of the order of 30 Å.

The total densities of states (Gordon et al, 1978) show the same trends with a slight decrease in the band width when the size is decreased, and some change in the density of states on the top side. This could then induce some change in the magnetic properties of small clusters. This is reasonable, as for very small sizes, the cluster atoms will behave more like free atoms. This is confirmed by the few experimental results existing on the electronic properties or magnetic ones of transition metal clusters (Ladas et al, 1978 ; Knight, 1977).

One can study also the relative stability of these clusters of various sizes by minimizing their cohesive energy as a function of the interatomic distance, as explained above. One finds always a contraction for the lattice, larger for smaller clusters sizes. This relaxation is the result of the balance between electronic and stain energy and leads to the relative stability of icosahedra with respect to FCC clusters for small sizes. This contraction is in agreement with recent results on Pt clusters using EXAFS (Moraweck el al, 1979).

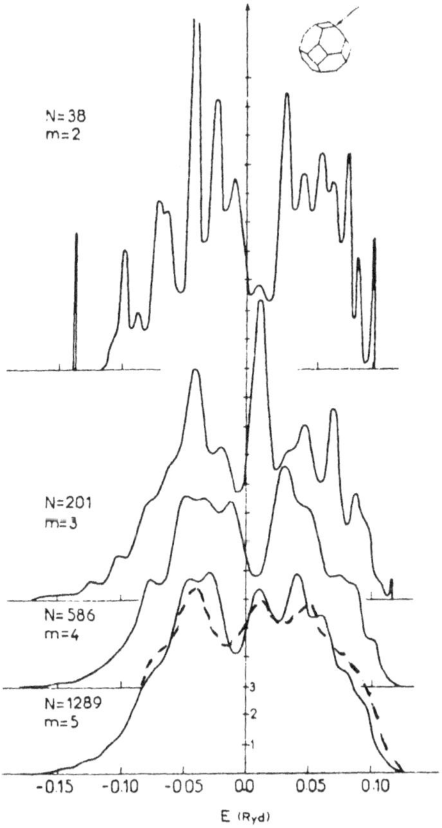

Fig. 2. LDS (—) on the edge between hexagonal and square faces for truncated octahedra of various sizes between 38 and 1289 atoms compared to the LDS (--) on the protruding edge of a 6(111) x (001) stepped surface which has the same nearest neighbour shell.

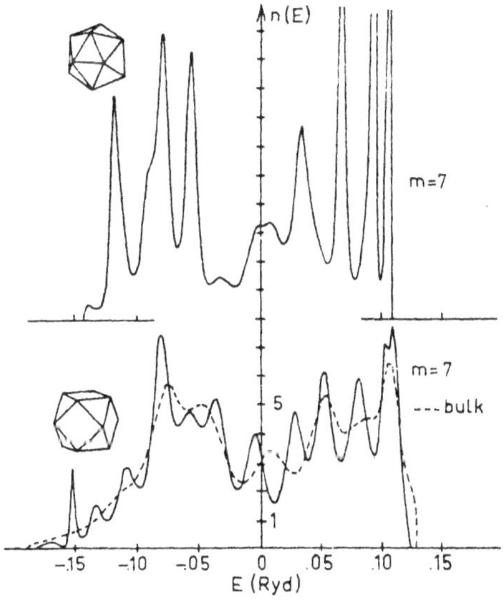

Fig. 3. Central atom LDS for icosahedra and cubooctahedra of 923 atoms.

3. ADSORPTION VERSUS ELECTRONIC STRUCTURE

Many attempts have been made to correlate the electronic factors of transition metals, such as their empty d orbitals, with their power to chemisorb, including for example the application of Pauling's valence bond theory or more recent quantum mechanical treatments of surfaces (Clark, 1974). But these approaches have some serious drawbacks, being generally qualitative models representing pictorially the complexity of the surfaces and of the chemisorptive bond.

However, any attempt at an understanding of adsorption and catalysis phenomena must take into account the reality of the electronic structure of the surface of the catalyst. The important changes occurring in the electronic properties when going from the bulk to surfaces or to clusters, such as the extra resonant surface states, will affect the chemisorption process.

When an adsorbate approaches a metal surface, there is an hybridization of its discrete electronic levels with the band of the metal leading to a broadening in virtual or resonance states. The chemisorption is a quite localized phenomena due to a strong perturbation which affects primarily the atoms of the substrate where adsorption took place and their close environment. This explains why some approaches use a simulation of the substrate by a very small cluster which enables to apply sophisticated quantum chemistry models to the chemisorptive bond. But, as we have seen above, the electronic properties of surfaces or clusters of the size of catalysers, i.e. a few dozen atoms, have their own characteristics, depending on their geometrical peculiarities, and are far from resembling the bulk or a very small cluster of a few atoms. One must then take into account in the description of adsorption, the main features of the electronic states of the surface substrate which have a determining influence as shown on two examples described below.

3.1 Strength of adsorption and surface states

When describing the adsorption of an atom A on a substrate, one can write the total hamiltonian as a sum of three therms

$$H = H_a + H_m + H_c$$

where H_a describes the free adsorbate, H_m the substrate and H_c the coupling between both. Depending on the strength of the coupling, one has two regimes for the LDS on the adsorbate : a regime of weak coupling giving an atomic character, i.e. a virtual bound state or a regime of strong coupling leading to a surface molecule.

The strength of the coupling will depend obviously on the effective over-lap between the adsorbate and the corresponding substrate site, on its coordination number and also on its local density of states. This last term will be particularly important when there exist resonant surface states close to the Fermi level.

Let us illustrate briefly this effect by an example : the adsorption of 4d atoms on (110) and (100) Mo. The surface resonant states existing on (100) Mo induce a strong binding and a surface molecule limit for the adsorption resulting from a resonance between these surface states and the adsorbate levels. Contrary to the case of (110) Mo whose LDS behaves like the bulk, the adsorption retains an atomic character (Fig. 4). The occurrence of resonant surface states thus increases the strength of the coupling and induces consequently a large anisotropy of the binding energy (Desjonquères et al, 1978)

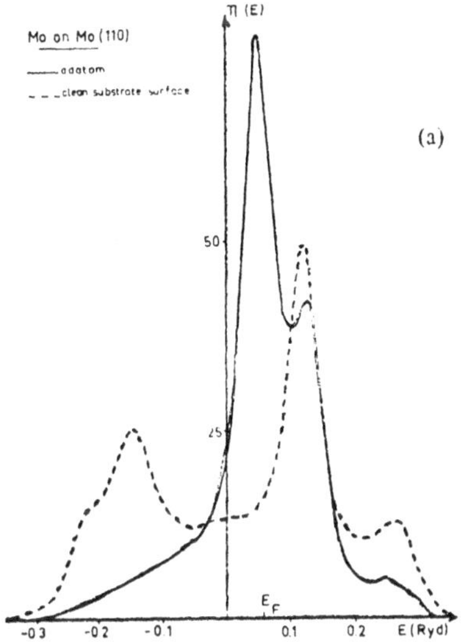

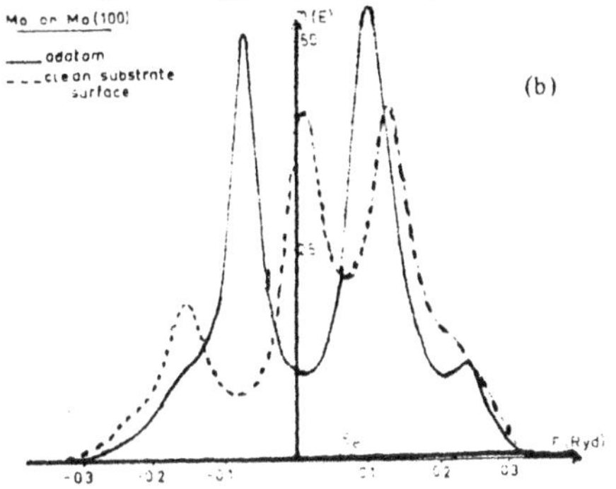

Fig. 4. LDS on a molybdenum adatom (—) compared with the surface LDS (--) of the clean substrate : a) (110) Mo, b) (100) Mo.

3.2 Adsorption and magnetism

The understanding of surface magnetic properties of transition metals or their charge when they participate in an adsorptive or a catalytic process still remains a challenging problem. However, magnetic techniques are often used to characterize the state of dispersion of supported metal catalysts, and to throw some light on the intermediate states produced in the adsorption process (Selwood, 1975). Any model description of the adsorption process must then permit to explain these magnetic properties.

For instance, the relative change in the magnetic properties of nickel particles when chemisorbing hydrogen and the correlative difference in the absolute quantities of chemisorbed hydrogen atoms at saturation coverage have been shown to depend on the surface structure (Martin et al, 1979) contrary to the initial heats of adsorption (Christmann et al, 1974). The lowering of the magnetic moment μ of nickel per atom of chemisorbed hydrogen is higher when the surface is made of dense planes, i.e. $\Delta\mu/\mu$ varies between 1.4 and 0.75 when the surface roughness increases. The surface stoichiometry of chemisrobed hydrogen at saturation, i.e. the ratio of adsorbed hydrogen to the number of surface nickel atoms H/Ni_S varies in the opposite way, between 0.6 and 1.1 when the surface roughness increases.

These features indicate the structure sensitivity of the hydrogen adsorption. We will show that they can be explained by the difference in the electronic properties of surface nickel atoms with the roughness of the surface.

One of the first question to ask is whether the surface states occurring for non dense cleavage planes will affect the magnetic properties of free nickel surfaces or clusters. We have been able to show that this is quite unlikely using a rigid exchange splitting model (Desjonquères and Cyrot-Lackmann, 1976). Our conclusions are in agreement with the most recent experiments, showing no noticeable change of the magnetic properties of surface nickel atoms from those of the bulk (Derouane et al, 1978).

When adsorbing a gaz such as H_2, O_2, Co.. on a nickel substrate, there will be a broadening of the adsorbate electronic levels and a displacement of the relative positions of the s and d bands of the nickel. The very small charge transfer between the adsorbate and the substrate is shown by the small change in work function when adsorbing (Christman et al, 1974). These shifts of the s and d bands will cause a reduction of the local density of states at the Fermi level on the active sites of nickel for adsorption and on its first neighbours. A simple reasoning using the Stoner criterion relating the existence of ferromagnetism to a high value of the density of states at the Fermi level will thus lead to a demagnetization of the active sites of nickel.

When H atoms are adsorbed, the s band will thus go up compared to the d band to achieve a s^0d^{10} configuration for active sites of nickel for adsorption instead of the usual $s^{0.6}d^{9.4}$ bulk ones. When the active sites have a dense geometry, such as that of the atoms of a (111) plane, the LDS is very similar to the bulk one (Fig. 1). Chemisorption of Hydrogen will lower the density of states and thus decrease rapidly the magnetic moment. On the contrary, when the active nickel sites have a smaller coordination number, such as on non dense planes, kinks.., there exist resonant surface states in their LDS (Fig. 2). The density of states decreases more rapidly but extends to higher energies in order to maintain charge neutrality. This leads to the same number of holes or magnetic moment (Fig. 5) as for a dense plane as we mention earlier. Chemisorption of H will then decrease the magnetic moment less rapidly than in the case of a dense plane. In the same way, the maximum surface stoichiometry for hydrogen adsorption which will correspond to the complete filling of the d band will be larger than for a dense surface plane. Let us point out that in this model, the

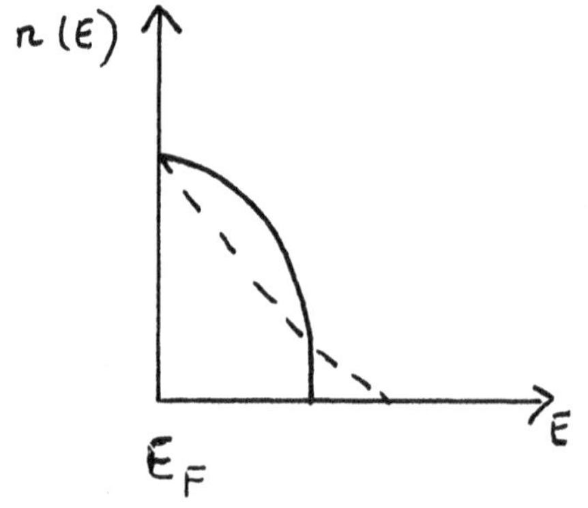

Fig. 5. Schematic top of the LDS of nickel surface atoms : — dense surface -- non dense surface.

product $\Delta\mu/\mu$ x H/Ni_S remains constant, which is roughly verified by experiments (Martin et al, 1979). Consequences of this model on the binding energy will be developped in a forthcoming paper.

ACKNOWLEDGMENTS

I would like to thank all the scientists who helped me to penetrate the strange and fascinating world of catalysis, in particular Drs. F. Gault, B. Imelik, M.V. Mathieu and especially G. Martin for stimulating discussions and for letting me know his results before publication. Discussions with Dr. M. Cyrot are also acknowledged.

REFERENCES

1. Balandin, A.A., 1969, Adv. in Catalysis 19, pp.
2. Christmann, K., Schober, O., Ertl, G. and Neumann, M., 1974, J. Chem. Phys. 11, pp. 4528-40.
3. Clark, A., 1974, The Chemisorptive bond, Academic Press.
4. Cyrot-Lackmann, F., Desjonquères, M.C. and Gordon, M.B., 1977, J. Physique 38, C2, pp.
5. Cyrot-Lackmann, F., 1970, J. Physique C1, pp. 67-74.
6. Derouane, E.G., Simoens, A., Colin, C., Martin, G.A., Dalmon, J.A. and Vedrine, J.C., 1978, J. Catal. 52, pp. 50-58.
7. Desjonquères, M.C. and Cyrot-Lackmann, F., 1975, Surf. Sci. 53, pp. 429-442.
8. Desjonquères, M.C. and Cyrot-Lackmann, F., 1976, Solid St. Comm. 18, pp. 1127-32 ; 1978, Solid St. Comm. 26, pp. 271-274.
9. Friedel, J., 1969, The Physics of Metals, ed. Ziman, J.M., pp. 340-408.
10. Friedel, J., 1976, J. Physique 37, pp. 883-894.
11. Gaspard, J.P. and Cyrot-Lackmann, F., 1973, J. Phys. C 6, pp 3077-96.
12. Gordon, M.B., Cyrot-Lackmann, F. and Desjonquères, M.C., 1978, Surf. Sci. 80, pp. 159-164.
13. Knight, W.D., 1977, J. Physique C2, 109-113.
14. Ladas, S., Dalla Betta, R.A. and Boudart, M., 1978, J. of Catal. 53, pp. 356-365.
15. Martin, G.A., Dutarbe, R. and Dalmon, J.A., 1979, To be published.
16. Moraweck, B., Clugnet, G. and Renonprez, A.J., 1979, Surf. Sci., To be published.
17. Renou, A. and Gillet, M., 1977, Thin Solid Films 41, pp. 15-28.

18. Selwood, P.W., 1975, Chemisorption and magnetization, Academic Press.
19. Solliard, C. and Buffat Ph., 1977, J. Physique 7 C2, pp. 167-170.
20. Somorjai, G.A., 1977, Adv. in Catal. 26, pp. 1-68.

BASIC PARAMETERS IN METAL-CATALYSED PROCESSES: THE "METHOD OF INVERSE TITRATION". x)

P. Heimbach, J. Kluth, H. Schenkluhn
Universität Essen - GHS, Fachbereich 8 (Organische Chemie), Postfach 68 43, D-4300 Essen 1 and
Max-Planck-Institut für Kohlenforschung, Postfach 01 13 25, D-4330 Mülheim-Ruhr, Deutschland

The formalisms of Ni-catalysed organic syntheses using alkenes, alkynes, 1, 3-dienes and 1, 3, 5-trienes or their heteroanalogous substrates are compared with those of enzyme chemistry. For the unevolved metalorganic systems we developed the "method of inverse titration" to get evidence for relevant actual "metal-fractions" in thecatalytic cycles. Depending on the real steady-state-concentrations of the active metal-complexes the amount of steering ligands has to be added. The concentration-control-maps give an insight into the relevant intermediates, their concentrations, correlations and coupling; this enables us to look at FMO-control in these catalytic processes leading to Metala-logy-Principle and its rules.

Starting point for nickel-catalyzed reactions was the observation of H. W. B. Reed, that butadiene can be cyclodimerized by Reppe-catalysts such as $(CO)_2NiL_2$ (1). G. Wilke and coworkers introduced carbonyl-free Ni-complexes and found, that without further ligands butadiene could be cyclotrimerized to three isomers of cyclododeca-1, 5, 9-triene (2, 3).

Our motivation for further investigations on nickel-catalysed organic syntheses was, to find out the scope and the kind of control in these processes. The following, formal cycloadditions of ethylene (or alkynes), butadiene and hexatriene have been achieved (4), e.g. see scheme 1.

In addition, depending on reaction-conditions and presence of further co-catalysts, open-chained oligomers and co-oligomers are catalytically formed, especially when heteroanalogous substrates are used (4, 6, 7): see scheme 2.

There are different types of investigations to elucidate the "black box" of catalytic processes as shown in SCHEME 3.

B. Pullman (Ed.): Catalysis in Chemistry and Biochemistry. Theory and Experiment. 227–239.

Scheme 1

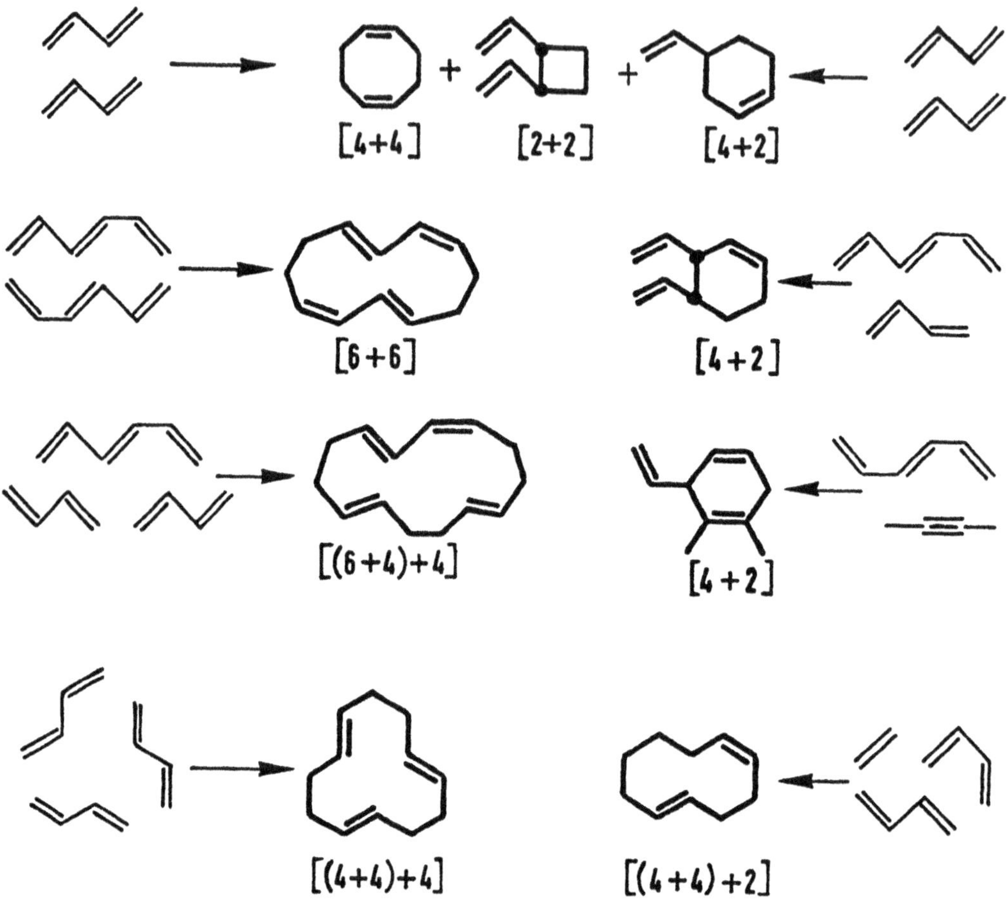

[ ] : symbol proposed for cycloadditions by (5). ( ) means coupled units in the first step. By these symbols and the name cycloaddition no comment on the reaction-mechanism is involved. These processes are stepwise metala-ringclosure-reactions.

Scheme 2

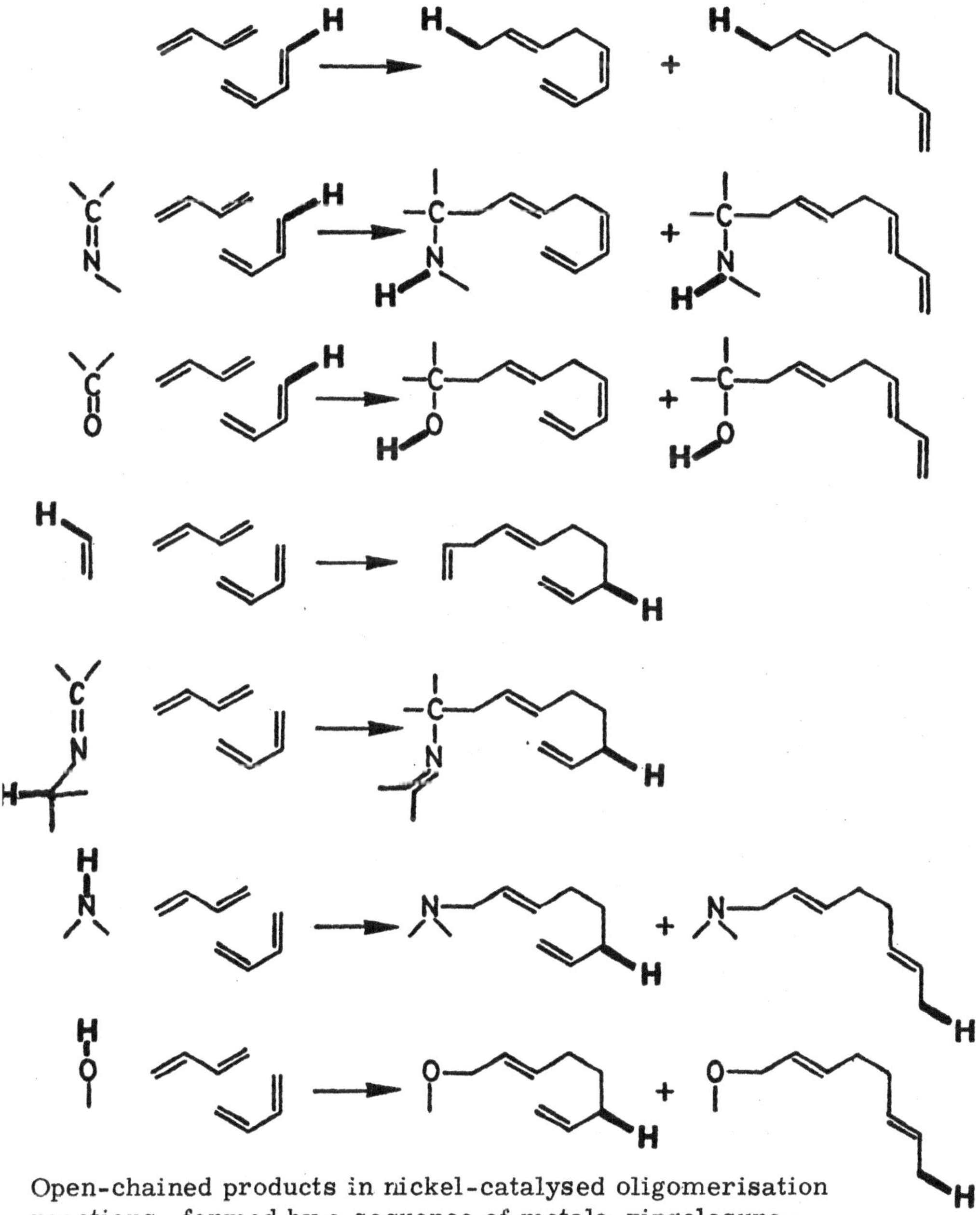

Open-chained products in nickel-catalysed oligomerisation reactions, formed by a sequence of metala-ringclosure-reactions and metalanalogous sigmatropic shifts.

Scheme 3

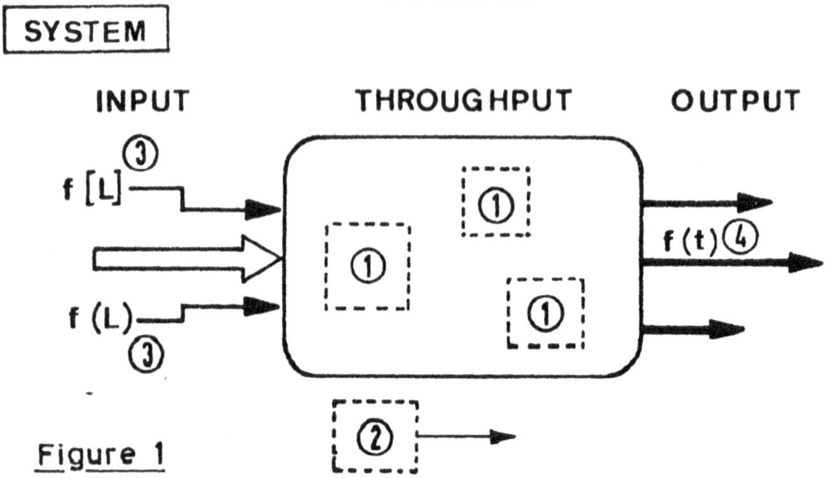

Figure 1

QUESTIONS

- ⇨ STRUCTURE ① ⟶ Isolation of intermediate-complexes
- ⇨ ELEMENTARY STEPS ② ⟶ Reactions of model-complexes and isolated intermediates
- ⇨ REACTION-SCHEME ③ ⟶ Kinetic and thermodynamic selectiv as an integral change of the produc stream by perturbation: f [L] and f (L)
- ④ ⟶ Roll of effectors by application of formalisms from enzyme chemistry f (t)

By gradually changing perturbation of the INPUT and controlling the effect on the THROUGHPUT by analysing the OUTPUT we increased stepwise the number of participating species in the catalytic processes and changed gradually the concentrations (symbol f [ ]) and stepwise the properties (symbol f ( )) of all compounds, which are involved in stoichiometry or only in processes (for isolation of intermediates see (8)).

Here we would like to focus on a comparison of enzyme-chemistry and metal-catalysis. In SCHEME 4 is shown, that the classical Michaelis-Menten formalism has to be extended, because there is more than one substrate bound to the active center in the relevant elementary step. In the given example of the system Ni/butadiene/tri-(ortho-bi-phenylyl) phosphite the sum of the squares of experimental errors for the given formalism is a minimum, when n is 2 (see figure 2 SCHEME 4). On the other hand we got with n = 2 concrete values for generalized $\bar{v}_{max}$ and $\bar{K}_M$ (figure 3 and 4, SCHEME 4). By kinetic investigations we could correlate lg $\bar{v}_{max}$ with lg k_{COD}/k_{VCH} linearly and got the information, that with increasing reactivity selectivity is also increased (figure 5, SCHEME 4). Here the original reactivity-selectivity-principle (R S P) is not valid (9) and this is typical for FMO-controlled processes with an early transition state (see later on SCHEME 8 and lit. (9 a)).

Because our unevolved metalorganic system is a complicated one (see SCHEME 5) with more than one series of dependent reaction cycles and multistep associations at different intermediates, we had to apply the "method of inverse titration" (for experimental procedure see (10)). In a classical titration process we change linearly the external amount of a base or an acid and look for the actual internal answer of the system (change of the pH-value). In the "method of inverse titration" we change logarithmically the external ratio lg $[L]_o/[Ni]_o$ and look for the internal answer of the system on the product-level, assuming that the distribution of the products (external dynamic) is connected with the "metal-fraction" (11) (internal dynamic of the system) in the product-determining step modified by kinetic factors. Figure 6 (SCHEME 5) shows the shape of a "titration-curve" in this method by a given K_D-value of 10^{-3} and "metal-fraction" $m/M_o = 1$. Figure 8 (SCHEME 5) shows the dependencies on changing K_D^o-values and variation of the "metal-fraction". The activator- or inhibitor-formalism also changes the shape and the position of the "titration-curves", but for simplicity this will not be discussed here.

Three different ligand-concentration-control-maps for the system Ni/butadiene/ligand are shown in SCHEME 6. The products formed are:

Scheme 4

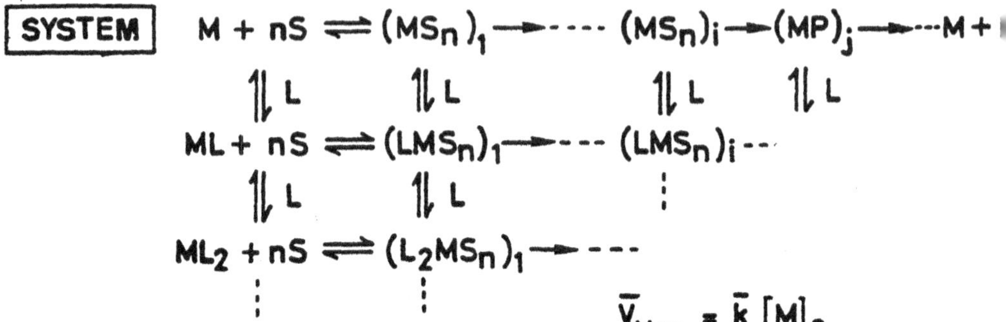

MICHAELIS MENTEN EQUATION

$$V = \frac{\bar{V}_{Max}\,[S]^n}{\bar{K}_M^n + [S]^n}$$

$$\bar{V}_{Max} = \bar{k}\,[M]_o$$

$$V = \frac{\bar{V}_{Max}}{2} \curvearrowright [S]^n = \bar{K}_M^n$$

Ni : ⁄⁄⁄ : TOPP (Benzene, 40°C)
1 : 200 : 1

TOPP =

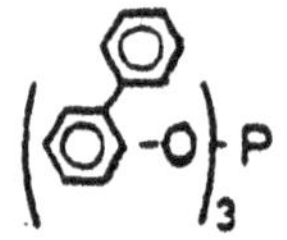

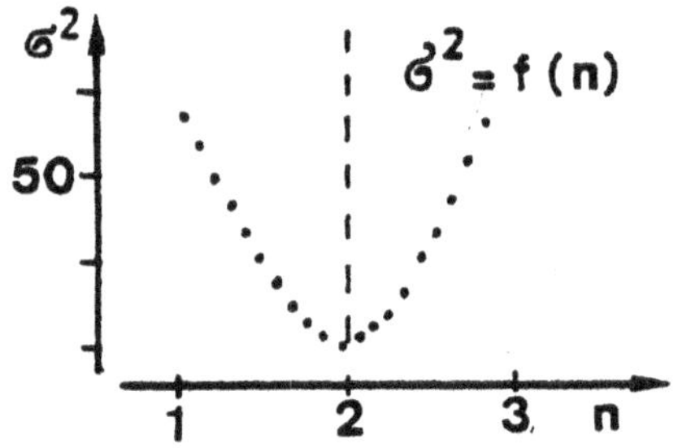

Figure 2

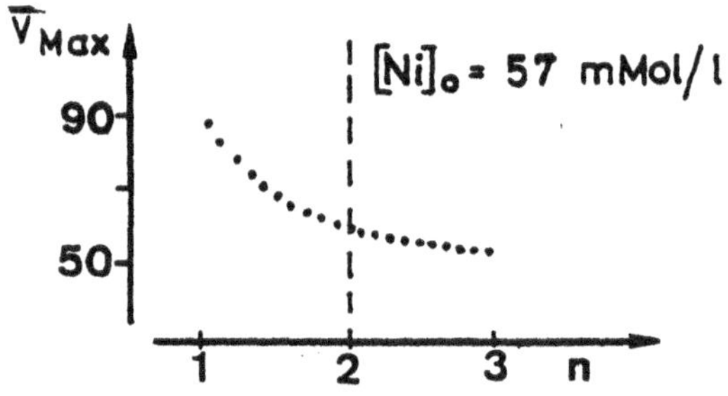

Figure 4

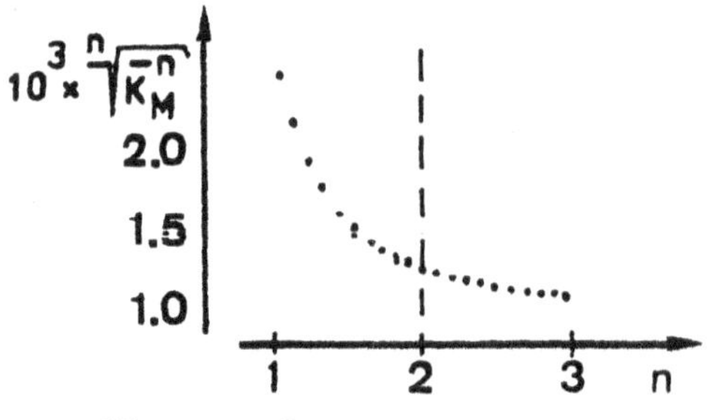

Figure 3

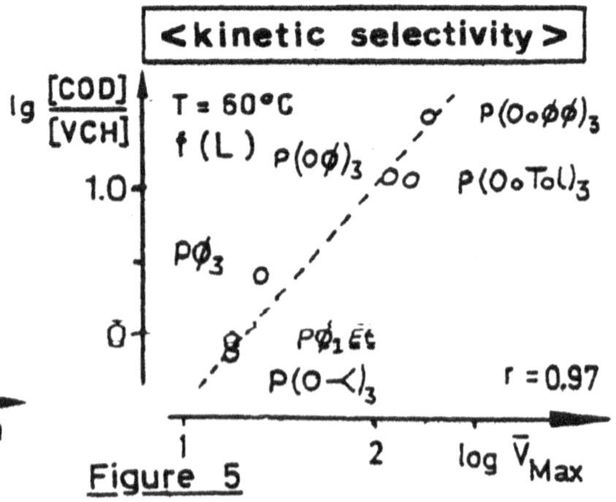

Figure 5

Scheme 5

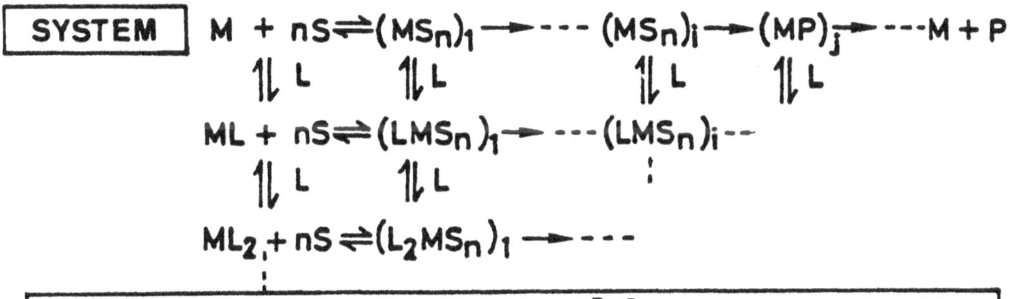

$$\frac{[M]_i}{[M]_o} = X_{M_i} = f\left(\mathrm{Lg}\,\frac{[L]_o}{[M]_o}\right) \qquad \frac{[P]_i}{\Sigma[P]_i} = f(X_{M_i}) = g\left(\mathrm{Lg}\,\frac{[L]_o}{[M]_o}\right)$$

INTERNAL DYNAMICS | **EXTERNAL DYNAMICS**

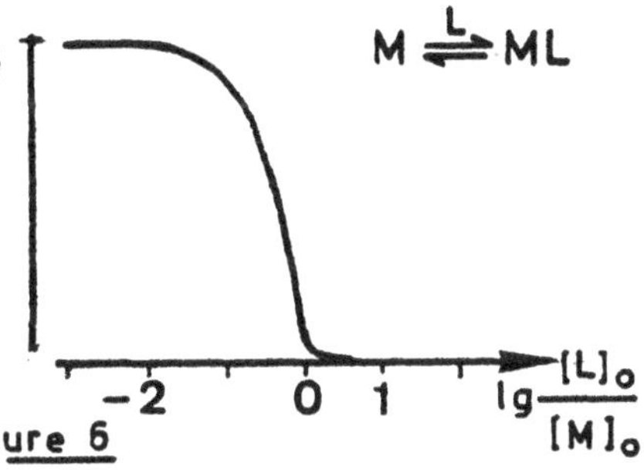

Figure 6

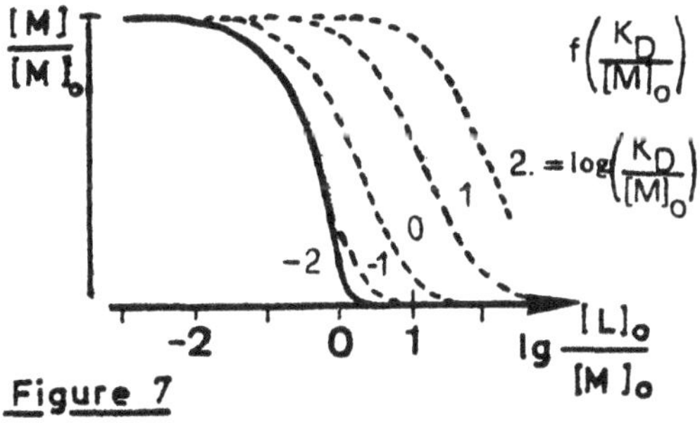

Figure 7

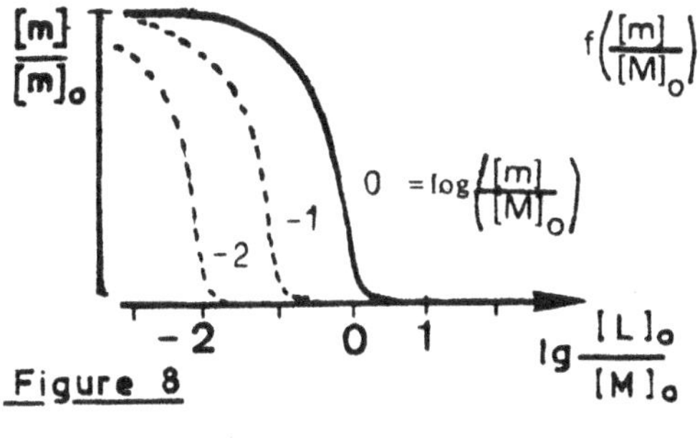

Figure 8

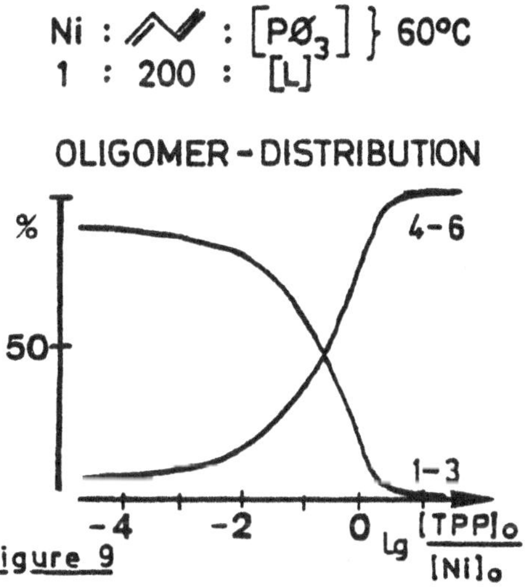

Figure 9

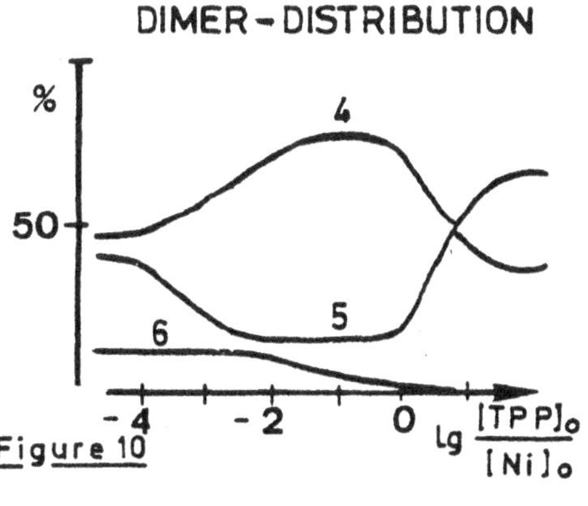

Figure 10

Scheme 6

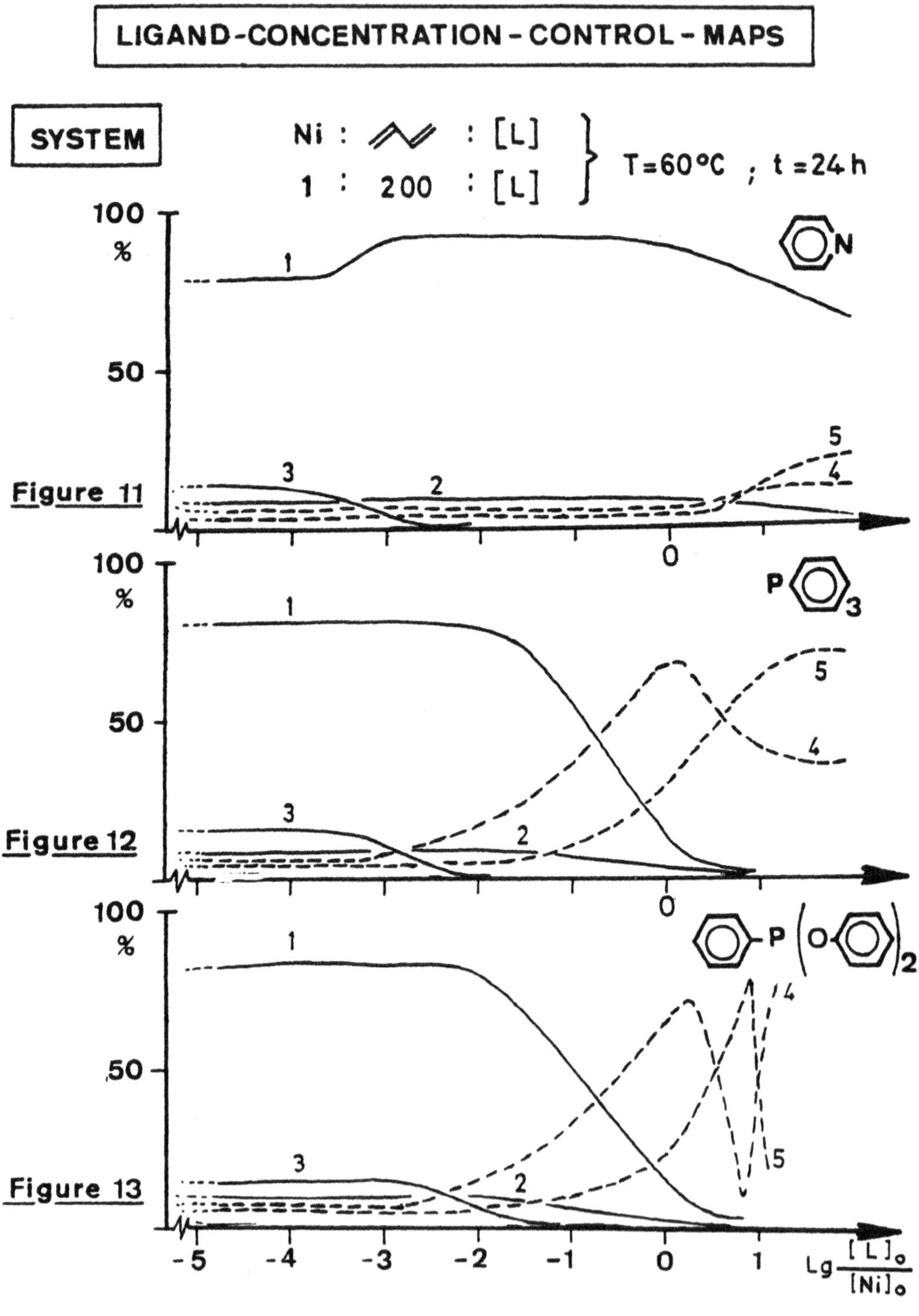
LIGAND-CONCENTRATION-CONTROL-MAPS
SYSTEM
Ni : [L]
1 : 200 : [L]
T=60°C ; t=24h
100
%
50
1
2
3
4
5
N
Figure 11
0
Figure 12
Figure 13
-5
-4
-3
-2
-1
0
1
Lg [L]o / [Ni]o

1

2

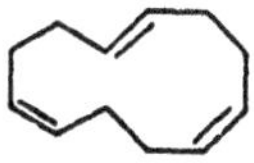
3

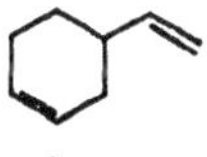
4

5

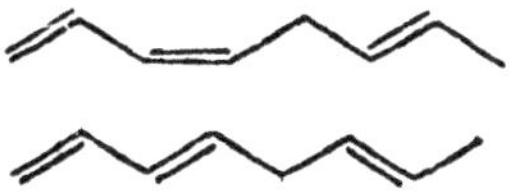
6

The most important result is, that we get evidence at least for two different "metal-fractions" in the catalytic cycles: the relative concentrations of these intermediates differ by a factor of about 10^2 - 10^4. The changing K-values of the ligand association processes and the variation of the low steady-state-concentrations by investigating the influence of these ligands can clearly be seen by comparison of figure 11 - 13 in SCHEME 6.

In addition in SCHEME 5 two other types of ligand-concentration-control-maps are shown. These maps can be derived from the original maps. In figure 9 (SCHEME 5) the "oligomer-map" gives the information, that intermediates in high steady-state-concentrations control the oligomerisation. On the other hand the isomer-distribution is controlled by different steady-state-concentrations. The consequences of these informations will be published in near future (12). One interesting application is, that for optimal selectivity the steering ligands have to be added with respect to the corresponding actual metal-fractions. In the system Ni/butadiene/morpholine/triphenyl-phosphane we can control the formation of n-octa-1, 3, 6-trienes (13) best by applying two different ratios of the additional ligands triphenylphosphane ($\lg [P(C_6H_5)_3]_o/[Ni]_o = -2,0$) and morpholine ($\lg [morpholine]_o/[Ni]_o = 1,5$)

In SCHEME 7 figure 14 the concentration-control-map of the four-component-system Ni/butadiene/propylene/triphenylphosphane is shown. Considering this map we are directly able to formulate a simplified reaction-scheme. The brackets { } mean evidence for complexes, out of which the corresponding dimers (DIM) or trimers (TRIM) are formed or that stop-complexes {STOP} are built up. The complexity of the {DIM} - and {TRIM} -intermediates, which are caused by donor/acceptor-perturbations in the substrates, the metals and the ligands will be discussed elsewhere (12).

For further interpretation of structure and reactivity of metal-organic intermediates theories (14) and models (15, 16) can be found in the literature (see SCHEME 8). Our group transferred the idea of the unifying principle in Organic Chemistry (17, 18) to Metalorganic Chemistry. The details of what we call "Metala-logy-Principle" will

Scheme 7

SYSTEM Ni , , PR_3 or $P(OR)_3$, CH_3

symbol M , S , L , S′

CO-OLIGOMERS

C_{11}

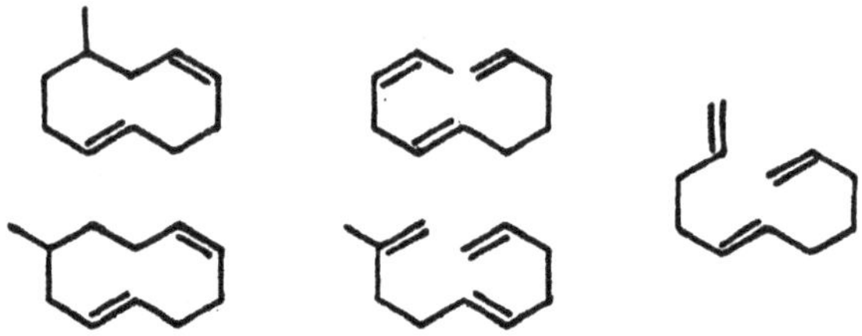

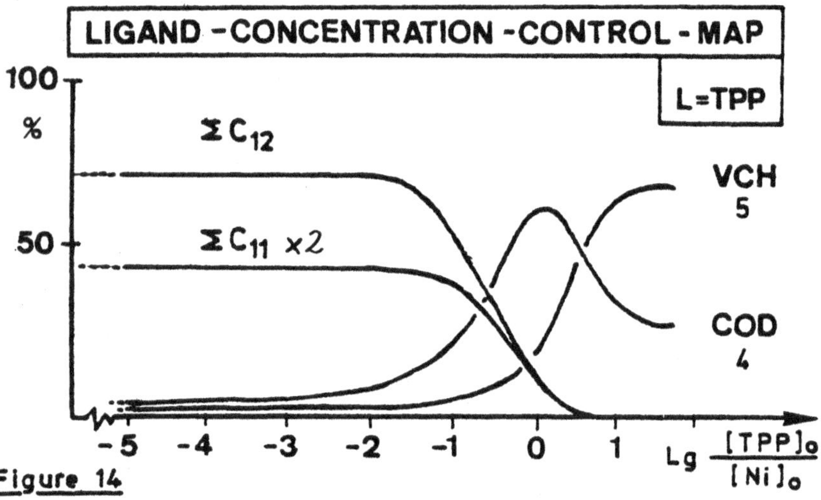

Figure 14

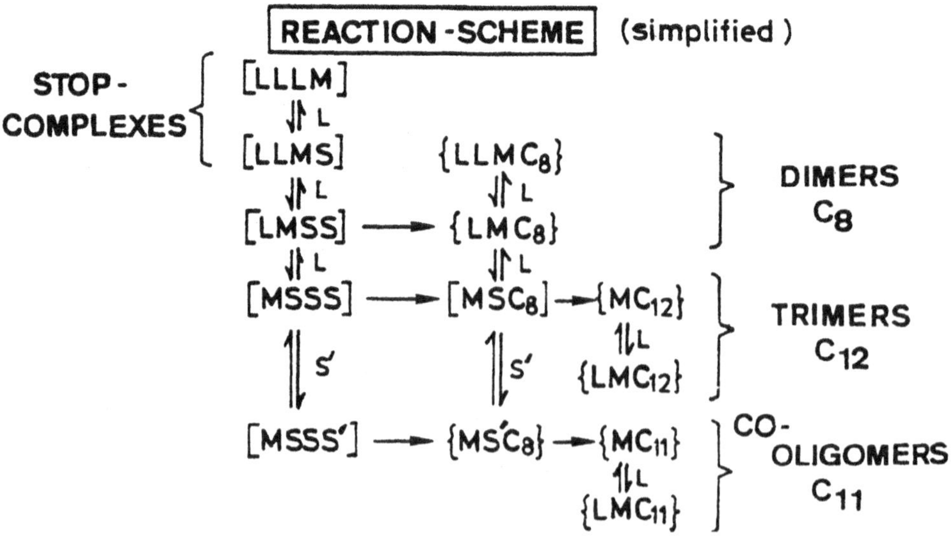

be published in extenso (12). Two examples for structure and reactivity of metalorganic compared with organic species in which the π-systems are coupled by a metal respectively by a σ-bond are seen in SCHEME 8.

This work was partly supported by Deutsche Forschungsgemeinschaft and the Fonds der Chemischen Industrie.

x) Control of Metal-Catalysed Reactions, Part IX; Part VIII: Sisak, A., Schenkluhn, H., Heimbach, P., Acta Chim. Acad. Sci. Hung. in preparation

1) Reed, H. W. B., J. Chem. Soc. London 1954, pp. 1931 - 1941

2) Wilke, G., et al., Angew. Chem. 75, pp. 10 - 20 (1963) Angew. Chem. Intern. Edit. 2, pp. 105 - 115 (1963)

3) Jolly, P. W., Wilke, G., "The Organic Chemistry of Nickel", Bd. 2, Academic Press, New York, San Francisco, London 1975

4) Heimbach, P., Angew. Chem. 85, pp. 1035 - 1049 (1973); Angew. Chem. Intern. Edit. 12, pp. 975 - 989 (1973)

5) Huisgen, R., Angew. Chem. 80, pp. 329 - 337 (1968); Angew. Chem. Intern. Edit. 7, 321 - 329 (1968)

6) Heimbach, P., Hugelin, B., Nabbefeld-Arnold, E., Reinehr, D., Roloff, A., Troxler, E., Angew. Chem. 89, pp. 260 - 261 (1977); Angew. Chem. Intern. Edit. 16, pp. 252 - 253 (1977)

7) Bandmann, H., Heimbach, P., Roloff, A., J. Chem. Res. (S) 1977, p. 261; J. Chem. Res. (M) 1977, pp. 3056 - 3069

8) a) Jolly, P. W., Wilke, G., "The Organic Chemistry of Nickel", Bd. 1, Academic Press, New York, London 1974
b) Barnett, B., Büssemeyer, B., Heimbach, P., Jolly, P. W., Krüger, C., Tkatchenko, J., Wilke, G., Tetrahedron Lett. 1972, pp. 1457 - 1460

9) Giese, B., Angew. Chem. 89, pp. 162 - 173 (1977); Angew. Chem. Intern. Edit. 16, pp. 125 - 136 (1977)

9 a) Heimbach, P., Roloff, A., Schenkluhn, H., Angew. Chem. 89, pp. 260 - 261 (1977); Angew. Chem. Intern. Edit 16, pp. 252 - 253 (1977)

10) Brille, F., Kluth, J., Schenkluhn, H., J. Molecular Catalysis

Scheme 8

PROBLEM : CATALYTICAL ELEMENTARY STEP

3 CENTER-FMO-RULES

Lit.[14)]

DONOR/ACCEPTOR-RULES

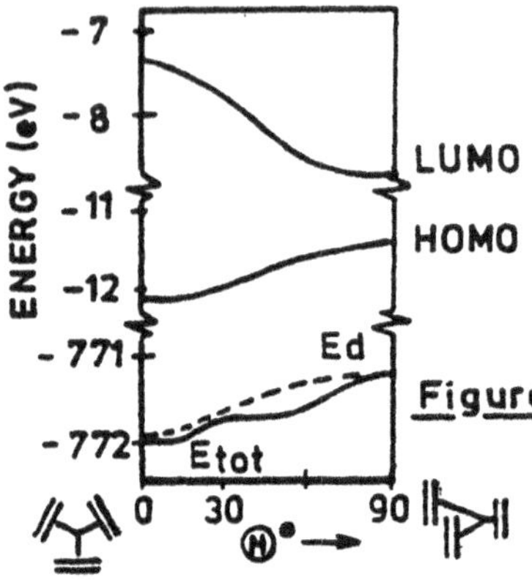

Figure 15

Lit.[15)]

METALA-LOGY-RULES

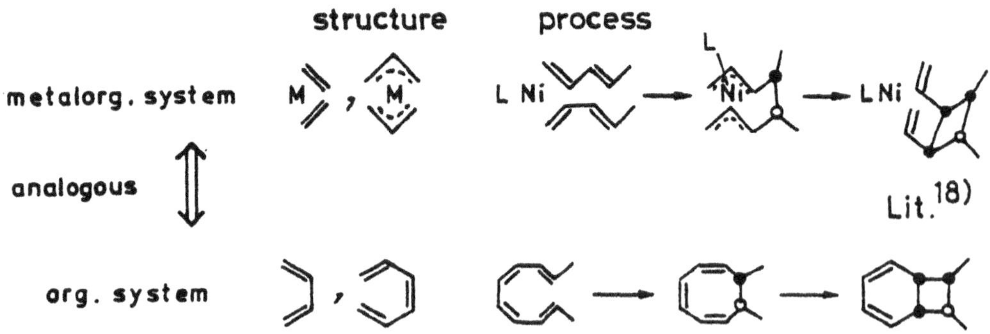

Lit.[18)]

Lit.[19)]

5, pp. 27 - 40 (1979)

11) The term "metal-fraction" is used in analogy to "mole-fraction" and (Lewis-)"acid-fraction".

12) Heimbach, P., Schenkluhn, H., in preparation

13) Heimbach, P., Nabbefeld-Arnold, E., Schenkluhn, H., in preparation

14) Fukui, K., "Theory of Orientation and Stereoselection", Springer-Verlag, Berlin, Heidelberg, New York, 1975

15) Rösch, N., Hoffmann, R., Inorg. Chem. 13, pp. 2656 - 2666 (1974)

16) Heilbronner, E., Bock, H., "Das HMO-Modell und seine Anwendung", Verlag Chemie, GmbH, Weinheim/Bergstr., 1968

17) Dewar, M. J. S., Dougherty, R. C., "The PMO-Theory of Organic Chemistry", Plenum Press, New York, London, 1975

18) Heimbach, P., Hey, H.-J., Angew. Chem. 82, p. 550 (1970) Angew. Chem. Intern. Edit. 9, p. 528 (1970)

19) a) Huisgen, R., Dahmen, A., Huber, H., J. Amer. Chem. Soc. 89, pp. 7130 - 7131 (1967); Tetrahedron Lett. 19, pp. 1461 - 1464 (1969)
b) Huisgen, R., Dahmen, A., Tetrahedron Lett. 19, pp. 1465 - 1469 (1969)

THE CHEMISORPTION OF ACETYLENE ON THE Pt(111) SURFACE. EXTENDED HUCKEL RESULTS

Angelo GAVEZZOTTI and Massimo SIMONETTA

Istituto di Chimica Fisica e Centro CNR,
Università di Milano, Milano, Italy

The chemisorption of acetylene on the Pt(111) surface has been studied by Extended Huckel molecular orbital calculations, with a new, more efficient basis set. The results are discussed and compared with those of previous calculations on the same subject. The results are in agreement with most of the available experimental evidence.

1. INTRODUCTION

Highly sophisticated experimental techniques are now in use in many laboratories for the study of the chemisorption of organic molecules on heavy-metal clean surfaces. None of them is however sufficient to give complete and unequivocal answers to the crucial problems of adsorption geometry and relative adsorption energies at various surface sites. On the other hand, quantum chemical calculations on this subject are still in an exploratory phase. One of the most challenging problems to date is that of the chemisorption of acetylene on the Pt(111) face, for which the recent literature has given contradictory interpretations; we would like to mention here three groups that are contributing on this particular subject, namely, Somorjai and coworkers (Lo, Chung, Kesmodel, Stair and Somorjai, 1977; Kesmodel, Baetzold and Somorjai, 1977; Kesmodel, Dubois and Somorjai, 1978); Ibach and coworkers (Ibach, Hopster and Sexton, 1977); and Demuth (Demuth, 1977,1979).They use combinations of Low Energy Electron Diffraction (LEED), Ultraviolet Photoelectron Spectroscopy (UPS), Electron Energy Loss Spectroscopy (ELS) and Temperature Programmed Thermal Desorption (TPTD) to try to unravel the complex dynamics of the chemisorption and subsequent evolution of chemisorbed species when acetylene is allowed to react with the Pt(111) surface. From their data, only a rough trend emerges with certainty, that is, acetylene may adsorb undissociated at low temperatures, although slightly distorted; at higher temperatures, fragments such as $H_2C{=}C$, $H_3C{-}C$, or even $H_3C{-}CH$, may be present on the surface. There is still considerable

B. Pullman (Ed.): Catalysis in Chemistry and Biochemistry. Theory and Experiment. 241–253.

contradiction as to which is the preferred fragment, and which is the preferred adsorption site.

On the theoretical side, ab initio calculations are at prese too expensive for clusters of heavy-metal atoms; a review on the use of semiempirical methods such as CNDO or EHT has been given (Simonetta and Gavezzotti, 1979). We have used previously EHT to study acetylene chemisorption (Gavezzotti and Simonetta, 1977) an fragmentation (Gavezzotti and Simonetta, 1979) on the Pt(111) surface, and for acetylene chemisorption on various faces of tungste (Gavezzotti, Ortoleva and Simonetta, 1979). An attempt was made to match the outcome of the calculations with that of experiment. Since then, new data have been acquired on the experimental side for platinum, but no final unanimous verdict has been reached. We present in this paper what we think is a significant contribution gives to the problem, since we use a substantially improved parame terization with respect to our previous work.

2. PARAMETERIZATION

It is well known that the results of semiempirical treatments are largely dependent on parameterization. Our previous basis set was based on an assembly of data from rather heterogeneous sources, and a single-zeta type function was used for the Pt 5d orbitals. Strongly repulsive curves were obtained for acetylene adsorption, and a general tendency was observed for the carbon atoms to locate themselves on top of Pt atoms whenever possible. To test the sensitivity of the geometrical predictions to the basis set, a new, less diffuse, double-zeta type function was particularly suited; we decided to adopt here the basis set proposed by Hoffmann (Summerville and Hoffmann, 1976). Moreover, for a number of metal atoms, this author has dealt succesfully with the problem of excessive charge transfer between metal and organic molecules.

Some sample binding energy curves, with our old and new parameterization, are shown in Figure 1. However, the new basis set is not enough as such to ensure reasonable binding energies. Followi a suggestion by Anders, Hansen and Bartell (1973), inner 5p orbitals are needed to provide the necessary repulsion at small internuclear separations; the amount of repulsion can be desumed from test calculations on metal-carbon binding energies. Figure 2 shows the effects of the inclusion of Pt 5p orbitals on the Pt-C binding energy curve; the difference between the curves with and without 5p orbitals gives the desired inner orbital repulsion, which is then added in each case to the total EHT energy for each C-metal contact. It may be noted that this repulsion may well be different in the case of a bare C atom and of a C atom bound to other atoms, such as in the acetylene molecule; test calculations show however that in this last case the correction obtained from the Pt-C curve (as described above) is only slightly overestimated. It is reasonable to assume that it is still less so in the case of a fragment

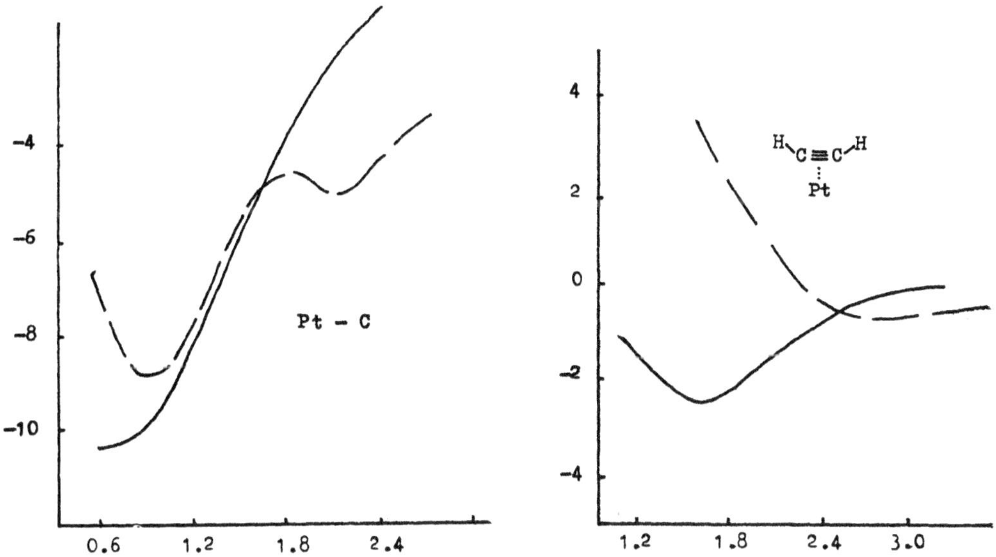

Figure 1. Binding energy (eV) versus Pt-C or acetylene-Pt distance with the old (broken line) and new (full line) basis sets. Distances in Å.

TABLE 1

Extended Huckel parameters

		VOIP, eV	Slater exponent[c]
Pt	5p[a]	-56.87	5.10
	5d[b]	-12.59	6.013(0.6334) 2.696(0.5513)
	6s[b]	-9.077	2.554
	6p[b]	-5.475	2.554
H	1s	-13.6	1.3
C	2s	-21.4	1.625
	2p	-11.4	1.625

[a]Gavezzotti and Simonetta (1977). [b]Summerville and Hoffmann (1976).
[c]For the 5d function, two exponents and the coefficients are given.

like H_2CC or H_3CC approaching the surface with the almost bare C atoms first.

Table 1 collects the Extended Huckel parameters. Table 2 has some interesting quantities, that are a measure of the goodness of the basis set and of the parameterization. The clusters A, B, C (Figure 3) represent 1-fold, 2-fold and 3-fold adsorption sites respectively. Their small-cluster analogs are 1, 2 and 3 (Figure 3). The binding energy per Pt atom in the clusters and the filled d-band width are roughly constant in the three large clusters. The charge shifts between atoms in each cluster are small compared with those obtained with the old parameterization.

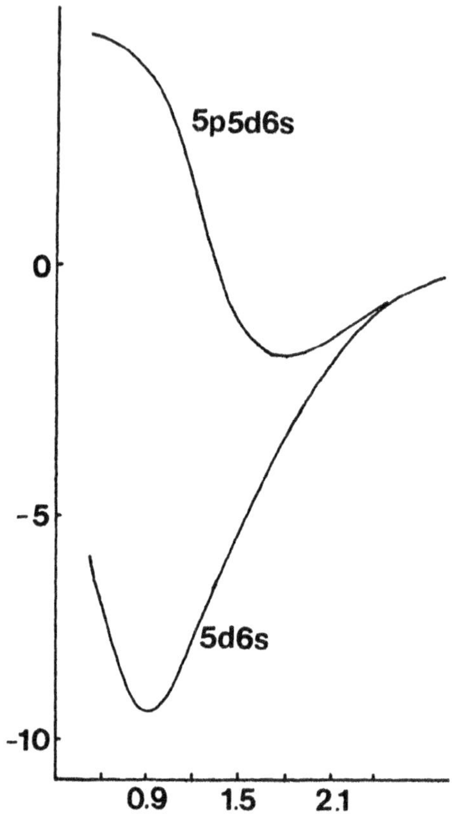

Figure 2. Binding energy (eV) versus bond distance (Å) for the Pt-C molecule. "New" parameters (as shown in Table 1).

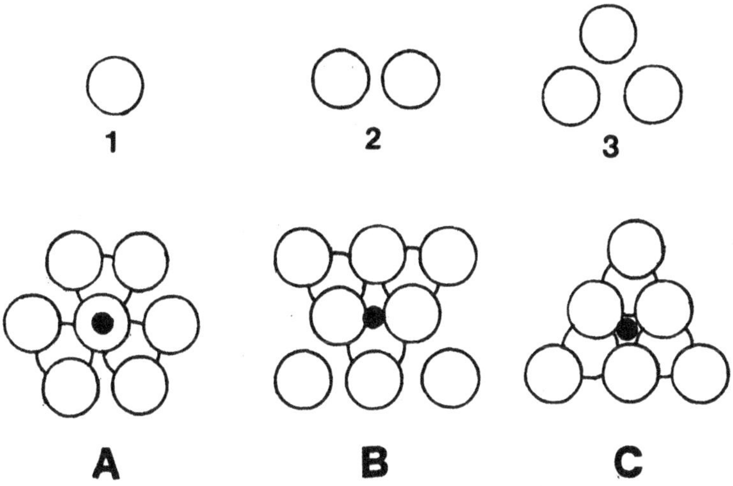

Figure 3. Small clusters (1, 2, 3) and large clusters (A, B, C) used to represent 1-fold, 2-fold and 3-fold adsorption sites on the Pt(111) surface. Atoms are always at the closest-contact distance proper of this surface.

TABLE 2

Properties of Pt_n clusters (energies in eV)

n	binding energy	filled d-band width	B.E./atom
2	0.194	-	0.097
3	0.495	-	0.165
9	3.202	1.79	0.356
10	4.026	1.86	0.403
11	4.064	1.88	0.369

Properties of the Pt-C molecule (energies in eV)

Pt basis set	R(Pt-C),Å	B.E.	bond overlap population	charge on C
5p5d6s	1.8	1.94	0.930	-0.45
5d6s	0.9	9.36	1.634	-1.35
5d6s6p	0.6	10.28	2.306	+0.12

3. CALCULATIONS FOR THE SMALL CLUSTERS

We consider as likely for chemisorption on Pt(111), after exposure to acetylene, the following three species:

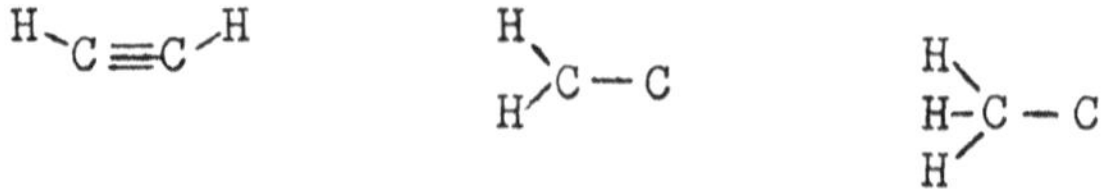

I II III

Back-bent acetylene (I) has a HCC angle of 140° and a C-C bond length of 1.3 Å; the reasons for this choice come from coordination chemistry (see for example Simonetta and Gavezzotti, 1978). The C-C distance is 1.3 Å also in II and III; C-H is 1.06 Å everywhere; HCH is 120° in II and 109.47° in III. The C-C bond was kept normal to the plane of the metal atoms for II and III, and parallel to this plane for I. The distance from the closest C atom, or from the C-C bond midpoint for I, to the surface plane was called z.

The interaction of these three fragments with clusters 1-3 (Figure 3) is instructive. A comparison between the adsorption energy curves obtained with the old and the new Pt parameters shows a large relative stabilization for the 3-fold adsorption site. Figure 4 shows the inner-repulsion corrected curves with the new basis set. The 3-fold coordination site is consistently favored, although by a small difference.

Another relevant point is that of the orientation of acetylene over 2-fold and 3-fold coordination sites. From the results for cluster 2 (see Figure 4), it is evident that a "di-sigma" arrangement is favored over the so-called "mu" arrangement, although this is at variance with the usual structures found in coordination compounds. It may be noted that in these compounds only heavily substituted acetylenes are found. On the other hand, the interaction with the three-atoms cluster is almost insensitive to the orientation of the incoming acetylene molecule (compare curves 3 and 3' in Figure 4; 3' is very slightly favored). Calculations on large clusters have consequently been performed only for the preferred orientations on small clusters. On the basis of previous experience, the orientation of acetylene over cluster A (1-fold site) and the orientation of H atoms in the fragments approaching the surface in an upright position has been considered to be irrelevant.

4. CALCULATIONS FOR THE LARGE CLUSTERS

The interaction of the three fragments with the three possible sites on the Pt(111) face, as represented by the three clusters A, B, C, is shown in Figure 5. Consistently with the results for the small clusters, the 3-fold coordination site C is always preferred. Table 3 collects some adsorption geometry and energy data.

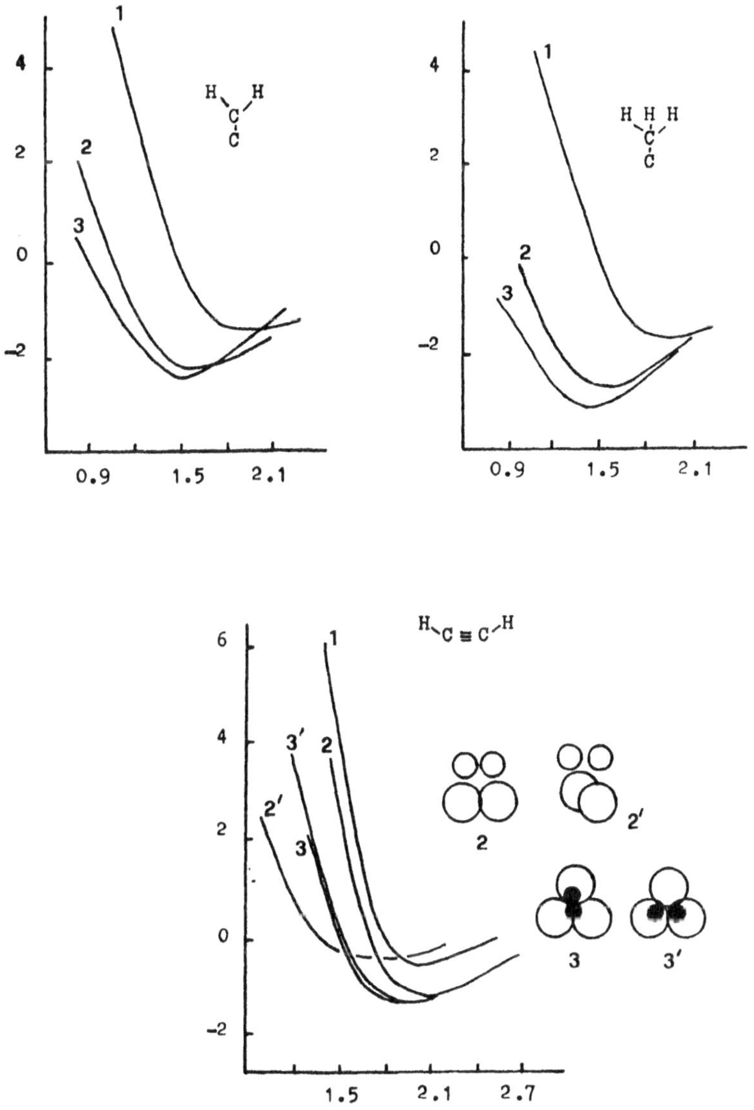

Figure 4. Binding energy (eV) versus z (Å) for I-III over clusters 1-3, Figure 3. Each curve is marked with a number that corresponds to the cluster to which it refers.

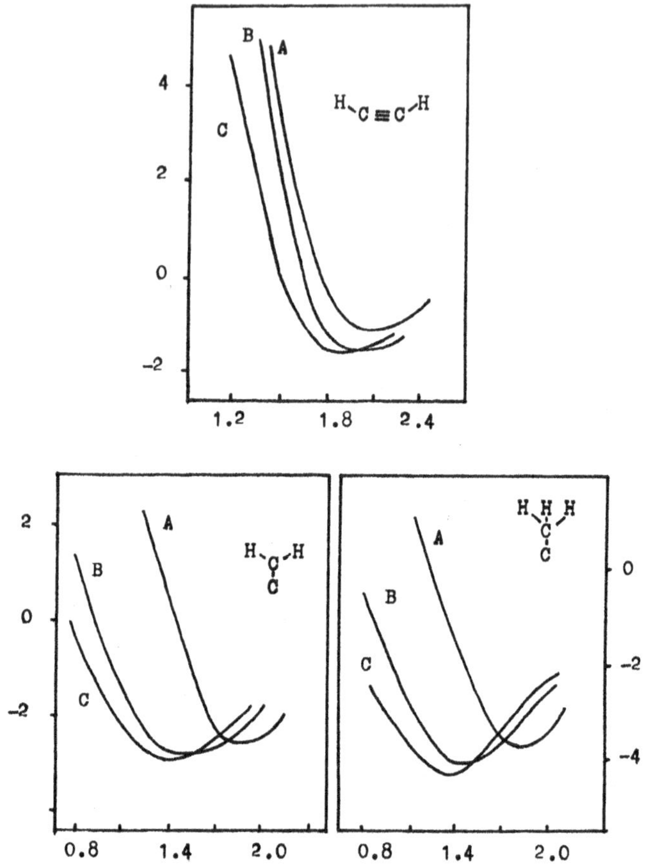

Figure 5. Binding energy (eV) versus z (Å) for I-III over clusters A-C, Figure 3. Each curve is marked with a letter that corresponds to the cluster to which it refers.

5. CATALYTIC EFFECT OF THE METAL

If the formation of II or III is considered likely over the metal surface, it is necessary to establish a path through which the rearrangements of the H atoms take place. The reaction:

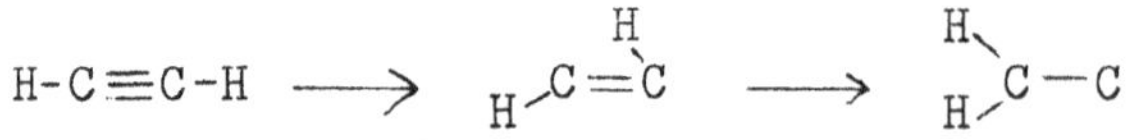
(1)

has been studied by ab initio calculations (Dykstra and Schaefer, 1978), and has been considered, together with the subsequent hydrogenation:

$$H_2C{-}C \; + \; H \longrightarrow H_3C{-}C \qquad (2)$$

in our previous calculations (Gavezzotti and Simonetta, 1979).

TABLE 3
Adsorption on clusters A–C, Figure 3. Charges for C atom nearest to the metal; the n's are Mulliken bond overlap populations.

	binding energy, eV	z, Å	Pt–C distance, Å	Charge on C	n(C–C)	n(C–Pt)
$HC{\equiv}CH$						
A	1.1	2.1	2.20	–0.06	1.56	0.15
B	1.6	2.0	2.13	–0.06	1.49	0.30
C	1.7	1.9	2.19	–0.04	1.45	0.31
$H_2C{-}C$						
A	2.6	1.9	1.90	+0.14	1.38	0.69
B	2.8	1.6	2.12	+0.02	1.37	0.39
C	3.0	1.4	2.13	+0.09	1.34	0.33
$H_3C{-}C$						
C – C = 1.3 A						
A	3.7	1.9	1.90	+0.54	1.08	0.79
B	4.0	1.4	1.97	+0.34	1.03	0.48
C	4.2	1.3	2.06	+0.21	1.02	0.41
C – C = 1.5 A						
A	3.4	1.8	1.80	+0.33	0.86	0.94
B	4.0	1.4	1.97	+0.15	0.83	0.52
C	4.2	1.3	2.06	+0.01	0.82	0.43

The former calculations showed that, while the metal surface had a strong catalytic effect on reaction (1), lowering the activation energy from 103 Kcal/mole in the gas phase to 24 on cluster A, the gas-phase hydrogenation of II to give III (reaction (2)) was essentially a zero-activation energy process. We have repeated the calculations with the new parameterization, using the same geometry for the species involved in reaction (1); the former results are on the whole confirmed (see Figure 6) with two major

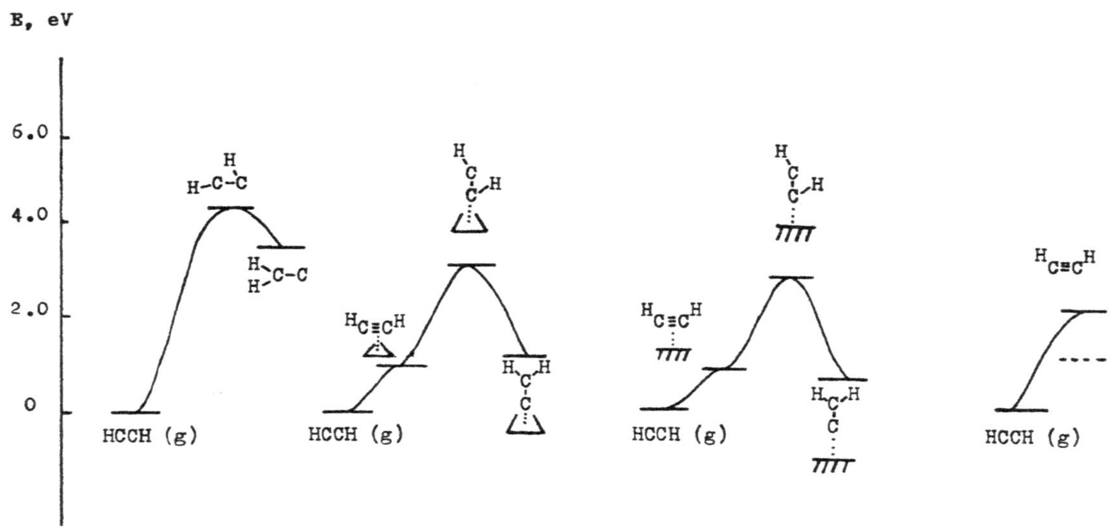

Figure 6. Schematic diagrams for the calculated activation energies for the rearrangement of acetylene to vinylidene. From the left: in the gas phase; over a three-atoms cluster, via back-bent acetylene; over cluster C, via chemisorbed back-bent acetylene. On the right: activation energy for back-bending of gaseous acetylene. The dotted level is the energy calculated for a 40° symmetrical deformation using the experimental bending force constant.

differences: 1) the surface (represented by cluster C) and a three atoms cluster have essentially the same catalytic effect, i.e. this effect does not seem to depend on cluster size; 2) the activation energy for the overall reaction is lowered from 103 to 65 Kcal/mole by the interaction with the metal, and the overall heat of reaction, that was largely negative in the former calculations due to the overestimation of the adsorption energies for all fragments, is now positive (about 1 eV). The surface reaction from back-bent acetylene to II has however a ΔH of -4 Kcal/mole and an activation energy of 46 Kcal/mole.

6. DISCUSSION

The calculated binding energy for back-bent acetylene is 39 Kcal/mole; an estimate of binding energy $>$ 25 Kcal/mole on Pt(100) has been made (Fischer and Kelemen, 1977) from experimental data. The calculated heats of adsorption of II and III (69 and 97 Kcal/mole, respectively) are for instance much larger than the experimental heat of adsorption of CO (about 30 Kcal/mole;Ertl, Neumann and Streit, 1977) but similar to the heat of adsorption of cyanide on Pt(100) (84 Kcal/mole; Bridge and Lambert, 1977). A BEBO estimate of the Pt-C bond energy (62.5 Kcal/mole; Weinberg, Deans and Merrill, 1974) can be compared with our calculated binding energy for the Pt-C molecule (46 Kcal/mole) or with our binding energies for II and III on site A (60 and 85 Kcal/mole, respectively). Calculated Pt-C distances are in the correct range to compare favorably with the sum of the metallic radius of Pt and of the covalent radius for C ($\sim$2.09 Å). The effects of the different bonding to the surface for I-III are also evident from an analysis of the C-C bond overlap populations (Mulliken, 1955): these quantities are best compared when the C-C distance is the same for all fragments. The overlap populations decrease from 1.45 for I to 1.34 for II and to 1.02 for III, the differences being partly due also to the different number of H atoms bound to a single carbon atom. The total Pt-C overlap populations decrease in the order III$>$II$>$I once the contributions of all the Pt-C interactions are taken into account, and increase on going from A to B to C sites. All this evidence is consistent with the binding energy differences (97, 69 and 39 Kcal/mole for III, II, I respectively) in indicating III bound to site C as the most stable species on the Pt(111) surface.

It is reasonable to consider that a bond lengthening may take place when II is hydrogenated to III, and a value of 1.5 Å for the C-C bond in III has been considered likely (Kesmodel, Dubois and Somorjai, 1978). Accordingly, the adsorption energy calculations have been repeated for III using R(C-C) = 1.5 Å. Aside from the obvious effects on the C-C bond overlap populations, no significant change was observed with respect to the calculations with R(C-C)= 1.3 Å (see Table 3), and the binding energy curves were practically undistinguishable from those shown in Figures 4 and 5.

The outcome of the present calculations will now be compared with that of our earlier work (Gavezzotti and Simonetta, 1977, 1979). Our former conclusions about the preferred adsorption sites for acetylene and the fragments II-III have to be revised in the light of the new results presented in this paper. A threefold adsorption site is now favored in all cases, although by small energ differences (0.1-0.2 eV, see Table 3) over the other adsorption sites. We have considered only one of the two possible threefold sites on the (111) face of fcc metals, namely, the one without a metal atom below the threefold hole. Our previous experience has shown however that the second layer of metal atoms has only a smal influence on the results of EHT calculations; this is not so for dynamical scattering calculations of LEED intensities (Kesmodel, Dubois and Somorjai, 1978), where the site we have chosen was found to be more favorable.

From the examination of the data in Table 3, our classificati of back-bent acetylene as a relatively loosely bound species survives the change of parameterization, and so do our previous inferences on the relative ease of formation of I-III on the surface. The catalytic effect of the metal has a trend that is qualitativel in agreement with the previously found one. Therefore, it is confirmed that at low temperatures back-bent acetylene (its activation energy of formation being small) is the preferred species, and lies on a threefold site; as the temperature is raised, the formation of II (catalyzed by the surface) and its subsequent easy hydrogenation to III are likely to take place (see Figure 6 for the energies of all these processes). Our present result that a threefold site is favored also for II-III is consistent with dynamical scattering LEED intensity analysis, and the calculated positive charge on carbon is consistent with the observed decrease in work functio after chemisorption (Kesmodel, Dubois and Somorjai, 1978).

The final point of debate is the preference for II (Demuth, 1979) or III (Kesmodel, Dubois and Somorjai, 1978) for the most stable species at higher temperatures; ELS results lend themselves to different interpretations. From our calculations, III should be favored in view of its larger binding energy; however, the relativ ease of formation of the two fragments must also depend on subtle differences in experimental conditions, including the amount of hydrogen present in the high-vacuum chamber (under this respect, also the formation of H_3CCH might be envisaged, as suggested by Ibach and coworkers, 1977). A possibility not yet explored to date is that of calculating the bending force constants and vibration frequencies for chemisorbed hydrocarbons by EHT, a method very efficient in obtaining the energies of valence angle distortions. Such calculations might well be of considerable help in the interpretation of ELS spectra. The experimetally found C-C distance of 1.5 Å is not in itself sufficient to discriminate between II and II since it is appropriate to both an ethane-like species (III) or a propylene-like species (like II on a threefold site).

REFERENCES

Anders,L.W., Hansen,R.S., and Bartell, L.S.: 1973,J.Chem.Phys.59, p.5277.

Bridge, M.E., and Lambert,R.M.: 1977,Surf.Sci.63,p.315.

Demuth,J.E.: 1977,Chem.Phys.Letters45,p.12.

Demuth,J.E.: 1979,Surf.Sci.80,p.367.

Dykstra,C.E., and Schaefer, H.F.III: 1978,J.Amer.Chem.Soc.100,p.1378.

Ertl,G., Neumann,M, and Streit,K.M.: 1977, Surf.Sci.64,p.393.

Fischer,T.E., and Kelemen, S.R.: 1977, Surf.Sci.69,p.485.

Gavezzotti,A., and Simonetta,M.: 1977,Chem.Phys.Letters 48,p.434.

Gavezzotti,A., and Simonetta,M.: 1979,Chem.Phys.Letters, in press.

Gavezzotti,A., Ortoleva,E., and Simonetta,M.: 1979,Theoret.Chim. Acta, submitted for publication.

Ibach,H., Hopster,M., and Sexton,B.: 1977,Appl.Surf.Sci.1,p.1.

Kesmodel,L.L., Baetzold,R.C., and Somorjai,G.A.: 1977,Surf.Sci. 66,p.299.

Kesmodel,L.L., Dubois,L.H., and Somorjai,G.A.: 1978,Chem.Phys. Letters, 56,p.267; J.Chem.Phys., in press.

Lo,W.J., Chung,Y.W., Kesmodel,L.L., Stair,P.C., and Somorjai,G.A.: 1977,Solid State Commun.22, p.335.

Mulliken,R.S.: 1955,J.Chem.Phys.23,p.1833.

Simonetta,M., and Gavezzotti,A.: 1978, in:The Chemistry of the Carbon-Carbon triple Bond, S.Patai, Editor, Wiley & Sons, New York.

Simonetta,M., and Gavezzotti,A.: 1979,Adv.in Quantum Chemistry, in press.

Summerville,R.H., and Hoffmann,R.: 1976,J.Amer.Chem.Soc.98,p.7240.

Weinberg,W.H., Deans,H.A., and Merrill,R.P.: 1974,Surf.Sci.41,p.312.

ASYMMETRIC HOMOGENEOUS HYDROGENATION OF PROCHIRAL OLEFINS AND CONTROL OF THE METAL CONFIGURATION IN LABILE OPTICALLY ACTIVE ORGANOTRANSITION METAL COMPLEXES

Henri Brunner
Universität Regensburg, Institut für Chemie,
Universitätsstr. 31, D-8400 Regensburg,
Federal Republic of Germany

Since 1966 the Wilkinson catalyst $RhCl[P(C_6H_5)_3]_3$ is used for the homogeneous hydrogenation of olefins mostly under conditions as mild as room temperature and hydrogen of atmospheric pressure *(1)*. In 1968 the groups of Horner and Knowles used optically active phosphines instead of triphenylphosphine for the hydrogenation of prochiral olefins and

Scheme 1

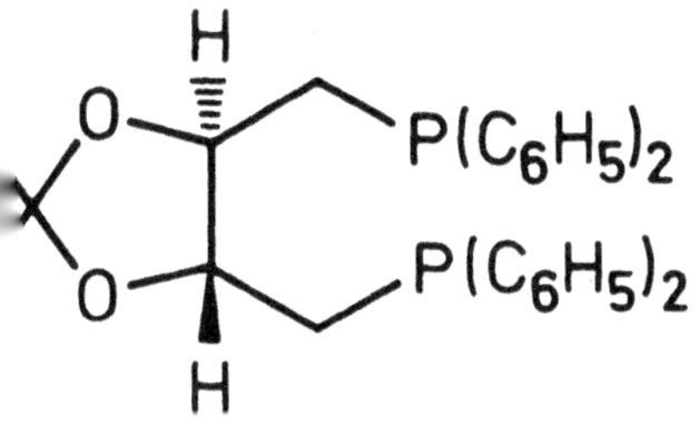

Diop

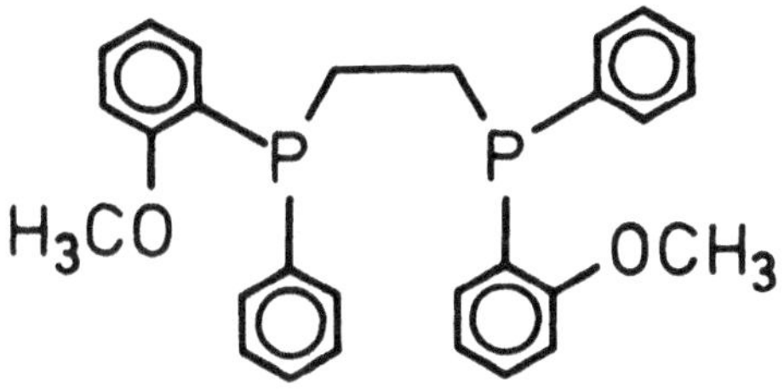

Dipamp

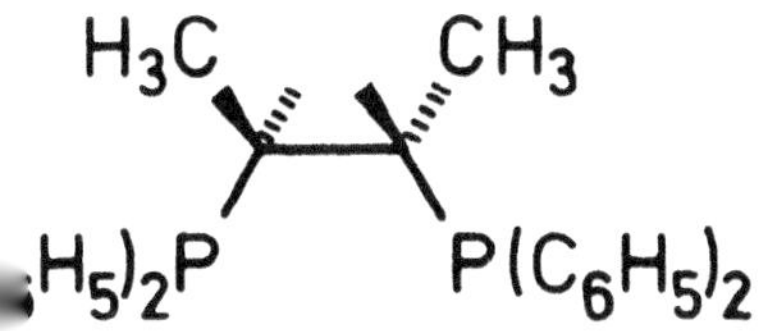

Chiraphos

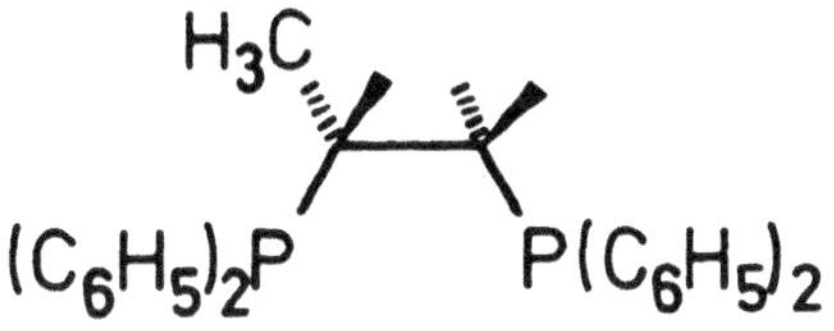

Prophos

B. Pullman (Ed.): Catalysis in Chemistry and Biochemistry. Theory and Experiment. 255–264.

Scheme 2

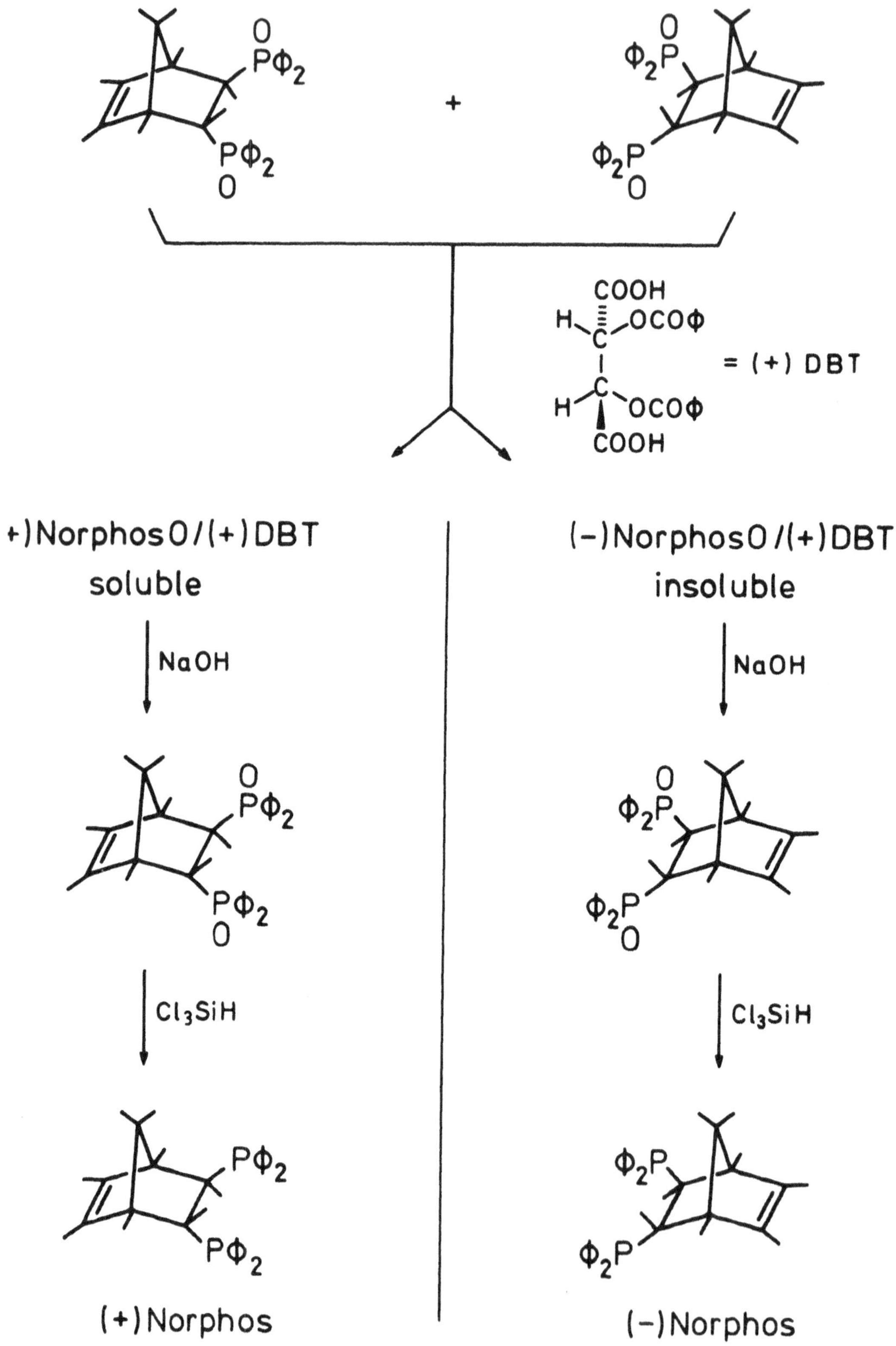

PΦ2
O
PΦ2
O
+
Φ2P
O
Φ2P
O
COOH
H
C
OCOΦ
C
H
OCOΦ
COOH
= (+) DBT
+)NorphosO/(+)DBT
soluble
(−)NorphosO/(+)DBT
insoluble
NaOH
NaOH
Cl3SiH
Cl3SiH
PΦ2
PΦ2
Φ2P
Φ2P
(+)Norphos
(−)Norphos

observed the formation of optically active hydrogenation products *(2,3)*. In the past decade progress in the field of catalytic asymmetric hydrogenation of prochiral olefins was mainly associated with the development of new optically active chelate ligands, four examples of which are shown in scheme 1 *(4-7)*.

We added the chelate ligand Norphos to this list of extremely effective co-catalysts, the synthesis of which is outlined in scheme 2. Starting material is the racemic mixture of NorphosO which can be obtained by Diels-Alder reaction of cyclopentadiene and the dioxide of trans-bis(diphenylphosphino)ethylene $(C_6H_5)_2PCH{=}CHP(C_6H_5)_2$ *(8)*, a commercially available and easily accessible compound. If the racemic mixture of NorphosO is treated with dibenzoyl-L-(+)-tartaric acid, abbreviated (+)DBT, half of the material, the less soluble diastereoisomer (-)NorphosO/(+)DBT, precipitates whereas the other half of the material, the more soluble diastereoisomer (+)NorphosO/(+)DBT, remains in solution. From both samples (+)DBT can be splitt off by treatment with NaOH. Reduction of (+) and (-)NorphosO by $SiHCl_3$ according to established procedures *(9)* gives (+) and (-)Norphos, respectively *(10)*. The surprising point of this procedure is the unexpected ease of resolution because phosphines usually are difficult to resolve.

In scheme 3 the catalytic hydrogenation of α[N-acetamino]-cinnamic acid is shown, the catalyst being $Rh_2Cl_2(COD\text{-}1,5)_2$/(-)Norphos. This optically active catalyst differentiates

Scheme 3

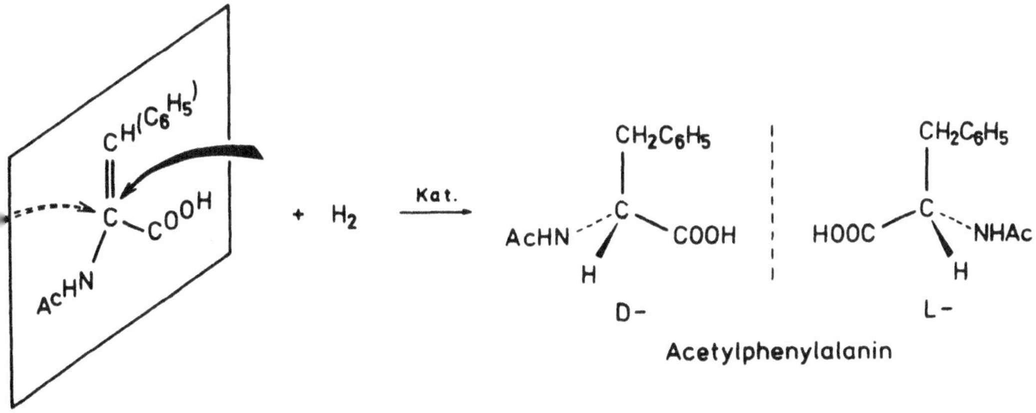

between front side and back side addition of hydrogen to the prochiral center to the extent of a formation of D- and L-acetyl-phenylalanine in a ratio of 2:98 in the solvent ethanol corresponding to an optical yield of 96 %. Similarly high optical inductions have also been observed with substrates other than α[N-acetamino]cinnamic acid.

Although the optical yields in favorable cases are close to enzyme stereospecificity, the whole field is still very empirical. For example, if a new optically active phosphine like Norphos has been prepared it is impossible to predict whether its rhodium complex will give a low, middle or high optical yield in an asymmetric hydrogenation according to scheme 3. Therefore, we attempted to improve this empirical situation by a control of the metal configuration in the catalyst according to the following concept. The catalytic species in the asymmetric hydrogenation with a Wilkinson type catalyst $RhClL_3$ is thought to be that of scheme 4 in which LL also can be a chelate ligand. In the catalyst the phosphines occupy two of the 6 octahedral positions. To a third position the Cl ligand or a solvent molecule may be coordinated. Three other positions cis to each other are needed to bind the olefin and the two hydrogen atoms in the oxidative addition of H_2. An important point is that in a structure as in scheme 4 and in similar structures the Rh atom is an asymmetric center.

Scheme 4

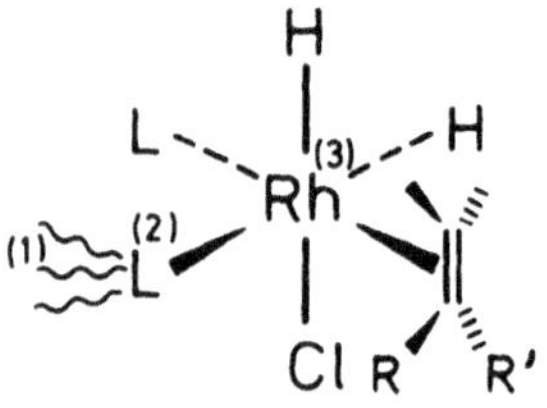

chirality (1) in the ligand side chain ~~~
(2) at the coordinating atom L
(3) at the coordination center Rh

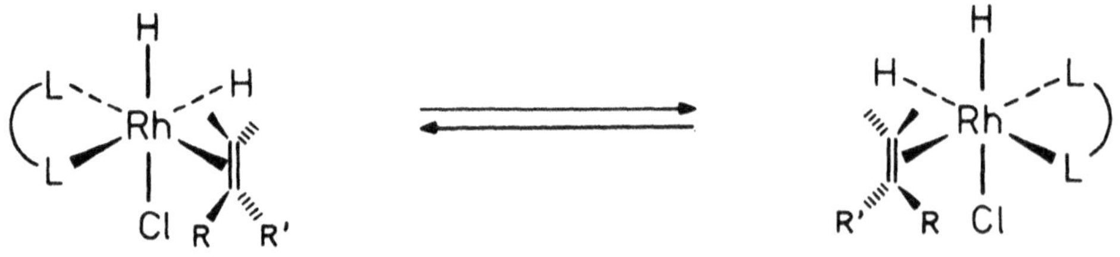

According to a rule of thumb an optical induction should become the larger the nearer the inducing chirality is to the prochiral center which is to be transformed into the new chiral center. For the catalytically active Rh complex of scheme 4 in case (1) the chirality in the side chain of the ligand L is far away from the prochiral center in the olefin. Consequently, the optical inductions for catalysts of type (1) usually are very low. In catalysts of the type (2) the coordinating atom of the ligand L, mostly phosphorus, is the inducing chiral center. Then, as in the Monsanto amino acid process *(5)* the observed optical inductions sometimes are very high. The rhodium atom [case (3)] is still nearer to the prochiral center. Therefore, if the rhodium atom would be the inducing chirality the effects would be expected to be even larger.

One possible approach to making the Rh atom the inducing asymmetric center would be to prepare a suitable optically active precursor which in situ can be transformed stereospecifically into the catalytically active species having a unique configuration at the Rh center. Since about 10 years we investigate the synthesis and properties of optically active transition metal complexes (reviews Lit. *11-17*), and such stereospecific processes have been found in our work, but it has also been demonstrated that the metal configuration in such complexes frequently is labile, especially when rapid dissociation reactions and reductive eliminations are involved. As this is the case in catalytic reactions, it is to be expected that a unique metal configuration is racemized after one or a few catalytic cycles. So, the approach to prepare an optically active rhodium complex which stereospecifically can be transformed into the optically active catalytic species cannot be the solution to the problem of control of the Rh configuration during the catalytic process. By the concept of asymmetric induction from an optically active ligand to the metal atom, however, it is possible to stabilize one of the metal configurations with respect to the other as exemplified by the model of the $C_5H_5Mo(CO)_2$ thioamidato complexes of schemes 5 and 6.

In scheme 5 the two possible isomers for a square pyramidal complex $C_5H_5Mo(CO)_2LX$ are shown carrying a C_5H_5 ligand on top of the pyramid, two carbonyls and an unsymmetrical chelate ligand LX at the basis. Both isomers contain an asymmetric carbon atom of configuration S´ in the amino side chain of the ligand. The only difference between them are the opposite Mo configurations R and S, respectively. Scheme 5 also shows that the two possible isomers RS´ and SS´ differ appreciably in their chemical shifts, especially in the area of the high intensity C_5H_5 singlets. So, it is easy to determine the diastereoisomer ratio RS´:SS´ by ^1H nmr integration. The two

Scheme 5

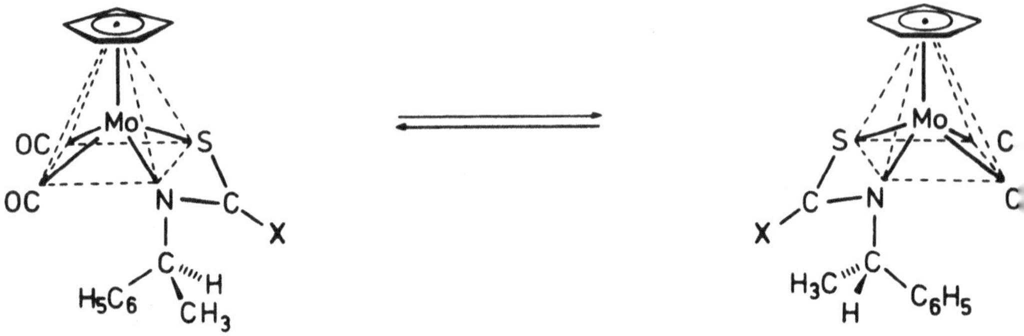

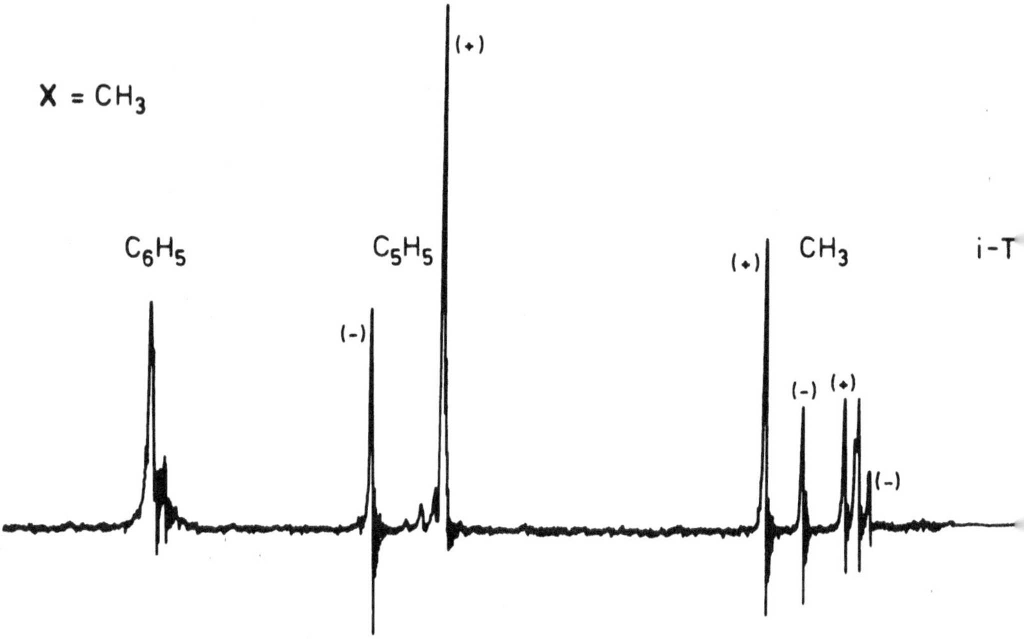

X	H	CH_3	C_6H_5	$C_{10}H_7$
(+)/(-)	50 : 50	69 : 31	77 : 23	87 : 13

diastereoisomers RS´ and SS´ can be separated by fractional crystallization or by preparative liquid chromatography to optical purity. On heating solutions of the $C_5H_5Mo(CO)_2$

thioamidato complexes to temperatures of about 50°C a rapid first order epimerization R⇄S with respect to the molybdenum atom takes place, probably by an intramolecular rearrangement, and the equilibrium mixture RS′:SS′ is obtained after ten half lives.

The diastereoisomer equilibria RS′:SS′ were investigated as a function of group X of the carboxylic acid used to prepare the corresponding thioamides (table in scheme 5). For the formic acid derivative X = H the equilibrium mixture corresponds to a 50:50 mixture of both isomers. That means, there is no optical induction from the stable S′ chirality at the asymmetric carbon atom on the formation of the two possible metal configurations R and S at equilibrium. For the acetic acid derivative X = CH_3 the equilibrium ratios RS′:SS′ is 69:31, identical with the spectrum in scheme 5. In this case there is an optical induction of the asymmetric carbon atom on the Mo equilibrium configuration of 38 %. For benzoic acid and 1-naphthoic acid with 77:23 and 87:13 the equilibrium ratios are even higher as shown in scheme 5, indicating equilibrium optical inductions of 54 and 74 %, respectively *(18, 19)*.

The table of scheme 6 summarizes these equilibrium inductions as a function of X in the first column and demonstrates the influence of a substituent variation at the inducing asymmetric center. In this variation the substituents hydrogen and phenyl at the asymmetric carbon atom were held constant and the substituent R was varied from methyl to ethyl to isopropyl. In the acetic acid series X = CH_3 the diastereoisomer ratio at equilibrium rises from R = CH_3 (69:31) to R = CH_2CH_3 (89:11) to R = $CH(CH_3)_2$ (98:2). This last underlined situation of an equilibrium induction of 96 % is represented in the spectrum of scheme 6. By fractional crystallization the disfavored isomer present in 2 % quantity at equilibrium can be removed completely. On heating the solution of the optically pure dominating isomer to 50°C, the signal of the less favored isomer appears until the equilibrium ratio of 98:2 is obtained. Compared to the acetic acid series, all the values for the corresponding benzoic acid series are a little larger rising to an optical purity of 98 % in the compound with R = isopropyl *(20)*.

As a result, the optical induction at equilibrium with respect to the metal configuration in $C_5H_5Mo(CO)_2LX$ complexes for the thioamidato ligand LX system can be varied from 0 to 98 %. By the same concept it should be possible by a substituent variation in an optically active ligand LL (scheme 4) to shift the equilibrium to one side, an effect identical to bringing the optical activity to the metal atom and keeping it there although there is a rapid equilibration with the other metal

Scheme 6

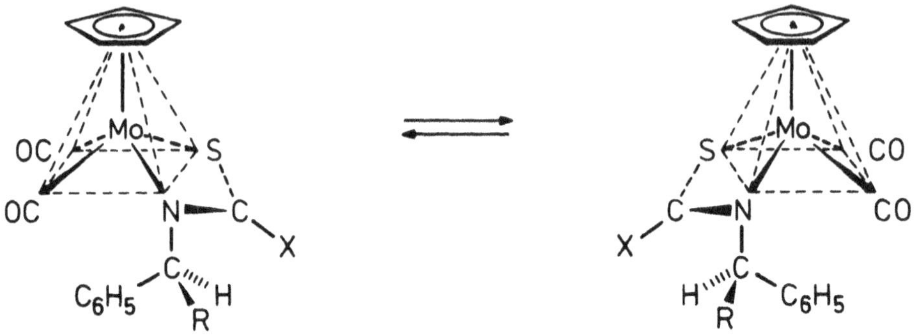

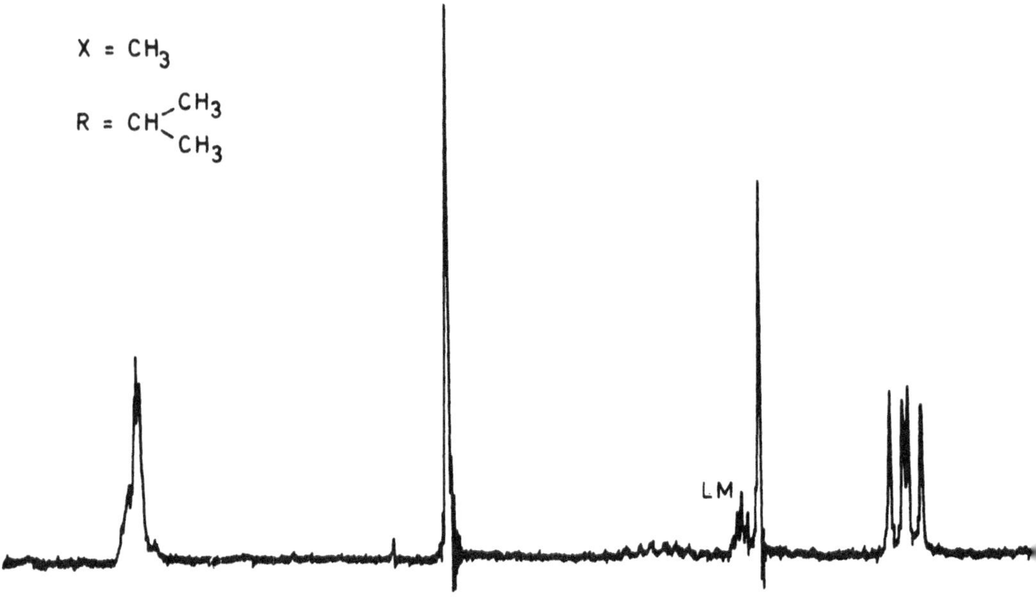

	R = CH_3	R = CH_2-CH_3	R = $CH(CH_3)_2$
X = H	50 : 50		
X = CH_3	69 : 31	89 : 11	98 : 2
X = C_6H_5	77 : 23	95 : 5	99 : 1
X = $C_{10}H_7$	87 : 13		

configuration *(16)*. This concept has also been veryfied with other bidendate ligands *(21)* and also with monodendate ligands *(22)*. Hopefully, the possibility to control the metal configuration in a catalyst similar to the example discussed in schemes 5 and 6, will also help to control the optical yields in asymmetric catalysis.

References

(1) Osborn, J.A., Jardine, F.H., Young, J.F., and Wilkinson, G.: 1966, J.Chem.Soc. A, pp. 1711.

(2) Horner, L., Siegel, H., and Büthe, H.: 1968, Angew.Chem. 80, pp. 1034; Angew.Chem.int.Ed.Engl. 7, pp. 942.

(3) Knowles, W.S. and Sabacky, M.J.: 1968, JCS Chem.Comm., pp. 1445.

(4) Kagan, H.B. and Dang, T.P.: 1972, J.Amer.Chem.Soc. 94, pp. 6429.

(5) Knowles, W.S., Sabacky, M.J., Vineyard, B.D., and Weinkauf, D.J.: 1975, J.Amer.Chem.Soc. 97, pp. 2567.

(6) Fryzuk, M.D. and Bosnich, B.: 1977, J.Amer.Chem.Soc. 99, pp. 6262.

(7) Fryzuk, M.D. and Bosnich, B.: 1978, J.Amer.Chem.Soc. 100, pp. 5491.

(8) Nesterova, N.P., Medved, T.Y., Polikarpov, Y.M., and Kabachnik, M.K.: 1974, Izv.Akad.Nauk SSSR Ser.Khim. 10, pp. 2210.

(9) Naumann, K., Zon, G., and Mislow, K.: 1969, J.Amer.Chem. Soc. 91, pp. 7013.

(10) Brunner, H. and Pieronczyk, W.: Angew.Chem., in press.

(11) Brunner, H.: 1971, Angew.Chem. 83, pp. 274; Angew.Chem. int.Ed.Engl. 10, pp. 249.

(12) Brunner, H.: 1974, Ann.N.Y.Acad.Sci. 239, pp. 213.

(13) Brunner, H.: 1975, Top.Curr.Chem. 56, pp. 67.

(14) Brunner, H.: 1977, Chemie in unserer Zeit 11, pp. 157.

(15) Brunner, H.: Acc.Chem.Res., in press.

(16) Brunner, H.: Adv.Organometal.Chem., in press.

(17) Brunner, H., in Koerner von Gustorf, E.A., Grevels, F.-W., and Fischler, I.: 1978, Organic Chemistry of Iron, Vol. 1, Academic Press, pp. 299.

(18) Brunner, H., Herrmann, W.A., and Wachter, J.: 1976, J.Organometal.Chem. 107, pp. C11.

(19) Brunner, H. and Wachter, J.: 1977, Chem.Ber. 110, pp. 721.

(20) Brunner, H. and Lukas, R.: Chem.Ber., in press.

(21) Brunner, H. and Rastogi, D.K.: Inorg.Chem., in press.

(22) Brunner, H., Doppelberger, J., Dreischl, P., and Möllenberg, T.: 1977, J.Organometal.Chem. 139, pp. 223.

HOMOGENEOUS AND HETEROGENEOUS HYDROGEN TRANSFER IN CARBINOLS BY GROUP VIII METAL CATALYSTS

Jochanan Blum
Department of Organic Chemistry, Hebrew University, Jerusalem

Ilan Pri-Bar and Ouri Buchman
Israel Atomic Energy Commission, Nuclear Research Center, Negev

ABSTRACT

The homogeneous and heterogeneous dehydrogenation, dehydration, reductive coupling and disproportionation of diaryl carbinols by group VIII metal catalysts were investigated. Conditions under which each of these catalyses take place independently have been defined using benzhydrol as substrate and $RuCl_2(PPh_3)_3$as catalyst. Kinetic measurements revealed that the various transformations of the carbinols are part of the same catalytic cycle. In aromatic hydrocarbon solvent, the active catalyst proved to be solvated $RuCl_2(PPh_3)_2$ while in DMSO it is $RuCl_2(PPh_3)(DMSO)_2$. The key intermediates are ruthenium alkoxides and ruthenium hydrides. Steric factors play a crucial part in catalyses by triphenylphosphine containing complexes but are of less significance in phosphine free heterogeneous systems.

The frequent application of alcohols as reagents and as solvents in catalytic processes and the existence of unavoidable side reactions in such systems[1], urged us to investigate the behavior of carbinols in the presence of some typical platinoid catalysts.

We have found that in the absence of external hydrogen acceptors complexes $RuCl_2(PPh_3)_3$, $Ru_2(CO)_6Cl_4$, $RhCl(PPh_3)_3$, $RhBr(PPh_3)_3$, $PdCl_2(PPh_3)_2$, $PtCl_2(PPh_3)_2$ and $Pt(PPh_3)_4$ promote transformations (1)-(5) in secondary alcohols.

Dehydrogenation	$R_2CHOH \rightarrow R_2CO + H_2$	(1)
Dehydration	$2R_2CHOH \rightarrow R_2CHOCHR_2 + H_2O$	(2)
Reductive coupling	$3R_2CHOH \rightarrow (R_2CH)_2 + R_2CO + 2H_2O$	(3)
Disproportionation	$2R_2CHOH \rightarrow R_2CH_2 + R_2CO + H_2O$	(4)
Hydrogenolysis	$R_2CHOH + R'_2CHOH \rightarrow R_2CH_2 + R'_2CO + H_2O$	(5)

Although in many cases the five processes proceed simultaneously, and even additional transformations occur[1c], we were able to work out conditions for diaryl carbinols under which each of reactions (1)-(5) takes place independently. E.g., at 180°C under exclusion of air, a

B. Pullman (Ed.): Catalysis in Chemistry and Biochemistry. Theory and Experiment. 265–272.

catalyst system containing $RuCl_2(PPh_3)_3$ in an aromatic hydrocarbon and small quantities of water promotes exclusively reductive coupling of benzhydrol (eq. 3). By exclusion of water or by application of an acid catalyst (e.g., $Ru_2(CO)_6Cl_4$)ether formation (eq. 2) becomes the main process. In the presence of a polar aromatic solvent (e.g., diphenyl ether), or upon addition of benzyl alcohol, reaction (4) and (5) predominate. Selective dehydrogenation of diaryl carbinol (eq. 1) takes place when a solution of $RuCl_2(PPh_3)_3$ in dimethylsulfoxide is used as catalyst.

Aliphatic carbinols do not undergo internal hydrogen transfer by $RuCl_2(PPh_3)_3$at 180°C. Secondary benzyl alcohols, however, are dehydrogenated under the conditions to give the corresponding ketones. Aromat tertiary carbinols that have β-hydrogen atoms are dehydrated and yield olefins (eq. 6).

$$RR'CHC(OH)(R)Ar \rightarrow RR'C{=}C(R)Ar + H_2O \qquad (6)$$

Examples of $RuCl_2(PPh_3)_3$-catalyzed transformations of some carbinols under comparable conditions are listed in Table I.

The high selectivity that could be achieved with $RuCl_2(PPh_3)_3$ and benzhydrol under appropriate conditions permitted the performance of detailed kinetic and mechanistic studies of which the essence will be given herewith.

Except for very high carbinol concentration, the kinetics proved to be first order in benzhydrol for both reactions (2) and (3). Betwee $5\text{-}15\times10^{-3}$ M a linear rate dependence on catalyst concentration has been observed. Modification of the electronic nature of the carbinol by introduction of substituents Cl, CH_3 and OCH_3 into the aromatic moietie indicated that electron releasing groups increase the rate while electr attracting substituents cause it to slow down. The Hammett ρ values of reactions (2) and (3) were calculated to be -2.2 and -6.4, respectively

The rates also prove to depend on the electronic nature of the ruthenium catalyst. The activities of $RuCl_2[(4\text{-}X\text{-}C_6H_4)_3P]_3$, for different substituents X, were shown to change in the order Cl > H > CH_3 > OCH_3. The corresponding Hammett ρ values for the reductive coupling and dehydration reactions were found to be +0.22 and +0.60. A secondary isotope effect of rate (H)/rate (D)=1.25 ± 0.15 was measure in (2) and (3) when benzhydrol-d_2 was employed.

Steric factors prove to have a substantial effect on the catalyses. By changing the degree of steric hindrance around the hydroxyl group in some aromatic carbinols, and by applying molecular models of carbinol-ruthenium adducts, we were able to demonstrate the existence of a direct correlation between the reaction rate and the steric factor Thus e.g., the initial rates of the reductive coupling of 9-fluorenol, 2,2'-dimethylbenzhydrol, benzhydrol, 10,11,-dihydrodibenzo[a,d]cyclohepten-5-ol and dibenzo[a,d]cyclohepten-5-ol in 1-methylnaphthalene were under comparable conditions (1.67 M carbinol, 11×10^{-3} $RuCl_2(PPh_3)_3$, 186°C

0.0, 0.8, 1.04, 18.5 and 43.0 mole x $liter^{-1}$ x min^{-1} x10^{3}, respectively.

TABLE I $RuCl_2(PPh_3)_3$-catalyzed Transformation of Some Carbinols*

Carbinol	Conversion %	Product distribution, mole %				
		Ar_2CH_2	olefin	ketone	$(Ar_2CH)_2$	ether
$C_6H_5CH(OH)CH_3$	50	25	-	75	-	-
$4\text{-}ClC_6H_4CH(OH)CH_3$	27	18	-	82	-	-
$C_6H_5CH(OH)C_2H_5$	10	-	-	100	-	-
$C_6H_5CH(OH)CH_2C_6H_5$	20	20		80	-	-
$C_6H_5CH(OH)C(CH_3)_3$	5	-	-	100	-	-
$(4\text{-}CH_3OC_6H_4)_2CHOH$	100	-	-	44	40	10
$(4\text{-}CH_3C_6H_4)_2CHOH$	90		-	41	41	18
$(C_6H_5)_2CHOH$	35	-	-	53	42	6
$(4\text{-}ClC_6H_4)_2CHOH$	26	3	-	50	40	3
$C_6H_5C(OH)(CH_3)_2$	97	-	92	3	-	-
$(C_6H_5)_2C(OH)CH_3$	100	-	100	-	-	-

* Reaction conditions: 0.5 mmole carbinol, 3×10^{-3} mmole catalyst in 0.3 ml 1-methyl-naphthalene; 3 h at 180° in a vacuum sealed ampoule.

Neither free radical scavengers (e.g., 2,5-di-tert-butylhydroquionone) nor free radical initiators (e.g., α,α-isoazobutironitrile) have any effect on the reaction rate. Small amounts of hydrogen chloride (as well as benzhydryl chloride) promote reactions (2) and (3) while triphenylphosphine and 2,6-dimethylpyridine serve as inhibitors.

As the catalyses take place under homogeneous conditions over a considerable range of temperatures in the same degree of selectivity, the thermodynamic values for reactions (2) and (3) could be obtained from Arrhenius plots of log rate against 1/T x 10^{3}. E_a (2), E_a (3), $\Delta H^{\ddagger}$ (2), and $\Delta H^{\ddagger}$(3), 200°C were found to be 34.5, 35.2, 35.2 and 34.3 kcal x $mole^{-1}$, respectively.

The kinetics of $RuCl_2(PPh_3)_3$-catalyzed dehydrogenation of benzhydrol in dimethylsulfoxide (eq. 1) differ considerably from those of the reductive coupling and dehydration reactions. For 1.0-3.3M benzhydrol (at 190°C) the reaction rate proved to be independent of the concentration of the carbinol. A linear relationship exists between the rate and catalyst concentration as long as $[RuCl_2(PPh_3)_3] < 3x10^{-3}$ M. Above this concentration the rate approaches asymptotically a limiting value. Modification of the ruthenium catalyst by introduction of substituents with different electronic properties in $RuCl_2[(4\text{-}X\text{-}C_6H_4)_3P]_3$ revealed a decrease in rate in the order $CH_3O > CH_3 > H > Cl$, (Hammett $\rho = -0.13$). A primary kinetic isotope effect rate(H)/rate(D)=4. has been observed for the dehydrogenation of Ph_2CHOH and Ph_2CDOD in DMSO. We have proven that the effect is not associated with any H-D exchange in the solvent.

Evaluation of our results led us to the conclusion that the $RuCl_2(PPh_3)_3$-catalyzed transformations (1)-(5) are parts of one and the same catalytic cycle shown in Scheme I.

The initial step, the dissociation of $RuCl_2(PPh_3)_3$ (eq. 7) that was

$$RuCl_2(PPh_3)_3 + \text{solv} \rightleftharpoons RuCl_2(PPh_3)_2\text{solv} + PPh_3 \quad (7)$$

suggested thirteen years ago[2], has been shown by ^{31}P NMR to go to completion at 190°C in the aromatic solvents. Similar measurements indicated the extrusion of 2-3 triphenylphosphine ligands from the starting catalyst in DMSO [eqs. (8) and (9)]

$$RuCl_2(PPh_3)_3 + 3DMSO \rightarrow RuCl_2(PPh_3)(DMSO)_3 + 2PPh_3 \quad (8)$$

$$RuCl_2(PPh_3)(DMSO)_3 + DMSO \rightleftharpoons RuCl_2(DMSO)_4 + PPh_3 \quad (9)$$

In catalyses (2) and (3) benzhydrol proved to undergo a slow non-oxidative addition to the active bis(triphenylphosphine) complex, $RuCl_2(PPh_3)_2$solv, forming a semicoordinative bond between the hydroxylic oxygen and the ruthenium atom. The resulting complex has been shown to lose another PPh_3 ligand and to undergo fast reductive elimination of HCl to give alkoxide $(Ph_2CHO\text{-})RuCl_2(PPh_3)$solv. which, in turn, may undergo fast β-hydride transfer to yield benzophenone and hydride $HRuCl_2(PPh_3)$solv [δ (C_6H_6) 36 ppm].

The benzhydryl cation, formed upon interaction of the hydrogen chloride and the starting carbinol, may either undergo rapid oxidative addition to $(Ph_2CHO\text{-})RuCl_2(PPh_3)$solv followed by formation of benzhydryl ether, or lead to intermediate $(Ph_2CH)Ru(H)Cl_2(PPh_3)$solv from which tetraphenylethylene is obtained. When the benzhydryl-ruthenium complex $(Ph_2CH)RuCl_2(PPh_3)$solv is trapped with HCl, diphenylmethane results rather than the tetraphenyl-compound. The competitive oxidative addition of HCl causes a decrease in the rate of reaction (2) and (3). The balance between the competing addition processes proved to be related to the stability and activity of the reacting carbo-cation.

The reaction of hydrogen chloride and other polar species are decelerated in DMSO owing to the formation of stable solvates. In this medium the active catalyst $RuCl_2(PPh_3)(DMSO)_2$ reacts with benzhydrol to give an alkoxide which undergoes β-hydride transfer

SCHEME I $RuCl_2(PPh_3)_3$-Catalyzed Transformations of Ph_2CHOH

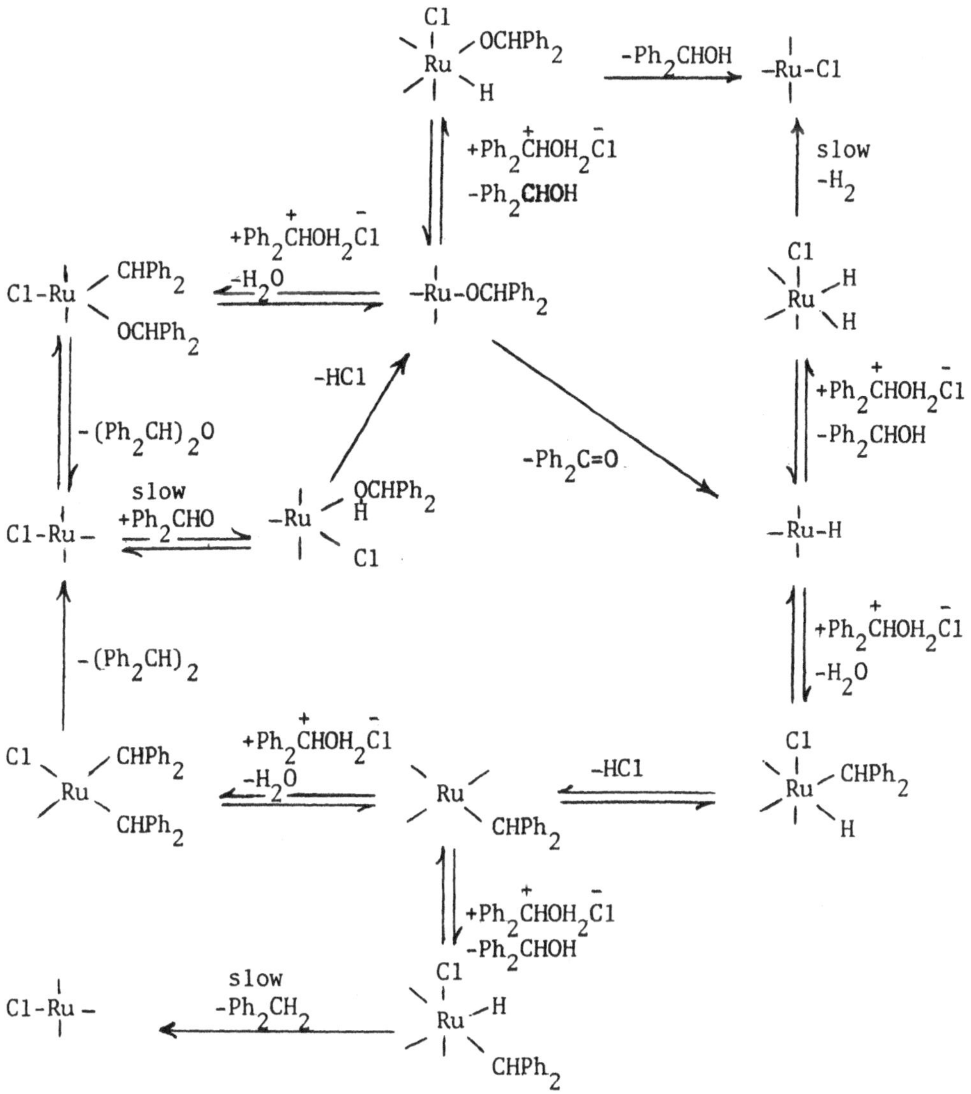

and elimination of benzophenone. The resulting ruthenium hydride may add hydrogen chloride (or benzhydryl chloride followed by elimination of benzhydrol) to give a dihydride. Slow evolution of molecular hydrogen is the rate determining step in process (1).

Full support in the proposed mechanisms of reactions (1)-(5) has been provided by a comprehensive study on $RuCl_2(PPh_3)_2$-catalyzed transformations of mixtures of alcohols. We have shown that carbinols

TABLE II Transformation of Benzhydrol by Homogeneous and Heterogeneous Catalysts under Comparable Conditions[a]

	Product distribution (relative yield, mole %)			
Catalyst	Ph_2CH_2	Ph_2CO	$(Ph_2CH)_2$	$(Ph_2CH)_2O$
$RuCl_2(PPh_3)_3$	4.4	47	46	2.6
Ru/C	28.7	51.6	18.7	1
Ru/C-NH_4Cl	9	43	38	10
$RhCl_2(PPh_3)_3$	1.8	22.5	23.7	52
Rh/C	9	32	32	27
$PdCl_2(PPh_3)_2$	3	10.2	10.2	76.6
Pd/C[b]	35	46	17.6	1.4
Pd/C-NH_4Cl	-	11.2	13.3	75.5
$PtCl_2(PPh_3)_2$	11	4.4	5.9	78.7
Pt/C	8.5	42.5	42.9	6.1
Pt/C-NH_4Cl	-	4.6	5.5	89.9

[a] Reaction System: 0.5 mmole carbinol, $3x10^{-3}$ mmole catalyst, 0.3 ml 1-methylnaphthalene. Homogeneous system-186°C, 4 h, Heterogeneous system-210°C, 24 h.

[b] cf., reference 3.

that do not form stable carbo-cations may serve as hydrogen donors and enhance hydrogenolysis but neither form symmetrical ethers nor undergo reductive coupling.

While triphenyl carbinol itself reacts only very slowly in 1-methylnaphthalene to give bis(triphenylmethyl) ether, a mixture of Ph_3COH and 2-butanol is transformed according to eq. (10) and eq. (11) in reasonable rate.

$$2Ph_3COH \longrightarrow (Ph_3C)_2O + H_2O \qquad (10)$$

$$Ph_3COH + EtCH(OH)Me \longrightarrow Ph_3CH + Et(CO)Me + H_2O \qquad (11)$$

Unhindered aliphatic alcohols (e.g., EtOH) yield in addition mixed ethers [eq. (12)].

$$EtOH + Ph_3COH \longrightarrow EtOCPh_3 + H_2O \qquad (12)$$

Benzyl alcohol that is completely inactive in oxygen-free solutions of $RuCl_2(PPh_3)_3$-1-methylnaphthalene is "activated" in the presence of benzhydrol. Thus, e.g., equimolar amounts of the two carbinols (1.6 M) and 11×10^{-3} M catalyst give after 4 h at 186°, 16.5 mole % of benzhydryl benzyl ether, 13.1 mole % of benzophenone, 11.7 mole % of diphenylmethane, 5.5 mole % of benzaldehyde, 3.0 mole % of benzhydryl ether 2.8 mole % of tetraphenylethane and 0.4 mole % of toluene.

In contrast to the homogeneous catalysts, pure powdered Ru, Rh, Pd and Pt are completely inactive in reactions (1)-(5). However, supported metals on carbon that were obtained from commercial sources have considerable activities. We have shown that the catalytic power of the supported metals is conditioned by the presence of minute amounts of chloride ions. Moreover, by addition of ammonium chloride it was possible to obtain heterogeneous systems that resemble in mode of reaction (though not in reactivity) to the corresponding homogeneous catalysts (see Table II).

It is remarkable that while sterically hindered carbinols are hardly affected in the presence of the bulky triphenylphosphine-containing complexes, (vide supra) they react smoothly in the presence of the supported metals (commercial grades). Heterogeneous catalysts prepared from chlorine free oxides and soot do not promote reactions (1)-(5) at all.

ACKNOWLEDGMENT. We are grateful to the U.S.-Israel Binational Science Foundation for financial support.

REFERENCES

1. Some examples from our laboratory: (a) Sasson, Y., Cohen, M., and Blum, J: 1973, Synthesis, pp. 359-360; (b) Sasson, Y., Blum, J., and Dunkelblum, E.: 1973, Tetrahedron Lett. pp. 3199-3202; (c) Sasson, Y., and Blum, J.: 1974, J.C.S. Chem. Comm. pp. 309-310; (d) Pickholtz, Y.,

Sasson, Y., and Blum, J.: 1974, Tetrahedron Lett. pp. 1263-1266;
(e) Sasson, Y., and Blum, J.: 1975, J. Org. Chem.: 1975, 40, pp. 1887-1896; (f) Albin, P., Blum, J., Dunkelblum, E., and Sasson, J.: 1975, J. Org. Chem. 40, p. 2402; (g) Sasson, Y., Zoran, A., and Blum, J.: 197[illegible] Proc. 45gh Annual Meeting, Isr. Chem. Soc. p. OR-27; (h) Sasson, Y., Albin, P., and Blum, J.: 1978, Proc. 45th Annual Meeting, Isr. Chem. So[illegible] p. OR-28.
2. Evans, D., Osborn, J.A., Jardine, F.H., and Wilkinson, G.: 1966, Nature 208, pp. 1203-1204.
3. Okada, H., Inoue, Y, and Hashimoto, H,: 1969, Kogyo Kagaku Zasski 72, pp. 2018-2023.

THE COOPERATE EFFECT OF ORGANOMAGNESIUM AND NICKEL COMPOUNDS IN THE CATALYTIC OLIGOMERIZATION OF 1.3-DIENES

H. Lehmkuhl, K. Mehler, A. Rufińska and D. Sekutowski
Max-Planck-Institut für Kohlenforschung,
Kaiser-Wilhelm-Platz 1, D-4330 Mülheim a.d. Ruhr

SUMMARY

Crotylmagnesium halide and nickelocene (molar ratio Mg/Ni ~100) catalyze the dimerization of butadiene to 3-methyl-1.4.6-heptatriene and the trimerization to 1-exo-methylene-2-methyl-4.6-divinyl-cyclohexane. Isoprenylmagnesiumhalide and nickelocene catalyze the dimerization of isoprene to a mixture of 3.3.6- and 2.3.6-trimethyl-1.4.6-heptatriene. On the basis of model reactions, the catalytic processes are suggested to occur through a sequence of Mg-C addition to C=C, exchange of allyl groups between Mg and Ni followed by displacement reactions in CpNi-π-allyl complexes or CpNi-alkyl intermediates by the 1.3-diene.

1. INTRODUCTION

The matrix for the catalytic cyclotrimerization of butadiene (1) to 1.5.9-cyclododecatriene[1] is the so-called "naked nickel". Wilke has defined as "naked" such Ni atoms which are surrounded only by ligands which can be displaced by butadiene[2]. Catalysts of this type can be produced by reaction of Ni(II) compounds with alkylaluminum as reducing agent in the presence of suitable ligands or by displacement of the allyl groups in bis-π-allylnickel or of 1.5-COD in bis-(1.5-cyclooctadiene)-nickel by butadiene.

Modification of the matrix by phosphine or phosphite ligands which cannot be displaced by butadiene result in a cyclodimerization reaction to give mainly 1.5-cyclooctadiene[3]. We have observed that, if crotylmagnesium halide is used instead of the organoaluminum for the preparation of the catalytic system, dimers and trimers of completely different structures are formed, viz. 3-methyl-1.4.6-heptatriene (2) and exo-methylene-methyl-divinylcyclohexane[4] (3). The ratio 2/3 depends on the reaction temperature and on the butadiene concentration.

B. Pullman (Ed.): Catalysis in Chemistry and Biochemistry. Theory and Experiment. 273–284.

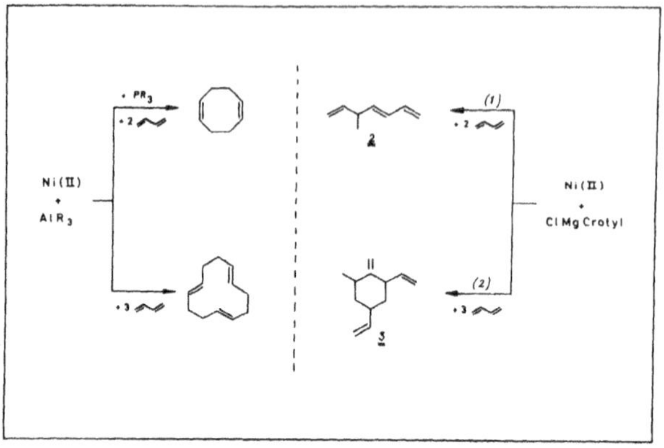

2. CATALYTIC DIMERIZATION OF BUTADIENE

One of the most effective precursors for the active catalyst is nickelocene (4). To elucidate the mechanism of the catalytic dimerization we have studied the system crotylMgCl (5) / $NiCp_2$ (4) / butadiene (1).

Nickelocene (4) reacts with crotylmagnesiumchloride (5) to give a 3/1 mixture of syn- (a) and anti- (b)-η^5-cyclopentadienyl-η^3-crotyl-nickel (6) in ~70% yield[5].

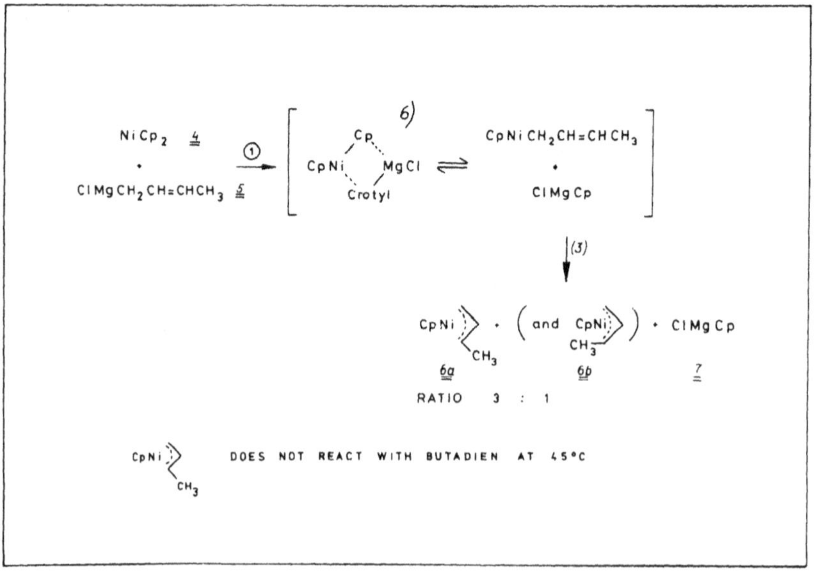

6 is a red-violet distillable oil, which does not react catalytically with butadiene in ether between 45 and 70°C.

The transfer of the second Cp-group in 4 from Ni to Mg by reaction with an excess of 5 (1/2 - 1/3) is much slower: the formation of bis-π-crotylnickel can be observed by ^1H-nmr, but during the reaction a mixture of 1- and 2-butene is formed which ultimately leads to consumption of all the crotyl groups.

Crotylmagnesiumhalide (5) reacts stoichiometrically in the absence of a Ni(II) compound: regioselective (~ 99%) metal→C-2 addition occurs to give 8 (ether, 96 h, 45°C, 45% yield), which reacts with further butadiene to give 9 (ether, 48 h, 85°C, 46% yield)[7].

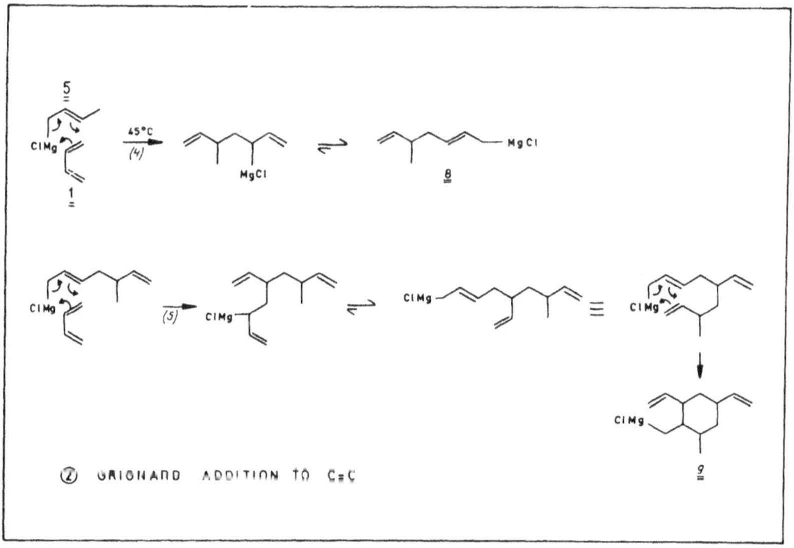

⑦ GRIGNARD ADDITION TO C=C

In the presence of 4 or 6 (Mg/Ni ~ 100/1) as well as other Ni(II) compounds such as $Ni(acac)_2$, $Ni(OEt)_2$ (but not bis(π-crotyl)-nickel) 2 and 3 are formed catalytically. Neither the rate of the reaction nor the regioselectivity of the C-C bond formation is changed in the presence of nickel and therefore we assume that the addition of the crotylmetal to 1 is associated with magnesium and not nickel. It is known from a comparison between the addition of crotyl compounds of Mg and Zn that the regioselectivity of the addition to butadiene is influenced drastically by both the metal and the substituents on the metal:

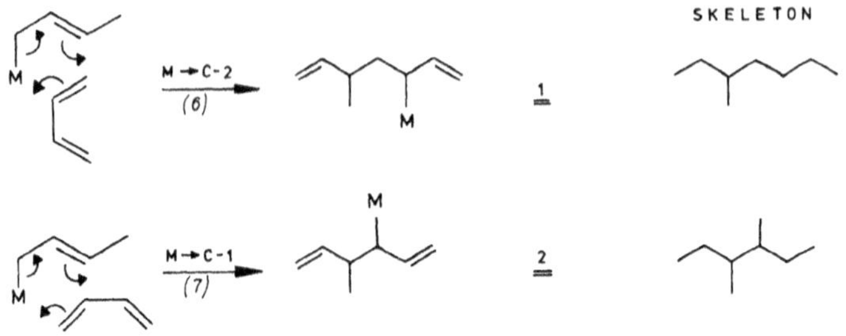

M	1 / 2
CROTYL Mg	98 / 2 [8]
Cl Mg	99 / 1
CROTYL Zn	52 / 48 [9]
Cl Zn	2 / 98
ClMg + Ni(II)	99 / 1

NO INFLUENCE OF Ni(II) ON

1. REGIOSELECTIVITY
2. RATE OF ADDITION

Two pathways for further reaction of 8 are possible:

a) exchange of allyl groups between 8 and 6 (eq. 8) or

b) addition of 8 to 1 to give 9 (eq. 5):

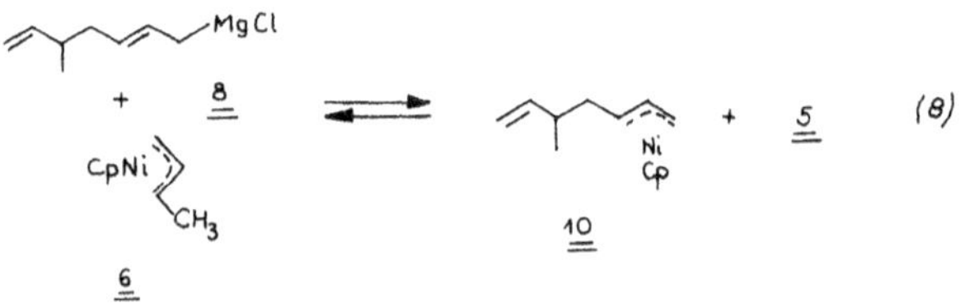

We have been able to demonstrate such an exchange reaction, e.g. between 5 and cyclopentadienyl-η^3-2-methylallylnickel 11, by ^{1}H-nmr spectroscopy (fig. 1):

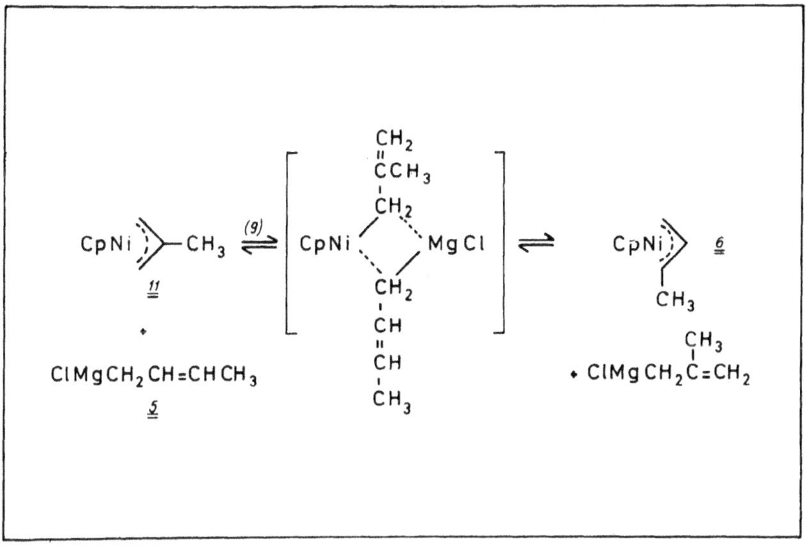

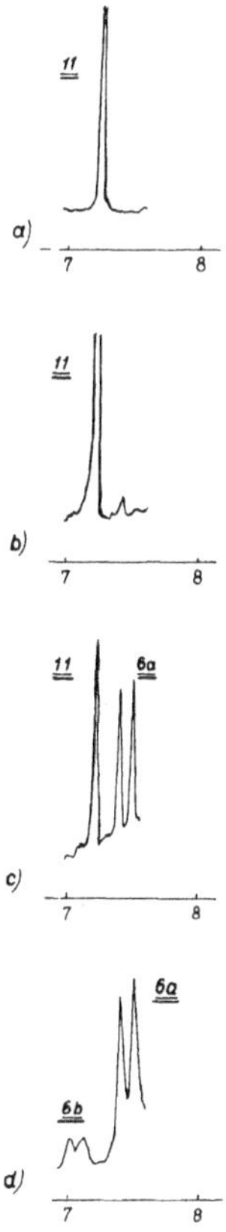

Figure 1:

^{1}H-nmr spectra of the syn protons of (a) 11, (d) 6 and a mixture of 11 and 5 at time 0 (b) and after 21 h (c) in D_8-THF.

In the next step 10 reacts with butadiene (1) by hydrogen transfer to give 3-methyl-1.4.6-heptatriene (2) and cyclopentadienyl-η^3-crotylnickel (6):

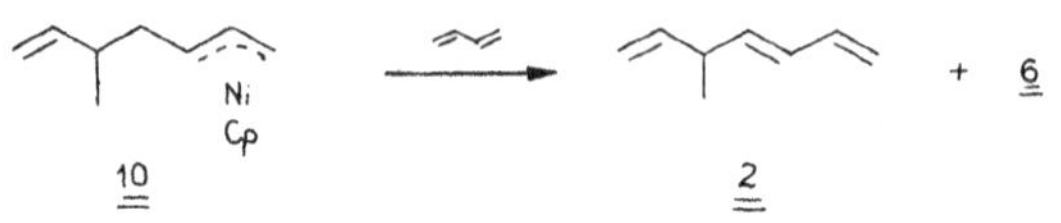

(10)

The reaction sequence (4), (8) and (10) describes a catalytic cycle for the dimerization of 1 to 2 (fig. 2)

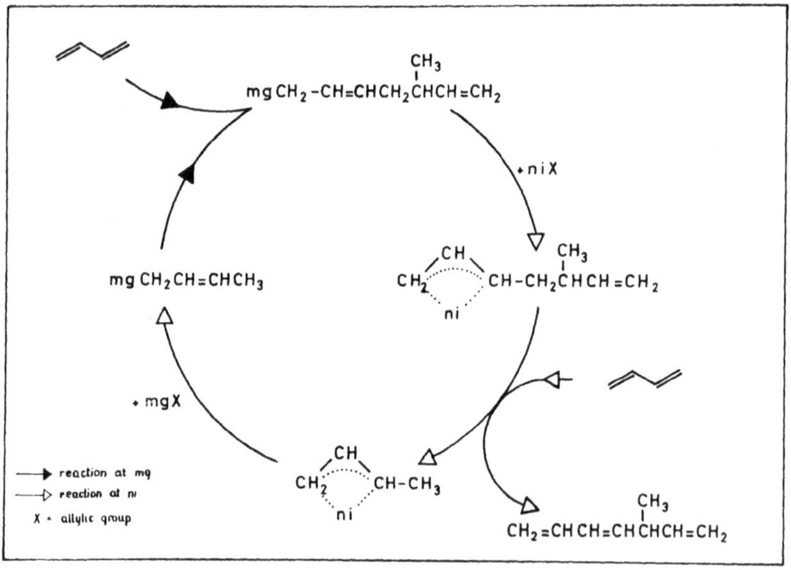

Figure 2: Catalytic cycle for the dimerization of butadiene to 2 by cooperation[10] between Mg and Ni compounds

3. CATALYTIC TRIMERIZATION OF BUTADIENE

If 8 adds to 1 before the exchange reaction (eq. 10) occurs, then the product 9 can react via a nickel-catalyzed reaction sequence involving hydrogen transfer to give 5 and 3.

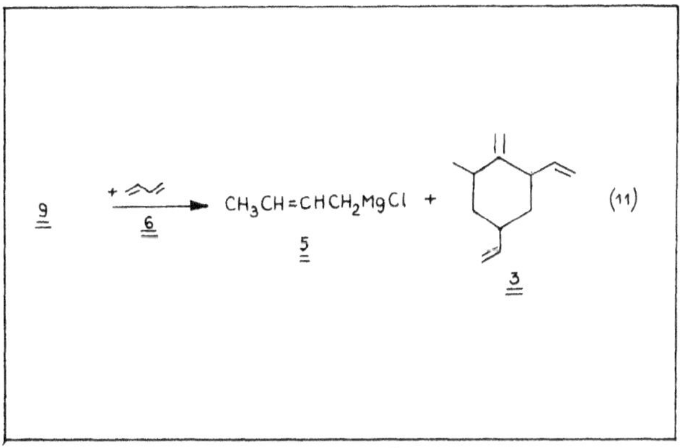

On the basis of model reactions, we assume that in the first reaction step an alkyl ⇄ crotyl exchange occurs between Mg amd Ni:

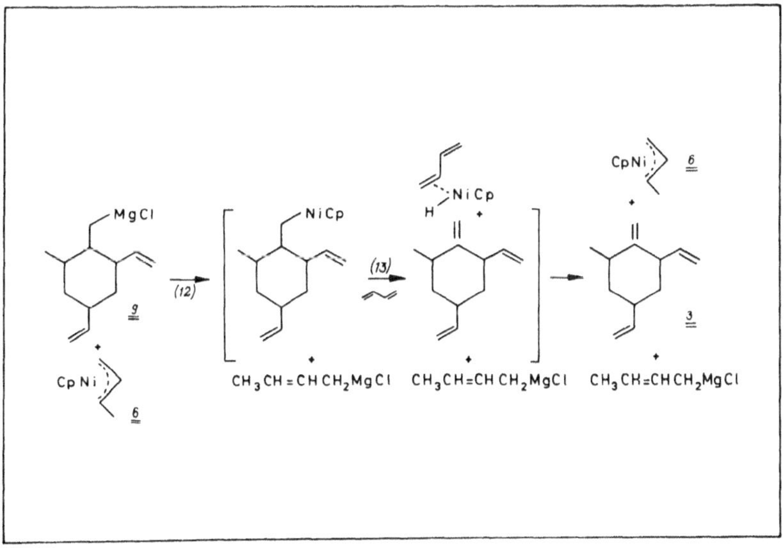

The resulting CpNi-alkyl complex is unstable and reacts via hydrogen transfer with butadiene forming CpNi-η^3-crotyl (6) and eliminating the alkene 3.

The last-mentioned reactions (12) and (13) were observed in the reaction between isopropyl- or isobutylmagnesium halides and nickelocene (4) and 1.3-dienes, e.g. 2.3-dimethylbutadiene.

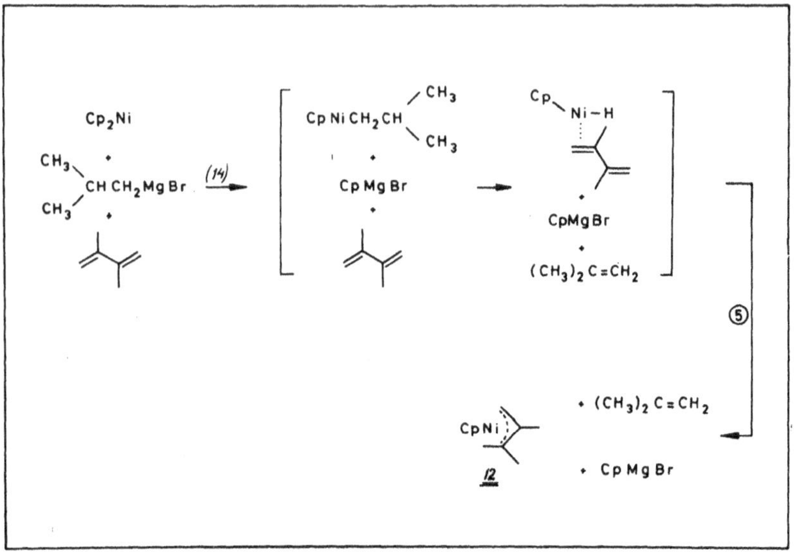

η^5-Cyclopentadienyl-η^3-1.1.3-trimethyl-allylnickel (12) was obtained in ~ 41% yield.

The reaction sequence (4), (5), (12) and (13) forms a catalytic cycle for the trimerization of butadiene to 3.

4. CATALYTIC DIMERIZATION OF ISOPRENE

The reaction of isoprene (13) in the presence of 3-methyl-2-butenyl-magnesiumchloride (14) and cyclopentadienyl-η^3-1.1-dimethyl-allylnickel (15), which can be prepared in an analogous manner to reaction (3) from 14 and 4 in 82% yield, gives a mixture of 3.3.6-trimethyl-(16) and 2.3.6-trimethyl-1.4.6-heptatriene (17) (eq. 15).

14 adds to isoprene (13) in a regioselective reaction (100% metal → C-3 addition)[7]. The skeleton according to 17 cannot be formed by this reaction. We have observed that in a mixture of 14 and cyclopentadienyl-η^3-1.1-dimethylallylnickel 15 (14:15 = 1/1 to 2/1) both 14 and 15 isomerize to 19 and 18 (20°C, THF), s. fig. 3. The isomerization of 14 is more rapid than that of 15 and becomes slower if the ratio 14/15 is altered to higher or lower values. 15 only isomerizes if heated

to 120 °C (t 1/2 ~ 80 h) to cyclopentadienyl-syn-η^3-1.2-dimethylallyl-nickel (18a) (eq. 16):

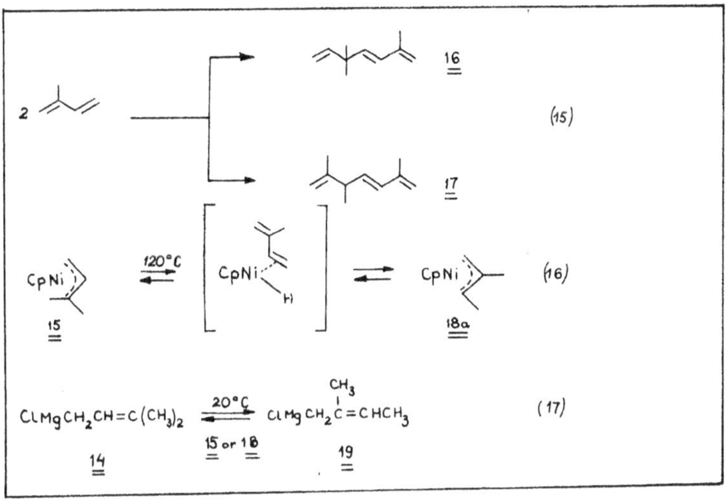

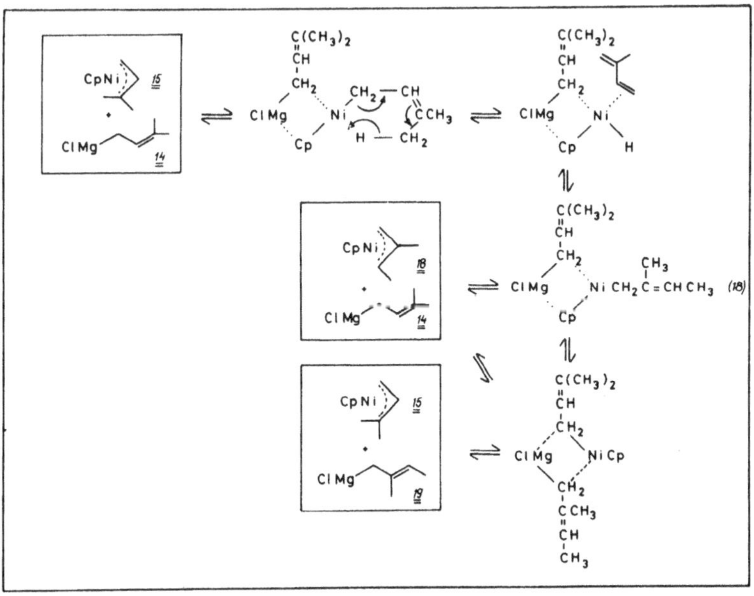

From these results we conclude that a complex between 14 and 15 is involved as an intermediate in these isomerizations. Two alternative mechanisms may be discussed: intermediate formation of a nickel-hydride-isoprene complex, s. eq. (18), or a direct hydrogen transfer within the allyl group from the CH_3 to C-3. Lack of direct interaction makes the latter improbable.

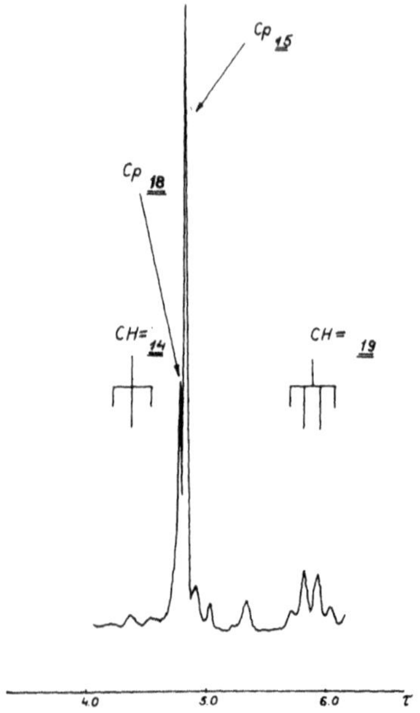

Figure 3:

Isomerization of 14→19 and 15→18 at 20°C after 114 h;
^{1}H-nmr spectrum in D_8-THF

14 / 19 ~ 15 : 85;
15 / 18 ~ 75 : 25.

14 or 19 add to isoprene in the familiar manner mainly by M→C-3 addition:

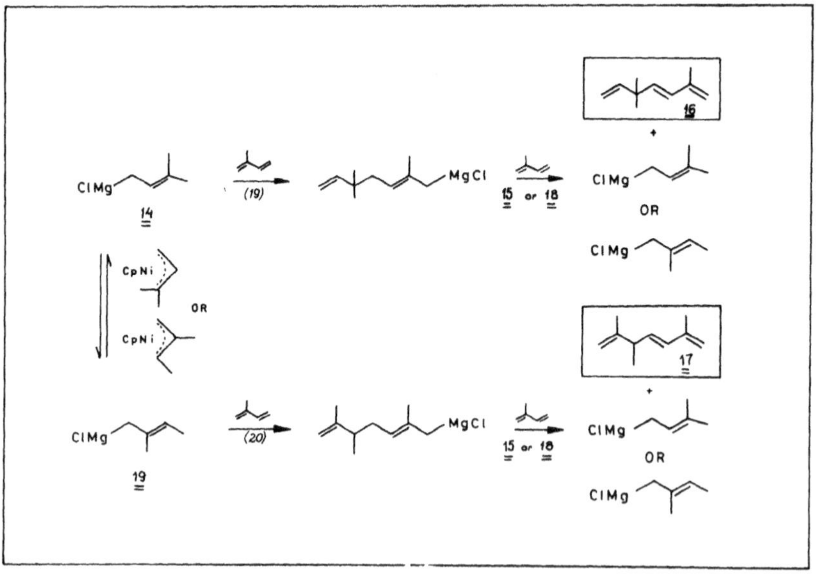

The other reaction steps can be assumed to be analogous to those discussed in the butadiene reaction.

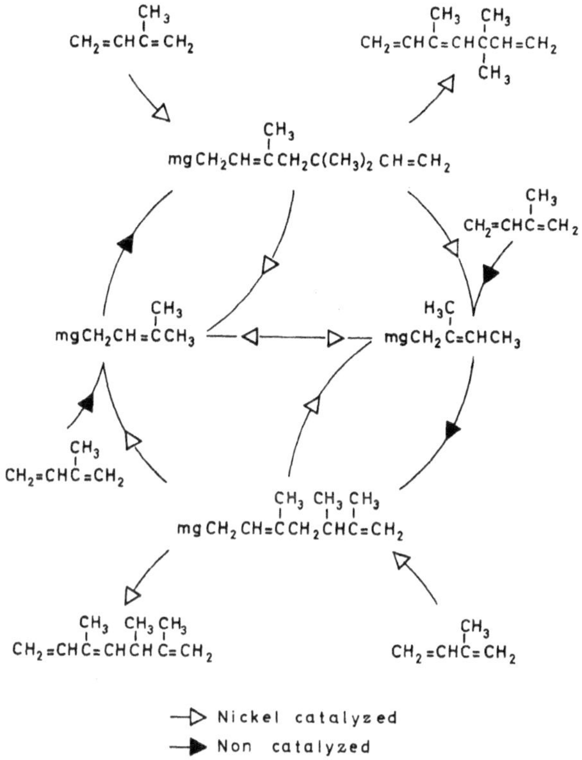

Figure 4:

The catalytic dimerization of isoprene to 16 and 17 involving both allyl-Mg and Ni compounds.

REFERENCES

1) Bogdanović, B., Heimbach, P., Kröner, M., Wilke, G., Hoffmann, E.G. and Brandt, J., Liebigs Ann. Chem. 727, pp. 143 (1969)

2) Jolly, P.W., and Wilke, G., The Organic Chemistry of Nickel, Vol. II, p. 134, Academic Press New York - San Francisco - London 1975

3) Brenner, W., Heimbach, P., Hey, H., Müller, E.W. and Wilke, G., Liebigs Ann. Chem. 727, p. 161 (1969)

4) Lehmkuhl, H., Chimia 31, p. 408 (1977)

5) McClellan, W.R., Hoehn, H.H., Cripps, H.N., Muetterties, E.L., and Howk, B.W., J. Amer. Chem. Soc. 83, p. 1601 (1961) have prepared CpNiη^3-allyl for the first time by this method.

6) Ustynyuk, Yu.A., Voevodskaya, T.I., Zharikova, N.A., and Ustynyuk, N.A., Dokl. Akad. Nauk SSSR 181, p. 372 (1968), Engl. 640.

7) Lehmkuhl, H., Reinehr, D., Mehler, K., Schomburg, G., Kötter, H., Henneberg, D., and Schroth, G., Liebigs Ann. Chem. 1978, p. 1449

8) Lehmkuhl, H., Janssen, E., unpublished results

9) Lehmkuhl, H., Doering, I., unpublished results

10) For a survey about activation of Grignard compounds by transition metal compounds see Felkin, H., and Swierczewski, G., Tetrahedron 31, p. 2735 (1975).

TRIMETHYLPHOSPHINE COBALTATES CONTAINING MONOOLEFIN AND DINITROGEN LIGANDS - MODEL COMPOUNDS FOR HOMOGENEOUS CATALYSIS

H.-F. Klein, R. Hammer, J. Wenninger, and J. Gross
Anorganisch-chemisches Institut der TU München

Properties and reactions of olefincobalt(0) complexes $Co(C{=}C)L_3$ (L = PMe_3; C=C = C_2H_4, C_3H_6, cyclopentene) are described. Reduction by potassium affords salts of the cobaltate anion $Co(C{=}C)L_3^-$ which upon protonation yield tbp complexes <u>cis</u>-$CoH(C{=}C)L_3$. Formal insertion of olefin into the Co-H bond is demonstrated by H/D exchange and by catalytic isomerization of olefins. Complexes <u>cis</u>-$CoCH_3(C{=}C)L_3$ are thermally stable and do not give an analogous insertion reaction. In olefin cobaltates dinitrogen replaces the olefin to give dinitrogen cobaltates L_3CoN_2M (M = K, 1/2 $Mg(THF)_4$). Electrophiles attack the ambidentate anion of these compounds, $L_3\overset{-}{Co}{=}\overset{+}{N}{=}\overset{-}{N}$, either at the cobaltate center (oxidative addition type) or at the terminal nitrogen (formation of diazenides $L_3Co{-}N{=}N{-}R$, R = $SiMe_3$, $SnMe_3$).

INTRODUCTION

The key process of a whole family of catalytic reactions is the formal insertion of a substrate molecule into a metal to ligand σ bond[1]. A metal center M, in solution protected by a number of ligands or solvent molecules,

B. Pullman (Ed.): Catalysis in Chemistry and Biochemistry. Theory and Experiment. 285–292.

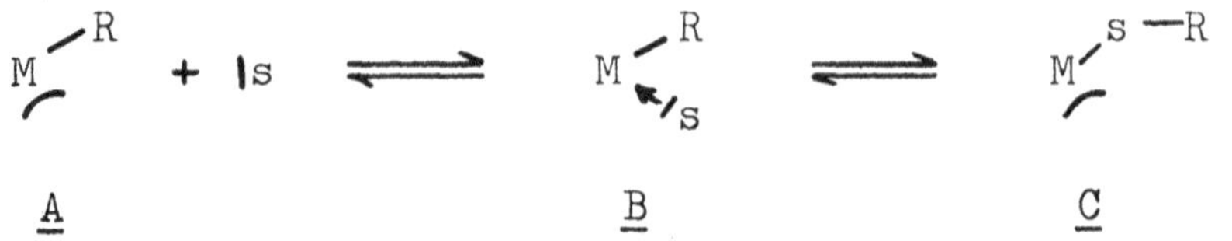

(1)

$$NiMeXL_2 + CO \rightleftharpoons NiMeX(CO)L_2 \rightleftharpoons Ni(COMe)XL_2 \qquad (2)$$

$L = PMe_3$

bears a functional group R. In the interesting cases of homogeneous catalysis R is a hydrogen or an alkyl rest. Substrate molecules (s = CO, olefin, etc.) reversibly occupy a vacant coordination site (and react with R whereby the vacant site is regenerated. We found the trimethylphosphine nickel and cobalt complex matrices useful for building model compounds because they can accommodate R and s in adjacent positions and simulate both steps of eq. (1). In the reversible carbonylation (2) we isolated thermally stable methylnickel[2] and acetylnickel compounds[3], type A and C, while the concentration of B in these equilibria is below the detection limit. For s = monoolefin, R = H, alkyl the trimethylphosphine cobalt matrix stabilizes model compounds of type B with A and C beyond detection (3). In this paper we describe some properties and reaction of olefincobalt complexes in the formal oxidation states Co^{I}, Co^{0}, and Co^{-I}.

Results and Discussion

Olefin Complexes

Coordination of monoolefins by a trimethylphosphinecobalt complex requires a high electron density on the metal. While Co^{I} halides[4] reversibly add olefins only at low

SCHEME I

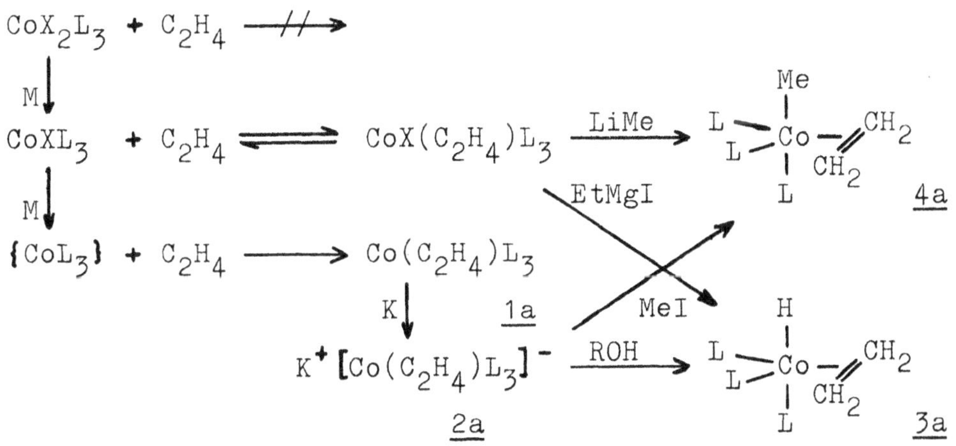

L = PMe_3; M = 1/2Mg; X = Cl,Br,I
olefin = propene: 1-4b, cyclopentene: 1-4c

temperatures the corresponding 16 electron species $CoRL_3$ (R = H, Me) decompose in the absence of an additional ligand but form stable olefin complexes 3a-c and 4a-c.
An appropriate reduction of $CoCl_2L_3$ yields paramagnetic 1a-c. These air sensitive compounds are highly soluble in hydrocarbons and remain monomers: μ_{eff}/μ_B = 1.85(9)[5]. At 20°C pentane solutions of 1a-c are green, only that of 1a turns blue below -10°C. From toluene glasses in liquid nitrogen well resolved e.s.r. spectra have been obtained displaying three sets of eight lines (I_{Co} = 7/2). Typical values are g_1 = 2.20, g_2 = 2.13, g_3 = 1.98; A_{Co1}, A_{Co2}, A_{Co3} = 30 - 60 Gauss.
Further reduction proceeds on a clean surface of potassium metal in ether solvents to give dark brown solutions of 2a-c. At -78°C orange precipitates are obtained which analyse as $K[Co(olefin)L_3]\cdot xOR_2$ (x = 0.5 - 1). 2a-c react with methanol and methyl iodide to give 3a-c and 4a-c, respectively. CD_3OD produces $CoH(C_2H_3D)L_3$ (ν_{CD} = 2100,

ν_{CoH} = 1850 cm^{-1}). 3a-c isomerize pentene-1 in bulk to give a mixture of all three isomers.[6]

These results can be explained by a σ-π process given in scheme II. ^{1}H and ^{31}P n.m.r. spectra of 3a[7] and 4a[8] at low temperatures (slow pseudorotation) place the H and CH_3 ligands in an axial position with the freely rotating olefin occupying an equatorial site. Thus a mobile *cis*-arrangement of R and olefin is provided as a necessary prerequisite of an insertion process. In particular this will include a special orientation which has been theoretically postulated[9]. The result shown in scheme II is that olefin inserts into the CoH bond but not into the CoC bond.

SCHEME II

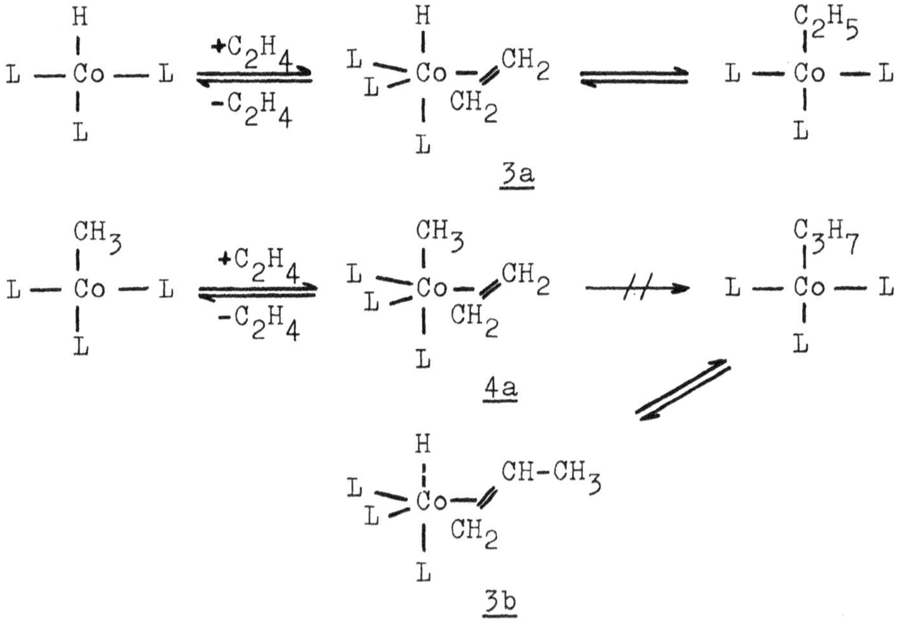

Thermolysis of 4a (80°C, toluene, 1 bar ethylene, 15 h) does not give 3b which is expected to arise from a beta-elimination of an intermediate $Co(C_3H_7)L_3$. In fact no CoH species is formed prior to the total disintegration of 4a.

Conclusion: For metal hydride olefin complexes evidence is accumulating[10-12] that this type of insertion reaction is quite general. However, there are virtually no reports supporting simple insertion reactions of olefins into transition metal to sp^3-carbon bonds which are not assisted by a Lewis acid cocatalyst. The conclusion drawn here is that the migration of an alkyl anion ligand onto a cis-coordinated olefin, or the corresponding insertion reaction, does not have an easy pathway. Therefore the mechanisms usually given for the oligomerization and polymerization of olefins have to be reconsidered. An alternative reaction scheme which has been recently proposed by M.L.H. GREEN et al.[13] has a better chance of being right.

Dinitrogen Complexes

In cobaltates 2b,c the olefin ligand is rapidly replaced by dinitrogen under ambient conditions[7] according to eq. (4), the reaction is not so fast with 2a but still complete. 6 is obtained in 70-80% yield as an orange solid which is stable up to 140°C. Dinitrogen also causes the reduction of trimethylphosphinecobalt halides by magnesium in tetrahydrofuran to proceed to the cobaltate stage which is usually not accessible with this metal. From the 4/1 ratio of $Me_3P/CoCl_2$ necessary for a 80% yield synthesis of 5, eq.(3), it is apparent that CoL_4[14] is an intermediate in this reaction. The subsequent reductive dis-

$$2\ CoCl_2 + 8\ L + 3\ Mg \xrightarrow{THF,\ N_2} L_3CoN_2\text{-}Mg(THF)_4\text{-}N_2CoL_3 \quad (3)$$

$$\underline{5}$$

$$\underline{2a\text{-}c} \xrightarrow{+\ N_2} L_3CoN_2K \quad (4)$$

$L = PMe_3$ $\underline{6}$

placement of the small phosphine ligand by dinitrogen is unusual because Me_3P easily substitutes the N_2 ligand in 6. This can be explained by assuming both an activation of dinitrogen on the magnesium surface and the magnesium chloride binding the excess phosphine.

The crystal and molecular structures of 5[15] and hexameric 6[7] reveal a novel association of end-on dinitrogen and an electropositive main group element. The stabilization of 5 by four THF donor molecules completing an octahedral coordination of the magnesium atom is essential as no other O or N donors could be introduced. Similarly a special role of ether and alkoxide donors has been demonstrated in the complex structure of a side-on NiN_2 compound[16]. However, in the structure of 6 the potassium ions are satisfied by N_2 donor centers only making up a double cube arrangement which reminds us of metal carbide structures. This fairly strong association is maintained even in ether solvents from which 6 is crystallized.[7]

The dinitrogen cobaltate anions in 5 and 6 are very electron rich and because of their bond lengths (d_{CoN} = 171 pm, d_{NN} = 118 pm) may be regarded as diazenide ions $C\bar{o}=\overset{+}{N}=N^-$. Accordingly reactions of electrophiles with 5 or 6 proceed either at the cobaltate center, i.e. as oxidative additions with complete loss of nitrogen as in (5)[7], or at the terminal nitrogen to give diazenides according to eq.(6).

$$L_3\overset{-}{Co}=\overset{+}{N}=\overset{-}{N} \xrightarrow[-\,Br^-]{+\,C_2H_5Br} \underline{3a} + N_2 \qquad (5)$$

$$L_3\overset{-}{Co}=\overset{+}{N}=\overset{-}{N} \xrightarrow[-X^-]{RX} L_3CoN=NR \xrightarrow{dec.} CoL_4 + Co_xN_yR_z \qquad (6)$$

R =	H,	$SiMe_3$,	$SnMe_3$
dec.:	-70	0	+45°C

The fixation of dinitrogen, e.g. according to eq.(4) and (6), proceeds in an extremely basic medium. There is little hope for a closed cycle reduction of dinitrogen to ammonia in this system because any NH species generated is rapidly deprotonated to give cobalt amides, imides or hydrazides, $Co_xN_yH_z$, in a complex oxidation reduction sequence which has not been elucidated. However, by suitable modification of the reaction conditions electrophilic attack can be turned regioselective, i.e. to occur at the terminal nitrogen according to eq.(6), giving diazenide intermediates which can be made sufficiently stable (R = $SnMe_3$) for a detailed study. We believe these cobalt diazenide derivatives to be the first step towards a catalytic synthesis of organonitrogen compounds. Further investigations along these lines are in progress.

Financial support of this work by the Fonds der chemischen Industrie and by the Deutsche Forschungsgemeinschaft is gratefully acknowledged.

(1) As described in most recent textbooks, e.g. Henrici-Olivé, G., and Olivé, S.: 1977, "Coordination and Catalysis", Verlag Chemie, Weinheim, New York, pp. 122-138.

(2) Klein, H.-F., and Karsch, H.H.: 1976, Chem. Ber. 109, pp. 2515-2523, and references cited therein.

(3) Klein, H.-F., and Karsch, H.H.: 1976, Chem. Ber. 109, pp. 2524-2532; Huttner, G., Orama, O., and Bejenke, V.: 1976, Chem. Ber. 109, pp. 2533-2536.

(4) Klein, H.-F., and Karsch, H.H.: 1975, Inorg. Chem. 14, pp. 473-477.

(5) Evans' n.m.r. method as described by Fritz, H.P., and Schwarzhans, K.-E.: 1964, J. Organomet. Chem. 1, pp. 208-211.

(6) Hammer, R.: 1977, Thesis Technische Universität München

(7) Klein, H.-F., Wenninger, J., and Schubert, U.: 1978, Z. Naturforsch. 33 b, pp. 1267-1273.

(8) ^{1}H n.m.r. (60 MHz, D_8-toluene, -60°C, TMS ext. τ= 10): τ $CoCH_3$ 11.0 d·t; $^3J(P_{eq}H)$ 12.4 Hz, $^3J(P_{ax}H)$ 1.3 Hz. $^{31}P\{^1H\}$-PFT n.m.r. (36.4 MHz, D_8-toluene, -90°C, H_3PO_4 ext. δ = 0 ppm): δP_{eq} 32.o ppm d(br), δP_{ax} -9.9 ppm t(br); J(PP) $\approx$40 Hz.

(9) Thorn, D.L., and Hoffmann, R.: 1978, J. Amer. Chem. Soc. 100, pp. 2079-2090.

(10) Byrne, J.W., Blaser, H.U., and Osborn, J.A.: 1975, J. Amer. Chem. Soc. 97, pp. 3871-3873.

(11) Green, M.L.H., and Mahtab, R.: 1979, J.C.S. Dalton 1979, pp. 262-265.

(12) Werner, H., and Feser, R.: 1979, Angew. Chem. 91, pp. 171-172.

(13) Green, M.L.H., and Mahtab, R.: 1978, J.C.S. Chem. Comm. 1978, pp. 604-606.

(14) Klein, H.-F.: 1971, Angew. Chem. Internat. Ed. Engl. 10, p. 343.

(15) Hammer, R., Klein, H.-F., Schubert, U., Frank, A., and Huttner, G.: 1976, Angew. Chem. Internat. Ed. Engl. 15, pp. 612-613.

(16) Jonas, K., Brauer, D.J., Krüger, C., Roberts, P.J., and Tsay, Y.-H.: 1976, J. Amer. Chem. Soc. 98, pp. 74-

CONFORMATIONAL TRANSITIONS OF THE MEMBRANE-BOUND CHOLINERGIC RECEPTOR

Jonathan B. Cohen and Norman D. Boyd
Department of Pharmacology, Harvard Medical School, Boston, MA 02115 USA

Abstract

Rapid-mixing and ultrafiltration techniques have been used to measure directly on the second time scale the kinetics and equilibria characterizing the binding of $[^3H]$-acetylcholine (AcCh) to membrane-bound nicotinic cholinergic receptors isolated from Torpedo electric tissue. The observed concentration dependence of the kinetics of ligand association, as well as the kinetics of ligand dissociation and of receptor reisomerization after removal of cholinergic agonist are analyzed in terms of a model consisting of a single binding site that exists in two conformations, each binding AcCh with different affinities. Rate and equilibrium constants are presented for the interaction of AcCh with those receptor conformations and for the conformational equilibria in the presence and absence of AcCh. The relationship between the receptor conformations defined by ligand binding and the process of channel activation and receptor desensitization are discussed.

The binding of acetylcholine (AcCh) or other cholinergic agonists by nicotinic cholinergic receptors results in increased permeability of the plasma membrane to cations. The cholinergic permeability response is not, however, the result of the equilibrium interaction of AcCh with its receptor. Upon release of AcCh by nerve stimulation, cholinergic receptors are exposed to millimolar concentrations of AcCh only for milliseconds, and the time course of the postsynaptic response is limited by the rapid loss of AcCh from the synaptic cleft due to hydrolysis and diffusion. When, on the other hand, a constant concentration of AcCh is maintained for seconds or longer, the initial permeability response is not maintained, but decreases progressively to a final equilibrium value. This phenomenon, known as pharmacological desensitization, was interpreted (1,2) to indicate the existence of two interconverting receptor conformations, one binding AcCh weakly and coupled to channel activation, and the second binding AcCh tightly but not coupled to channel activation.

B. Pullman (Ed.): Catalysis in Chemistry and Biochemistry. Theory and Experiment. 293–304.

Definition of the conformations of the cholinergic receptor associated with channel activation and with desensitization depends upon the analysis of both the kinetics of binding of AcCh and the kinetics of the permeability response. Studies of this sort can be carried out *in vitro* because it is possible to isolate from *Torpedo* electric tissue a membrane fraction highly enriched in nicotinic receptors (3). These highly specialized membranes, in which the nicotinic receptor constitutes about 50% of the protein, remain functional in the sense that cholinergic agonists increase their permeability to alkali cations (4-7) while the prolonged exposure to agonists results in desensitization (5,8). These membranes have been utilized in numerous studies of the mechanism of cholinergic permeability control (for recent reviews, see 9,10).

Analysis of the interaction of cholinergic agonists with the *Torpedo* receptor-rich membranes revealed an apparent paradox. Agonists such as AcCh or carbamylcholine (Carb) were bound with extremely high affinity at equilibrium. For AcCh, the equilibrium dissociation constant, K_{eq} = 10nM and for Carb, K_{eq} = 100 nM (reviewed in 10). However, the agonist concentrations necessary to stimulate the permeability response of those vesicles were orders of magnitude higher. The concentration of AcCh necessary for a half maximal permeability (flux) response was 2,000 nM, and that of Carb was 30,000 nM (4-6). One possible explanation for these results is that the membrane-bound *Torpedo* receptor exists in several interconvertible conformations. The agonist interacts initially with a receptor conformation characterized by a low affinity for the agonist and coupled to the permeability response, while the equilibrium binding achieved only after prolonged exposure to AcCh reflects the existence of an agonist-stabilized high affinity receptor conformation which is not associated with increased membrane permeability. Studies of the kinetics of binding of radioactive α-neurotoxins in the presence of cholinergic ligands (11-14) as well as spectroscopic studies utilizing extrinsic (15-17) and intrinsic (14) fluorescence probes have provided evidence for the existence of such conformational transitions. However, the data obtained in these indirect studies have been interpreted in terms of differing models for the interaction of agonists with the membrane-bound receptor.

In order to characterize directly the interaction of AcCh with its receptor, we have developed procedures to measure the kinetics of interaction of $[^3H]$AcCh with the *Torpedo* membranes, and in this report we summarize these studies. *Torpedo* receptor-rich membranes were isolated according to the procedure of Sobel et al. (18). All binding studies were carried out in *Torpedo* physiological saline (250 mM NaCl, 5 mM KCl, 3 mM $CaCl_2$, 2 mM $MgCl_2$, 5 mM NaP_i, pH 7) in the presence of 0.1 mM diisopropylphosphofluoridate to inhibit any remaining acetylcholinesterase. The ligand binding kinetics were determined by a binding assay combining rapid-mixing and ultrafiltration techniques. Reaction rates and amplitudes at times greater than 5 seconds after mixing were determined as described previously (7). To study binding kinetics occurring more rapidly, an automated rapid-mixing ultrafiltration apparatus was constructed that is similar in principle to the quenched flow apparatus described by Fersht and Jakes (19). Mixing was achieved by the use of a pneumatic ram and an eight jet tangential mixer (20). The reaction mixture was collected in an incubation tube and then filtered at the desired time by positive pressure. The temporal resolution of the apparatus is limited by the time necessary to filter the sample volume (about 0.1 sec).

1. RATE OF ASSOCIATION OF [^{3}H]-ACCH

The equilibrium binding of [^{3}H]-AcCh to *Torpedo* receptor-rich membranes determined by ultrafiltration was well-defined by a single class of binding sites characterized by K_{eq} = 14 ± 2 nM (*T. marmorata*) or K_{eq} = 8 ± 2 nM (*T. californica*). Preliminary experiments established that a reaction mixture 25nM in AcCh binding sites and 100 nM

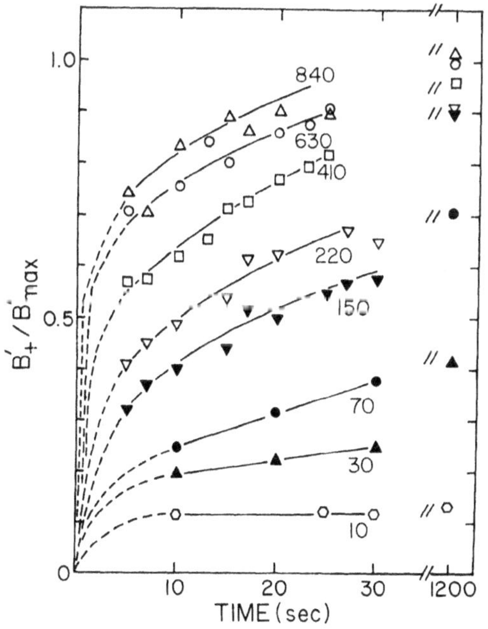

Figure 1. Kinetics of binding of [^{3}H]-AcCh to membrane-bound cholinergic receptor of *Torpedo marmorata*: dependence of the slow component on AcCh concentration. The receptor concentration in the reaction mixture was 80 nM, and that of [^{3}H]-AcCh (nM) as indicated.

in [^{3}H]-AcCh attained equilibrium only minutes after mixing (10). In a first series of experiments the slow phase of the association kinetics was determined for [^{3}H]-AcCh concentrations varying between 5 nM and 800 nM (Fig. 1). It is striking that at low ligand concentrations all the binding occurred within the first 10 sec, while at intermediate concentrations (70 nM to 200 nM) the amplitude of the slow phase of the binding was larger than that of the rapid phase. At still higher ligand concentrations the amplitude of the rapid phase was further increased while that of the slow phase now decreased. At the highest concentrations studied as at the lowest, most of the binding occurred rapidly.

For the slow phase of the association kinetics plots of the logarithm of $B'_\infty - B'_t$ against time were found to be linear, indicatin a first order process and permitting the determination of the rate constant (k_s) and the amplitude (B'_s) of the slow phase from the slope and intercept respectively. In addition, the amount of binding occurring rapidly (B'_f) is equal to $B'_\infty - B'_s$. The dependence of B'_f and of k_s on the concentration of AcCh are presented in Fig. 2. The

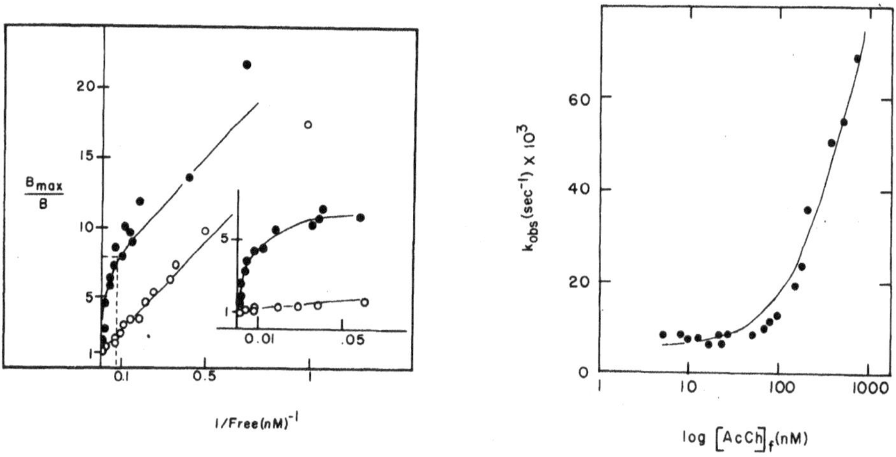

Figure 2. Concentration dependence of the amplitude of the rapid phase of AcCh binding (B'_f, O) in comparison with that of the equilibrium AcCh binding (B'_{eq}, O) (left), and of k_s, the observed rate constant of the slow binding (right). For each concentration of AcCh, k_s and B'_f were determined from a plot of logarithm of $B'_\infty - B'_t$ against time where B'_∞ and B'_t were obtained from experiments similar to those described in Fig. 1. The theoretical curves were calculated according to the reaction cycle and parameters of Table 1.

concentration dependence of B'_f was characterized formally by two classes of binding sites: 20% of the sites bound AcCh with K_{Hi} = 2 nM, and 80% of the sites bound AcCh with K_{Lo} = 1000 nM. This is in sharp contrast to the binding at equilibrium which was characterized by a single class of sites binding [^{3}H]-AcCh with K_{eq} = 14 nM. The rate constant k_s was essentially independent of ligand concentration as AcCh varied between 5 nM and 50 nM and then increased an order of magnitude as the ligand concentration increased to 500 nM. Furthermore, it was found that for a given concentration of AcCh, k_s did not change as receptor concentration was varied between 5 nM and 100 nM.

2. DISSOCIATION AND RECOVERY RATES

The rate of dissociation of AcCh-receptor complexes was measured as the rate of exchange of [^{3}H]-AcCh for non-radioactive AcCh. Membranes preequilibrated with [^{3}H]-AcCh were mixed with a solution containing excess non-radioactive AcCh. At 23°C, k_{dis} the rate of dissociation of [^{3}H]-AcCh was equal to 0.15 sec^{-1}, and at 4°C it was a factor of 3 slower (Fig. 3). Reversibility of the ligand associated conformational isomerization was determined in the following manner. A concentrated membrane suspension was equilibrated with carbamylcholine at a concentration sufficient to occupy at least 80% of the sites, and the solution was then diluted several hundred fold so that at equilibrium only 10% of the sites would be occupied by carbamylcholine. If the agonist-induced conformational transition was reversible, the unliganded receptor must reisomerize to that conformation equilibrium

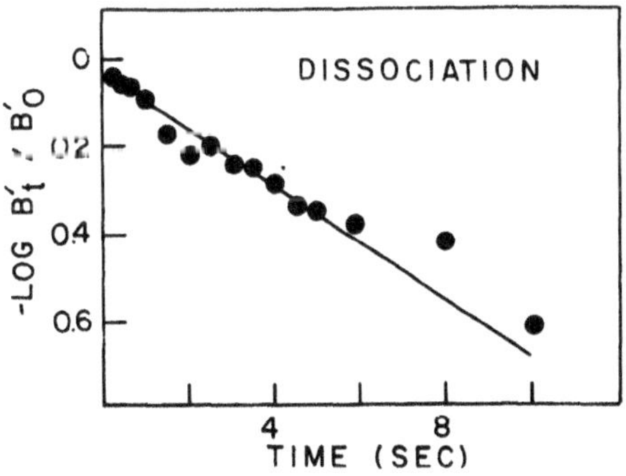

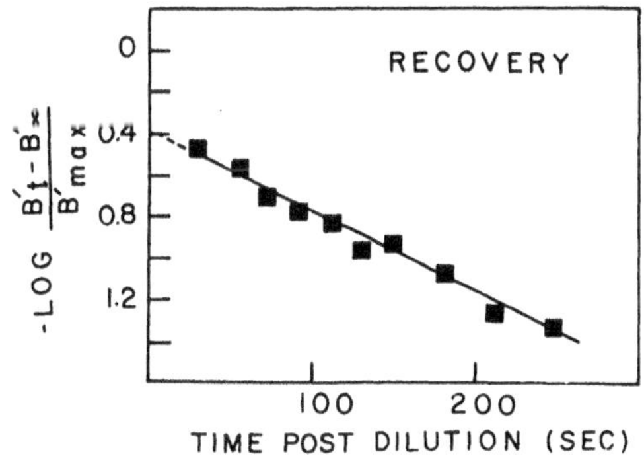

Figure 3. (Left) Kinetics of Dissociation of [^{3}H]-AcCh from membrane-bound Torpedo receptor at equilibrium (23°C).

Figure 4. (Right) Reversibility of agonist-induced conformational perturbation after removal of agonist by dilution (23°C).

appropriate for the unliganded receptor. The rate of recovery was measured by titrating with [^{3}H]-AcCh the number of high affinity receptor sites remaining at any time after dilution. At 23°C the time course of recovery was characterized by a rate, k_{rec} = .010 sec^{-1}, while at 4°C k_{rec} was equal to .003 sec^{-1} (Fig. 4). At each temperature k_{rec} was an order of magnitude slower than k_{dis} but was equal to k_s, the rate of receptor isomerization to the high affinity conformation when incubated with the lowest concentrations of [^{3}H]-AcCh.

3. PLAUSIBLE REACTION MECHANISM

The observed association kinetics establish the existence of a ligand induced conformational isomerization. These experiments alone can be accounted for qualitatively by the existence of two independent receptor pools: one binding AcCh rapidly with high affinity; a second binding AcCh weakly initially and then isomerizing to a high affinity receptor conformation. However, such a simple model is ruled out by the fact that the rate of dissociation of [^{3}H]-AcCh-receptor complexes was an order of magnitude greater than the rate of reisomerization of the receptor to its low affinity conformation upon dilution and dissociation of the agonist from the agonist stabilized high affinity conformation.

The simplest model that has the potential to account for the observed association, dissociation, and recovery kinetics is a reversible cyclic model, similar to that originally proposed by Katz and Thesleff (1). In Table 1 we define the parameters characterizing a cycle where a single binding site exists in two interconverting conformations defined by different ligand affinities. The relaxation times characteristic of such a cycle have been reviewed recently with reference to the kinetics of binding of a hapten with immunoglobulin A (20). On the basis of our data, it was appropriate to assume that the ligand binding steps were in rapid equilibrium relative to the receptor isomerization rates. With this simplification the cycle was specified by 5 independent parameters: three equilibrium constants and a rate for each conformational isomerization. In fact, since the overall equilibrium constant characterizing the equilibrium binding of AcCh was known independently, only two equilibrium constants in the reaction cycle remained independent. The reaction cycle was solved numerically by the use of a modified Euler's method for the appropriate concentrations of AcCh and receptor. For most of our experimental conditions, neither AcCh nor receptor was in excess, and no simplification of the reaction kinetics based upon buffered concentrations was appropriate.

The parameters listed in Table 1 were used to generate the solid curves in Fig. 2. It can be seen that the model does reproduce the data well. Several comments are appropriate. The concentration

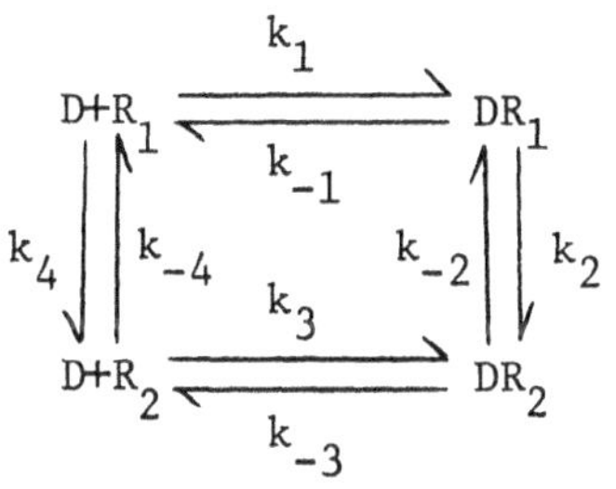

$$K_1 = \frac{D \cdot R_1}{DR_1}; \quad K_3 = \frac{D \cdot R_2}{DR_2}$$

$$K_2 = DR_1/DR_2; \quad K_4 = R_1/R_2$$

$$K_1 \cdot K_2 = K_3 \cdot K_4$$

$$K_{eq} = \frac{D(R_1+R_2)}{DR_1 + DR_2} = \frac{K_1(1+K_4)}{1+K_2}$$

	$K_{eq} = 15$ nM		
$K_1 = 1000$ nM	$K_2 = 0.013$	$K_3 = 2.4$ nM	$K_4 = 5.6$
	$k_{+2} = 0.18\ s^{-1}$	$k_{+3} = 7\times10^7 M^{-1} s^{-1}$[a]	$k_{+4} = 0.0005\ s^{-1}$
	$k_{-2} = 0.0024\ s^{-1}$	$k_{-3} = 0.15\ s^{-1}$[a]	$k_{-4} = 0.0028\ s^{-1}$

[a]Parameters determined for binding of $[^3H]$-AcCh to Torpedo californica membranes, 23°C.

Table 1. Binding of $[^3H]$-AcCh by receptor-rich membranes of Torpedo marmorata, 4°C.

dependence of B'_f is essentially determined by the equilibrium constants K_1, K_3 and K_4 and not by underlying rate constants. This insensitivity to the rate constants is expected since the ligand binding steps are in rapid equilibrium relative to the conformational transitions. The concentration dependence of k_s is sensitive not only to the rate constants $k_{\pm4}$ and $k_{\pm2}$, but also to the equilibrium constant K_1. Comparison of the rate constants $k_{\pm2}$ and $k_{\pm4}$ reveals a selective effect of AcCh on the rate constant of the transition from the low to high affinity receptor conformation. In the absence of AcCh k_{+4} is equal to 0.0005 sec^{-1} while k_{+2}, the rate constant for the transition of the liganded receptor, is over three hundred fold greater. On the other hand, the rate constants k_{-2} and k_{-4} differ by less than a factor of two. While the simple model does reproduce the data to a first approximation, systematic deviations do exist in the fit of the rate data. The observed rate constant k_s increases more rapidly with increasing AcCh concentration than can be accounted for by the simple model. Because of such a deviation, we feel it is not useful to emphasize at this time the uncertainty of the parameters in Table 1 based upon a least square goodness of fit. Rather, we will extend the studies to higher concentrations of AcCh to permit further refinement of the model.

4. BIMOLECULAR ASSOCIATION RATES

The results of the previous sections establish that 20% of the AcCh binding sites were present in a receptor conformation (R_2) binding AcCh with high affinity (K_3 = 2nM). Furthermore, that high affinity binding was in rapid equilibrium when compared with the rates of the receptor conformational transitions. In order to provide further evidence in support of the model and to further define the rates and equilibria, we measured the rate of binding of [^{3}H]-AcCh to that receptor conformation. At low concentrations of [^{3}H]-AcCh, the initial rapid binding is to the pre-existing high affinity receptor conformation and not to the low affinity conformation. An automated rapid-mixing ultrafiltration rig was used to measure the kinetics of binding of [^{3}H]-AcCh with a resolution of 0.2 sec. In Fig. 5 is shown a typical experiment where the rate of binding of [^{3}H]-AcCh was measured for a reaction mixture 50nM in total receptor sites and 4nM or 8nM in [^{3}H]-AcCh. The determination of the exact reaction times is rather complicated and will be discussed separately (N.D. Boyd & J.B. Cohen, manuscript in preparation), but it is possible to determine from such data the rate of binding of [^{3}H]-AcCh under conditions of either excess receptor or excess AcCh. In Fig. 6 is shown the dependence of that rate constant on the concentration of [^{3}H]-AcCh. For a

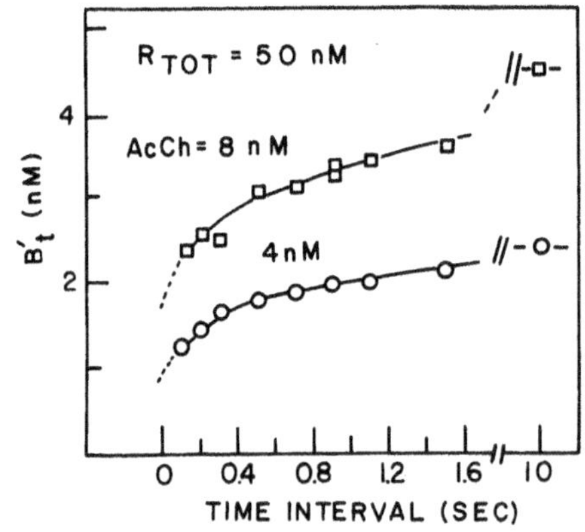

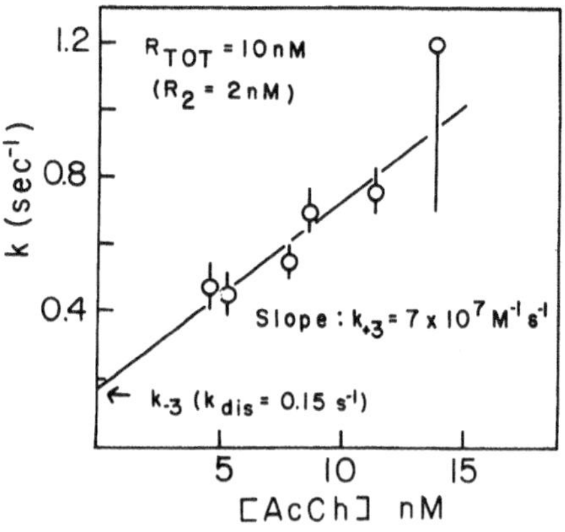

Figure 5. (Left) Kinetics of binding of [^{3}H]-AcCh to membrane-bound cholinergic receptor of *Torpedo californica* (23°C). In the presence of low concentrations of AcCh, the rapid binding measured on the sub-second time scale measures occupancy of preexisting high affinity receptor (R_2 = 10 nM in this experiment).

Figure 6. (Right) Concentration dependence of k the rate constant characterizing the binding of AcCh to the high affinity receptor conformation R_2. Reaction rates were determined from experiments similar to those shown in Fig. 5.

bimolecular association process in the presence of excess AcCh, $k = k_+ [AcCh] + k_-$ (see, for example 22). The ordinate intercept in Fig. 6 defines the unimolecular dissociation rate (k_{-3} in terms of the nomenclature of Table 1) and the slope defines the bimolecular association rate (k_{+3}). In Figure 6 we have also indicated k_{dis}, the independently determined rate of dissociation of [^{3}H]-AcCh-receptor complexes at equilibrium (Fig. 3). The value of k_{-3} determined from the association kinetics is the same as k_{dis} and equal to 0.15 sec^{-1} at 23°C; $k_{+3} = 7 \times 10^7\ M^{-1}\ sec^{-1}$. The equilibrium constant calculated from these rates, k_{-3}/k_3 = 2.1nM, is in good agreement with K_3 determined from the rates and amplitudes of the slow phase of the ligand binding (Fig. 2 and Table 1). The rate constants characterizing the binding of AcCh by the high affinity receptor conformation are included in Table 1.

In addition to the high affinity receptor conformation, R_2, the concentration dependence of B'_f indicated the existence of a receptor conformation (R_1) binding AcCh with a dissociation constant K_1 = 1000nM. The kinetics of binding of AcCh associated with that low affinity binding can also be measured by the ultrafiltration technique because it turns out that they are not particularly rapid. For example, in the presence of 100 nM AcCh and 25 nM receptor at 23°C, 50% of the sites were occupied during the rapid phase of the binding. While the binding to R_2 occurred more rapidly than could be measured, the binding associated with R_1 was characterized by a half time of 3 sec. We will present elsewhere an analysis of this intermediate phase of the ligand association kinetics. It is interesting to point out, however, that if the binding is due to a simple bimolecular association as suggested by the reaction scheme in Table 1, the bimolecular rate constant is about $2 \times 10^6\ M^{-1}\ sec^{-1}$, a factor of thirty slower than k_{+3}. An alternate interpretation that must be considered is that K_1 is an apparent constant characterizing a bimolecular association coupled to a conformational transition and the observed binding kinetics reflect the underlying rates and equilibria rather than a bimolecular association rate.

5. CONCLUSION

In this report we have demonstrated that it is possible to measure quantitatively on the second and subsecond time scale the interaction of the neurotransmitter AcCh with its membrane-bound cholinergic receptor. Two kinetically distinct components of the AcCh association kinetics have been studied. First, analyses of the concentration dependence of the reaction rates and amplitude of the slowest phase of the association kinetics provided clear evidence that in the absence of AcCh the cholinergic receptor exists in two interconverting conformations distinguished from each other by the affinity with which each binds AcCh. Secondly, the rapid binding occurring at low AcCh concentrations has been shown to be due to direct ligand binding to a preexisting high affinity receptor conformation. Also, although we have not presented data here, the full AcCh binding kinetics includes an intermediate phase associated with the binding of AcCh to a preexisting low affinity receptor conformation.

The concentration dependence of the rates and amplitudes of the slow phase of the association can be determined with great precision. As a result the values of K_3, K_4, and $k_{\pm 4}$ are very well defined in terms of the model in Table 1. One strong prediction of this model is that K_4 as well as $k_{\pm 4}$ are characteristic of the receptor and independent of the particular agonist. We confirmed this by also measuring the kinetics of binding of $[^3H]$-Carb and have found for both $[^3H]$-AcCh and $[^3H]$-Carb that $K_4 = 5 \pm 0.5$ and that the sum of k_{+4} and k_{-4} is equal to $0.0028 \pm 0.0004\ sec^{-1}$. Previous studies where α-neurotoxins have been used to monitor the binding of cholinergic ligands have been interpreted in terms of cyclic models where K_4 varies from 2 to 10 (12-14). It appears that the use of the indirect method provides only a qualitative characterization of the reaction cycle. However, the results reported here do agree with parameters deduced from the kinetics of binding of a fluorescent cholinergic ligand (16).

The kinetic parameters characterizing the binding of AcCh to the high affinity conformation (k_{+3} and k_{-3}) were found to be fully consistent with the value of K_3 determined from the analysis of the amplitude of the slow phase of the binding. It is striking that a receptor conformation exists that binds a molecule as simple as AcCh extremely high affinity (K_3 = 2 nM) at a rate that is almost diffusion controlled ($k_{+3} = 7 \times 10^7\ M^{-1}\ S^{-1}$). Comparable information is available for only one other cholinergic ligand, a fluorescent cholinergic agonist, for which the rate constants are almost identical (16).

The available data have been fit quite well by a simple model consisting of one binding site and two conformations (Table 1). In fact, it may appear surprising that such a simple model is adequate, and we have indicated that systematic deviations do exist between calculated and experimental rates (Fig. 2). The deviation can be readily accounted for by the use of a model consisting of two interacting binding sites, a model consistent with the fact that the isolated cholinergic receptor is known to contain two AcCh binding sites per functional unit (for a review, see 9). However, until data are obtained at higher AcCh concentrations it is not appropriate to introduce further parameters to improve the fit.

A second limitation of the simple model concerns the characterization of the transient low affinity receptor conformation. As we have indicated the binding characterized by K_1 is either due to a slow bimolecular association or to a composite process involving ligand binding and conformational transitions. It is noteworthy that the dissociation constant characterizing the transient low affinity binding (K_1 = 1000 nM) is very close to the concentration of AcCh (2000 nM) causing a half maximal flux response from the Torpedo vesicles (4). However, the binding of AcCh that causes channel activation must do so by a change of receptor structure (conformation). Also, the cholinergic permeability response occurs within a fraction of a millisecond in isolated muscle fibers, and even in the Torpedo vesicles the response occurs at least within a fraction of a second (R. Neubig & J. Cohen, unpublished observations). These facts make it likely that the transient low affinity binding that we report here is probably not a simple direct bimolecular association. Further studies will be necessary to relate quantitatively those conformations determined by ligand binding and those characterized by ion transport. The studies presented here, however, establish that it is possible to characterize by direct methods the kinetics of interaction of AcCh with its membrane-bound receptor. Such studies provide a description of the dynamic flexibility of the cholinergic receptor and serve as a basis for interpreting its various functional states.

ACKNOWLEDGEMENTS

This research was supported by U.S. Public Health Service Grants NS-12408, NS-00155 and by Training Grant MH-14275.

REFERENCES

1. Katz, B. and Thesleff, S.: 1957, J. Physiol. 138, pp. 63-80.
2. Rang, H.P. and Ritter, J.M.: 1970, Mol. Pharmacol. 6, pp. 357-382.
3. Cohen, J.B., Weber, M., Huchet, M. and Changeux, J-P.: 1972, FEBS Letters 26, pp. 43-47.
4. Popot, J.L., Sugiyama, H. and Changeux, J-P.: 1976, J. Mol. Biol. 106, pp. 469-483.
5. Bernhardt, J. and Neumann, E.: 1978, Proc. Natl. Acad. Sci. USA 75, pp. 3756-3760.
6. Miller, D.L., Moore, H.P.H., Hartig, P.R. and Raftery, M.A.: 1978, Biochem. Biophys. Res. Commun. 85, pp. 632-640.
7. Neubig, R.R., Krodel, E.K., Boyd, N.D. and Cohen, J.B.: 1979, Proc. Natl. Acad. Sci. USA 76, pp. 690-694.
8. Sugiyama, H., Popot, J.L. and Changeux, J-P.: 1976, J. Mol. Biol. 106, pp. 485-496.
9. Heidmann, T. and Changeux, J-P.: 1978, Ann. Rev. Biochem. 47, pp. 371-411.
10. Cohen, J.B.: 1978, In: Molecular Sepcialization and Symmetry in Membrane Function (A.K. Solomon and M. Karnovsky, Eds.), Harvard University Press, Cambridge, MA, pp. 99-128.

11. Weber, M., David-Pfeuty, T. and Changeux, J-P.: 1975, Proc. Natl. Acad. Sci. 72, pp. 3443-3447.
12. Weiland, G., Georgia, B., Lappi, S., Chignell, C. and Taylor, P.: 1977, J. Biol. Chem. 252, pp. 7648-7656.
13. Quast, U., Schimerlik, M., Lee, T., Witzemann, V., Blanchard, S. and Raftery, M.A.: 1978, Biochem. 17, pp. 2405-2414.
14. Barrantes, F.J.: 1978, J. Mol. Biol. 124, pp. 1-26.
15. Grunhagen, H., Iwatsubo, M. and Changeux, J-P.: 1977, Eur. J. Biochem. 80, pp. 225-242.
16. Heidmann, T., Iwatsubo, M., and Changeux, J-P.: 1977, C.R. Acad. Sci. Paris 284, pp. 771-774.
17. Quast, U., Schimerlik, M. and Raftery, M.A.: 1978, Biochem. Biophys. Res. Commun. 81, pp. 955-964.
18. Sobel, A., Weber, M. and Changeux, J-P.: 1977, Eur. J. Biochem. 80, pp. 215-224.
19. Fersht, A.R. and Jakes, R.: 1975, Biochemistry 14, pp. 3350-3356.
20. Gibson, Q.H. and Milnes, L.: 1964, Biochem. J. 91, pp. 161-171.
21. Lancet, D. and Pecht, I.: 1976, Proc. Natl. Acad. Sci. USA 73, pp. 3549-3553.
22. Bernasconi, C.F.: 1976, Relaxation Kinetics, Academic Press, New York.

THE CATALYTIC EFFECT OF HYDROGEN BONDING UPON EPOXIDE RING-OPENING

Peter Politzer and Vernon M. Estes

Chemistry Department
University of New Orleans
New Orleans, LA, 70122, U.S.A.

I. INTRODUCTION

A particular type of catalysis that has been attracting increasing attention in recent years is that due to a properly situated hydroxyl group which facilitates the opening of an epoxide ring by means of intra-molecular hydrogen bonding. This has been proposed, for several molecules of biochemical interest, as a means of explaining their high degrees of reactivity.

It is of course well known that epoxide ring-opening reactions are accelerated by the presence of acids [1]; the mechanism is believed to involve protonation of the epoxide oxygen, with a consequent weakening of the C-O bonds and a simultaneously increased positive character on the part of the ring carbons, making them more attractive to a nucleophile [1-4]. Delocalization of charge in the transition state is presumably also a factor in this mechanism. It does not seem unreasonable that hydrogen bonding to the epoxide oxygen might have an effect similar to that of protonation, but to a lesser extent.

The purpose of the present investigation was to examine this proposed hydrogen-bonding catalysis quantitatively, and to learn more about its properties and effects. First, however, some of the evidence for its presence, in several different systems, shall be reviewed briefly.

B. Pullman (Ed.): Catalysis in Chemistry and Biochemistry. Theory and Experiment. 305–321.

II. PRESUMED EXAMPLES OF CATALYSIS BY INTRA-MOLECULAR HYDROGEN BONDING

A. Triptolide, Tripdiolide:

Triptolide (Ia) and tripdiolide (Ib) are naturally-occurring compounds that have been shown to possess significant anti-leukemic activity [5,6]. Their NMR spectra indicate strong hydrogen bonding between the hydroxyl group on C(14) and the epoxide

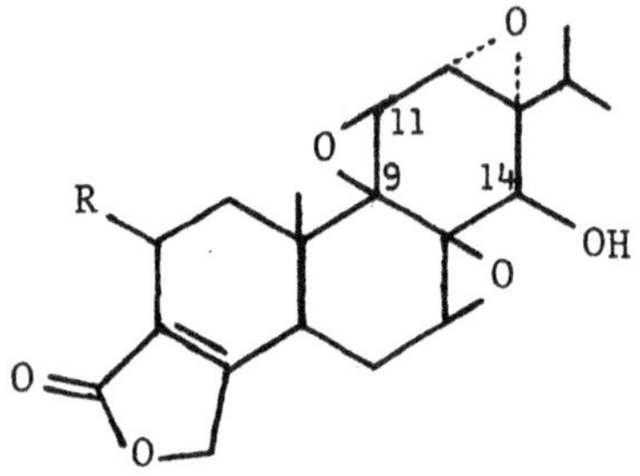

Ia: R = H

Ib: R = OH

at positions 9 and 11. It has been proposed that this hydrogen bonding catalyzes the opening of the C(9)-C(11) epoxide ring, which is believed to be a critical step in the biological mode of action of these systems [6]. Consistent with this proposal is the observation that no anti-leukemic activity is shown by the α-isomer of Ia, in which the 14-OH is trans to the C(9)-C(11) epoxide ring, nor by the related compound in which the 14-OH has been replaced by a ketone group. It has been hypothesized that the anti-tumor mechanism involves a ring-opening interaction at C(9) with nucleophilic thiol groups on growth-regulating enzymes [6]. Indeed, the reaction of Ia with propanethiol was found by NMR to disrupt the hydrogen bonding between the 14-OH and the C(9)-C(11) epoxide. This reaction proceeded more than 20 times more rapidly for Ia than for its α-isomer under the same conditions.

B. Hydroxy-Epoxycholestanes:

Intra-molecular hydrogen bonding is believed to catalyze the opening of the epoxide rings in two isomeric hydroxy-epoxy-cholestanes, II and III [7,8]. The reactions of these compounds with the nucleophilic azide ion, N_3^-, led to the opening of the epoxide rings and formation of the respective azides at C(6) in II and C(4) in III. These reactions proceeded significantly more

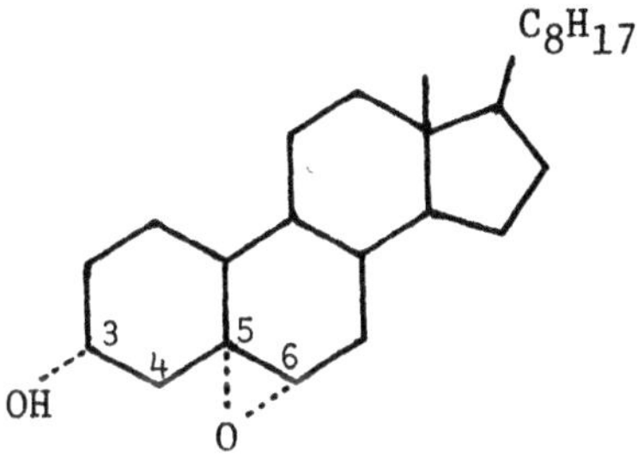

II: 3α-hydroxy-5α,6α-epoxy-cholestane

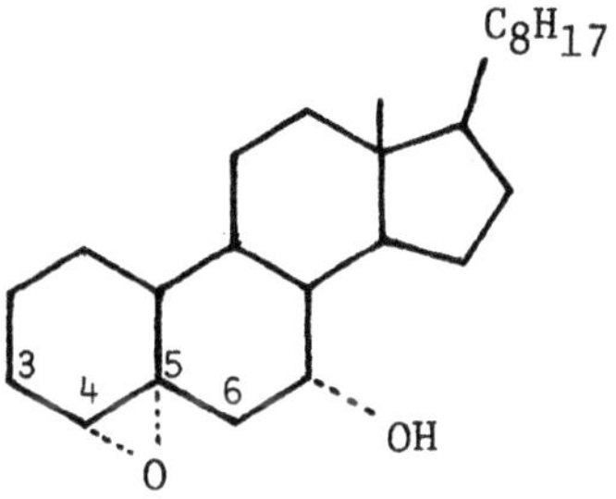

III: 4α,5α-epoxy-7α-hydroxy-cholestane

rapidly than the corresponding ones using modified forms of II and III in which the hydroxyl group was either not present or was oriented in such a manner as to preclude hydrogen bonding. It was proposed that the reactivities of II and III are increased by the presence of weak hydroxyl-epoxide hydrogen bonds which permit a delocalization (to the hydroxyl group) of the negative charge which would otherwise build up on the epoxide oxygen in the transition state [7,8]. It was also conjectured that a second effect of the hydrogen bonding might be to make the epoxide carbon atoms more positive and thus more reactive toward a nucleophile [8].

C. Diol Epoxides of Benzo[a]pyrene:

Considerable evidence now indicates that one or both of two metabolically-produced isomeric diol epoxides, IV and V, are carcinogenically-active forms of benzo[a]pyrene [9-12]. Both IV and V interact with nucleophilic sites on nucleic acids, the epoxide ring opening and a covalent bond forming to carbon 10 [9,11-13]. They are known to react with both DNA and RNA [12,14]. While both IV and V exhibit a high degree of

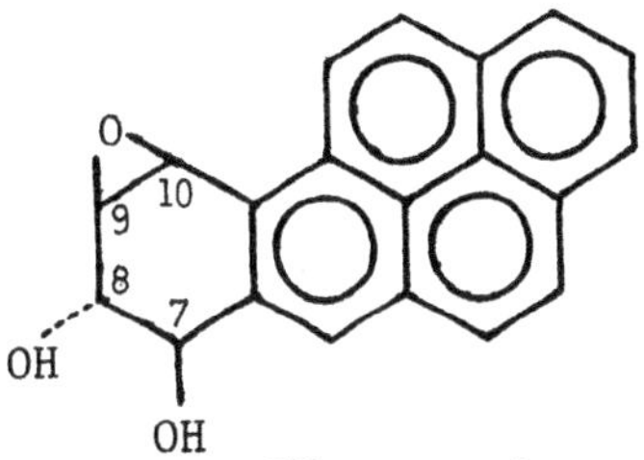

IV: syn isomer

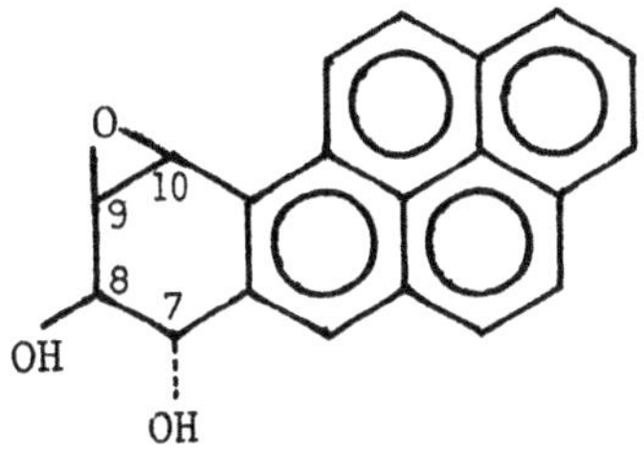

V: anti isomer

activity toward nucleophiles, this is especially true of the *syn* isomer, IV [15-17]. For example, the reaction of IV with sodium 4-nitrothiophenolate has a rate constant 163 times as great as that of V -- and about 500 times as great as that of the corresponding epoxide, VI, that lacks the two hydroxyl groups [15,16]. These and other such observations have been interpreted as reflecting a catalytic effect due to hydrogen bonding between the 7-OH and the epoxide oxygen in the *syn* form, IV [18-20]. Such intra-molecular hydrogen bonding is considered unlikely in the case of the *anti* isomer, V.

VI

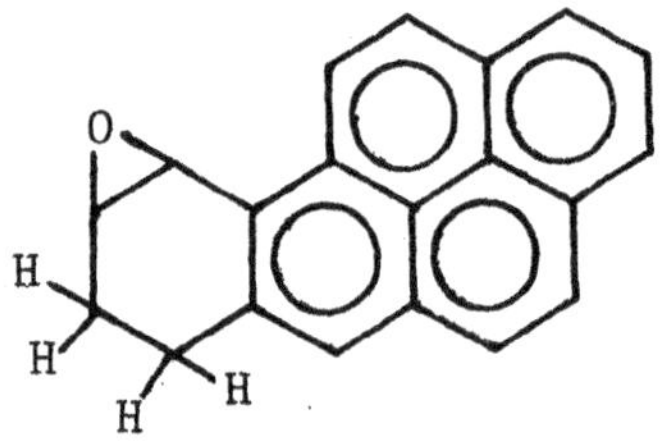

D. Picrotoxinin:

Picrotoxinin, VII, an analeptic drug, contains an epoxide ring at positions 12 and 13 which is known to be very resistant to nucleophilic attack, a fact that has been ascribed to shielding of the epoxide ring by the lactone groups [21]. Dalzell *et al* have suggested, however, that a very reactive epoxide can be

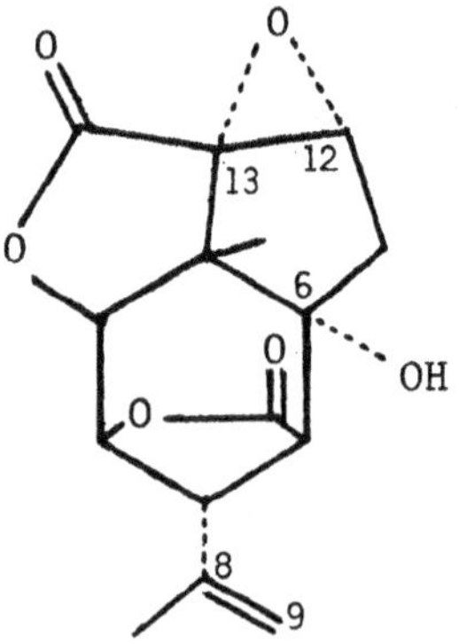

VII

formed at the C(8)-C(9) double bond, its reactivity being greatly enhanced by hydrogen bonding with the hydroxyl group on C(6) [21]. Although this epoxide was apparently not isolated, its existence was inferred from subsequent reaction products. Interaction with such nucleophiles as diethylamine, which resulted in amination at C(9), was found to proceed much more readily with the hydroxyl group at position 6 than when it was replaced by an acetate group, which eliminated the possibility of hydrogen bonding. It should be pointed out, however, that the hydrogen bonding to which Dalzell *et al* ascribe such a marked catalytic effect would involve the somewhat unlikely formation of a seven-membered ring, in competition with the formation of a presumably more stable six-membered ring involving the C(12)-C(13) epoxide.

III. PREVIOUS WORK

We have earlier made a preliminary study of hydrogen-bonding catalysis, taking two diol epoxides of benzene, VIII and IX, as model systems [22]. These are of course closely related to one of the examples discussed in section II, since they correspond to the terminal, non-aromatic rings in the *syn* and *anti* 7,8-diol-9,10-epoxides of benzo[a]pyrene, structures IV and V.

Using an *ab initio* SCF molecular-orbital procedure (Gaussian 70, with an STO-5G basis set [23]), we computed optimized structures for VIII and IX, to seek possible indications of the existence of hydrogen bonding. The final structure obtained for the *syn* isomer, VIII, fully supported the idea of a hydrogen bond between the 4-OH hydrogen and the epoxide oxygen [22]. The hydrogen atom was found to preferentially point toward that oxygen; when the hydrogen was made to rotate away by 90°, the energy of the system increased by 1.7 kcal/mole. The computed O···O and O···H distances were also quite consistent with the presence of a weak intra-molecular hydrogen bond in VIII.

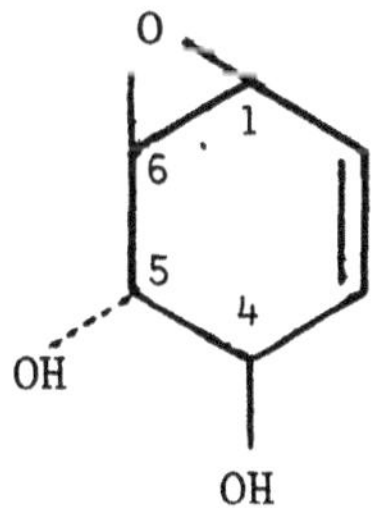

VIII: *syn* isomer

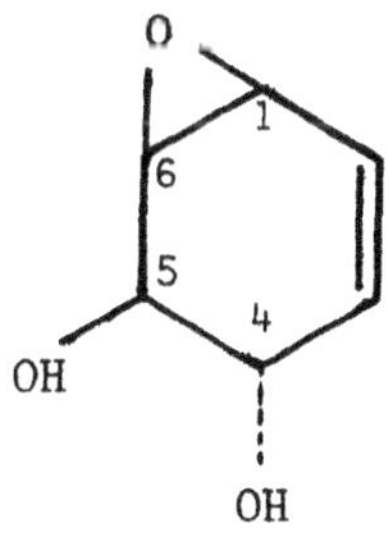

IX: *anti* isomer

On the other hand, there were no indications whatsoever of hydrogen bonding in the _anti_ isomer, IX [22]. The 5-OH hydrogen, which would have to be involved, preferred to point _away_ from the epoxide oxygen and to be at a relatively large distance from it, and resisted all efforts to impose orientations more consistent with hydrogen bonding. (Similar conclusions regarding hydrogen bonding in the _syn_ and _anti_ isomers have been reached, on the basis of computations, by Lavery and Pullman for VIII and IX [24] and by Yeh _et al_ for IV and V [25].)

The effect of the hydrogen bonding upon the strengths of the C-O bonds in VIII was assessed by calculating the energy required to stretch each one individually by a given amount (0.050 A), holding all other bond lengths constant. This was done for both the optimum _syn_ structure and also for the less stable one in which the hydrogen bonding had been disrupted by moving the hydrogen away from the epoxide oxygen. The results showed very clearly that the presence of the hydrogen bond does indeed weaken both C-O bonds, and they further indicated that this effect is more pronounced for the C(1)-O bond than for the C(6)-O [22]. This was confirmed by computing the respective force constants. Thus, this study fully supported the concept of hydrogen bonding as a catalytic agent for epoxide ring opening, and even predicted that the ring would open preferentially at the position corresponding to the actual site of nucleophilic attack in the diol epoxides of benzo[a]pyrene, C(10) in structures IV and V (see section II-C).

There is another point of interest that should be mentioned. The opening of a C-O bond in the oxides of benzene and naphthalene and in the diol epoxides of benzo[a]pyrene, in the absence of acid catalysts, has sometimes been described as going through a zwitterionic structure, X, in which negative charge has accumulated on the oxygen and positive charge on the carbon [19,26-29].

X

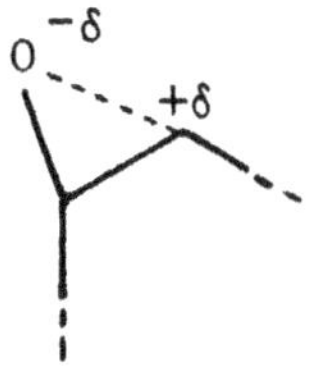

Our results for the _syn_ diol epoxide of benzene, VIII, did not support this picture. When a C-O bond was stretched, the calculated charges (Mulliken population analysis [30]) on the carbon and oxygen changed very little from their values in the equilibrium structure (+0.02 and -0.24, respectively). What slight changes did occur were such as to decrease the charge separation rather than to increase it [22]. Similar observations have been made by Kaufman *et al* in calculations on various benzo[a]pyrene diol epoxides [31].

IV. PRESENT RESULTS

The preliminary study of the benzene diol epoxides that was described in section III did clearly show a catalytic effect of hydrogen bonding upon epoxide ring opening, through the weakening of the C-O bonds. It also indicated that these bonds opened without buildups of charge on the atoms involved. In order to test these initial conclusions, we have now examined the same points for the case of ethylene oxide with a hydrogen-bonded water molecule, $C_2H_4O \cdot H_2O$. This is a smaller system than the benzene diol epoxides, which makes possible a more thorough study. Of course the hydrogen bonding here is inter-molecular, whereas it was intra-molecular in all of the examples considered in section II; a possible consequence of this difference shall be discussed later.

Since our purpose is to investigate the effect of hydrogen bonding, it is first necessary to study ring opening in the case of ethylene oxide alone, with nothing else present, so as to have a point of reference. Furthermore, in view of the well-established catalytic effectiveness of protic acids in epoxide ring-opening, it seemed useful to include in this investigation a second reference point, the protonated epoxide. It is anticipated that the system of particular interest, $C_2H_4O \cdot H_2O$, will fall somewhere between the two extremes represented by C_2H_4O alone and $C_2H_4OH^+$. For each of these three systems, we shall assess the strengths of the C-O bonds, and also the charge separation, if any, that accompanies the opening of these bonds. This should reveal how the effect of hydrogen bonding on these two properties compares with that of outright protonation.

The results for ethylene oxide and for its protonated form shall be presented and discussed first. All of the studies being reported here were carried out computationally, using the *ab initio* SCF GAUSSIAN 70 program [23] with an STO-5G basis set.

A. Ethylene Oxide:

We have already earlier computed a structure for ethylene oxide [3], which is in very good agreement with the experimental one. Some key features of this structure, and also the calculated atomic charges, are given in Table I. It is the starting point for the present work.

Table I. Properties of Calculated Ethylene Oxide Structure [3].

Property	Calculated Value	Experimental Value [32]
C-O distance	1.43 A	1.435 A
C-C distance	1.49 A	1.470 A
C-O-C angle	63°	
H-C-H angle	115°	116.3°
Angle between H-C-H plane and C-C axis	25°	21.9°
Atomic charges:		
C	-0.05	
O	-0.21	
H	+0.08	

We have used two methods of measuring the strengths of the C-O bonds. First, we computed the energy required to stretch one of them by a fixed amount, 0.05 A, while holding all other bond lengths constant. Second, we calculated an "effective ring-distortion force constant" [22]. Holding the C-C and one C-O distances constant, the total energy was computed for two non-equilibrium values of the other C-O bond length, greater and smaller by 0.05 A than the optimum value, R_e. These two energies, together with that for the equilibrium structure, were used to obtain a parabolic E vs. R_{C-O} relationship. A force constant was then determined by evaluating $(\partial^2E/\partial R^2_{C-O})_{R_e}$. (It should be noted that the process by which this force constant is obtained involves a distortion of all of the bond angles of the epoxide ring, in addition to a stretching of one C-O bond, so that it is not simply a stretching force constant.)

An important feature of the present work is that at each non-equilibrium value of the C-O distance, some re-optimization of the structure was carried out. In this first case, ethylene oxide alone, this re-optimization consisted of varying the O-C-H angles, thereby avoiding an artificial degree of strain at each carbon atom and giving a more realistic result. No such re-optimization had been performed in the benzene diol epoxide calculations.

The energy required to stretch a C-O bond by 0.05 A was found to be 1.2 kcal/mole; the computed force constant was 7.7 mdynes/A. The significance of these values is that they provide a reference point for those to be computed for the other two systems being considered here, $C_2H_4OH^+$ and $C_2H_4O \cdot H_2O$. Of more immediate interest, however, is the variation of atomic charges as one C-O bond is stretched. This is shown in Table II. There is clearly no tendency whatsoever for increased charge separation on the carbon and oxygen that are involved; instead, their charges actually approach each other to a small extent. This exactly parallels the situation found in the case of the *syn* diol epoxide of benzene, as discussed in section III.

Table II. Variation of Calculated Atomic Charges in Ethylene Oxide as One C-O Bond is Stretched.

Bond Lengths (A)		Atomic Charges				
C_1-O	C_2-O	C_1	C_2	O	H(C_1)	H(C_2)
1.431	1.431	-0.047	-0.047	-0.212	+0.076	+0.076
(equilibrium)						
1.431	1.480	-0.043	-0.058	-0.205	+0.075	+0.078
1.431	1.681	-0.031	-0.088	-0.192	+0.073	+0.083
1.431	2.180	-0.035	-0.104	-0.167	+0.060	+0.093

B. Protonated Ethylene Oxide:

An optimized structure for protonated ethylene oxide has also been computed previously [3], and is summarized in Table III. The C-O bonds are significantly longer than in the unprotonated molecule, suggesting that they have been weakened by the interaction with the proton. A second effect of this interaction is the more positive character that is now shown by the carbons. In both ways, therefore, the proton is facilitating a ring-opening interaction with a nucleophile.

As before, the stretching of one of the C-O bonds was accompanied by a partial re-optimization of the structure. This took the form of varying the position of the proton and of the methylene hydrogens. The energy required to stretch the C-O bond by 0.05 A is 0.6 kcal/mole, and the ring-distortion force constant is 4.7 mdynes/A. Both are lower than for ethylene oxide

Table III. Properties of Calculated Protonated Ethylene Oxide Structure [3].

Property	Calculated Value
C-O distance	1.48 A
C-C distance	1.49 A
$O-H^+$ distance	1.00 A
Out-of-plane angle of $O-H^+$ *	58°
H-C-H angle	117°
Angle between H-C-H plane and C-C axis	17°
Atomic charges:	
C	+0.04
O	-0.18
H^+	+0.38
H (<u>cis</u> to H^+)	+0.17
H (<u>trans</u> to H^+)	+0.19

* Angle by which the $O-H^+$ axis comes out of the C-C-O plane.

alone, which indicates that the C-O bonds are weaker in the protonated structure. This demonstrates the catalytic effect of protonation upon epoxide ring-opening.

The variation in atomic charges that occurs as the C-O bond is stretched is now quite different from what it was in the non-protonated case, as can be seen in Table IV. The carbon atom that is involved becomes distinctly more positive as the bond is stretched. The oxygen becomes more negative, but to a lesser extent; it appears that some of its electronic charge is being passed on to the proton. (Indeed, the net sum of the C_2, O and H^+ charges changes by only 0.038 over the whole range of C_2-O distances in Table IV.) Some negative charge also goes to the hydrogens on C_1. In fact, the overall pattern of charge migration revealed in Table IV can be described as a movement of

Table IV. Variation of Calculated Atomic Charges in Protonated Ethylene Oxide as One C-O Bond is Stretched.

Bond Lengths (A)		Atomic Charges					
C_1-O	C_2-O	C_1	C_2	O	$H(C_1)^*$	$H(C_2)^*$	H^+
1.483	1.483 (equilibrium)	+0.044	+0.044	-0.178	+0.170 +0.187	+0.170 +0.187	+0.377
1.483	1.534	+0.040	+0.047	-0.173	+0.169 +0.186	+0.173 +0.190	+0.368
1.483	1.737	+0.028	+0.086	-0.180	+0.165 +0.182	+0.184 +0.202	+0.332
1.483	1.984	-0.004	+0.162	-0.207	+0.155 +0.167	+0.208 +0.219	+0.301
1.483	2.234	-0.032	+0.215	-0.211	+0.148 +0.156	+0.219 +0.227	+0.277

*In the columns $H(C_1)$ and $H(C_2)$, the first of each pair of numbers refers to the hydrogen that is <u>cis</u> to the proton, the second refers to the hydrogen that is <u>trans</u>.

electronic charge from the C_2-methylene group to the remainder of the system. At least to some extent, the situation resembles the one depicted in structure X.

C. Ethylene Oxide with a Hydrogen-Bonded Water Molecule:

As part of the present investigation, we have computed an optimum structure for the system $C_2H_4O \cdot H_2O$, ethylene oxide with a hydrogen-bonded water molecule. Considerable effort was made to determine the best relative orientations and separation of the two molecules. In addition, the positions of the methylene hydrogens were allowed to vary. Our final structure is shown in Figure 1. The water molecule is in the plane perpendicular to the C-C bond at its midpoint; the hydrogen bond is seen to point toward a lone pair on the epoxide oxygen, a tendency that has been noted previously [33]. The O-H and O-O distances, 1.95 A and 2.93 A, respectively, are fully consistent with hydrogen bonding [34]. The energy of the interaction

$$C_2H_4O + H_2O \longrightarrow C_2H_4O \cdot H_2O$$

was found to be $\Delta E = -3.8$ kcal/mole, which suggests that this hydrogen bond may be stronger than that in the *syn* diol epoxide of benzene, VIII, where disrupting the hydrogen bonding increased

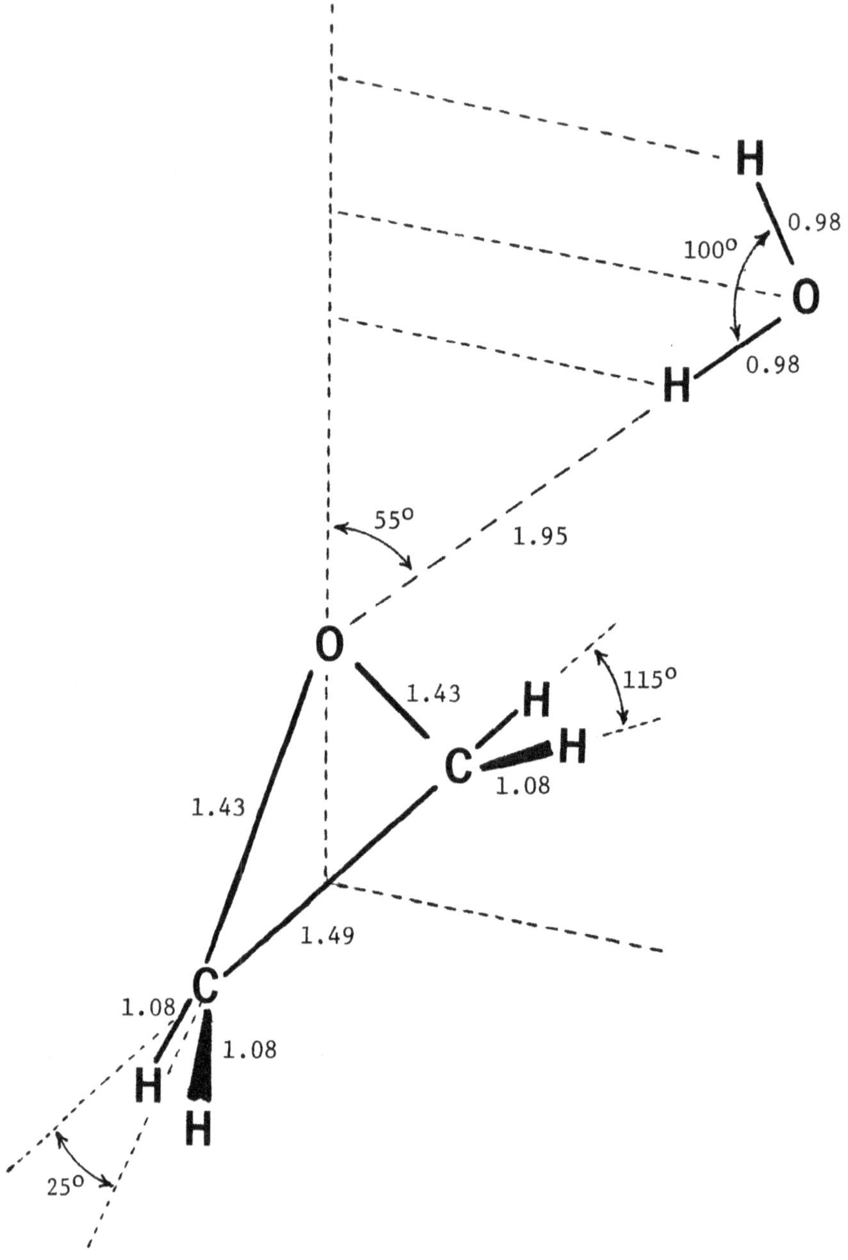

Figure 1. Calculated structure for $C_2H_4O \cdot H_2O$ complex. The water molecule is in the plane perpendicular to the C-C bond at its midpoint. All distances are in Angstroms.

the energy of the system by just 1.7 kcal/mole (section III). In the structure shown in Figure 1, the bond lengths within the C_2H_4O and H_2O portions are the same as in the free molecules. In particular, we found no tendency for the C-O bond lengths to change, or for the bridging hydrogen to move away from its own oxygen and toward that of the epoxide.

As before, stretching a C-O bond was accompanied by re-optimization of the overall structure. This included varying the orientation of the water molecule relative to the distorted epoxide ring, and also the positions of the methylene hydrogens. In addition, the bridging hydrogen was always given the opportunity to move closer to the epoxide oxygen, which it refused to do.

The values obtained for the two quantities being used as measures of bond strength are perhaps somewhat surprising. The energy required to stretch a C-O bond by 0.05 A is 1.5 kcal/mole, while the ring-distortion force constant is 8.7 mdynes/A. These are quite similar to the results that were obtained for ethylene oxide alone; indeed they are even slightly higher. Thus, the presence of a hydrogen-bonded water molecule very clearly does _not_ weaken the epoxide C-O bonds.

As far as the variation of atomic charges with C-O bond stretching is concerned, there really is very little of it. The carbon in question becomes somewhat more negative, but otherwise the changes are extremely small, as can be seen in Table V; the pattern is very much the same as in the cases of ethylene oxide alone and the _syn_ diol epoxide of benzene.

V. DISCUSSION

With regard to the question of whether a separation of charge accompanies the opening of an epoxide C-O bond (see structure X), our results present a consistent picture. When the epoxide oxygen has been protonated, then some degree of charge separation (and delocalization to the proton) does occur. When the epoxide is unprotonated, however, there is no tendency whatsoever for charge separation; instead, the carbon and oxygen charges approach each other to some extent. This picture is _not_ changed significantly by hydrogen bonding to the epoxide oxygen.

One further point of interest is related to the atomic charges in the various equilibrium structures. The charges in the C_2H_4O portion of the hydrogen-bonded complex are really quite similar to the corresponding ones in free ethylene oxide (compare the first lines in Tables II and V). In particular, a hydrogen-bonded water molecule does not produce the marked increase in positive character on the part of the carbons, indeed of the whole methylene groups, that occurs upon protonation (Table IV). Accordingly,

Table V. Variation of Calculated Atomic Charges in Ethylene Oxide with Hydrogen-Bonded Water Molecule as One C-O Bond is Stretched.

Bond Lengths (A)		Atomic Charges†				
C_1-O	C_2-O	C_1	C_2	O	$H(C_1)^*$	$H(C_2)^*$
1.432 (equilibrium)	1.432	-0.043	-0.043	-0.222	+0.084 +0.081	+0.083 +0.081
1.432	1.482	-0.039	-0.054	-0.219	+0.083 +0.080	+0.086 +0.084
1.432	1.682	-0.026	-0.081	-0.216	+0.080 +0.078	+0.095 +0.092
1.432	1.932	-0.021	-0.084	-0.219	+0.076 +0.075	+0.100 +0.097
1.432	2.182	-0.036	-0.106	-0.167	+0.069 +0.066	+0.098 +0.097

†The charges in the water molecule remain essentially unchanged for the whole range of C_2-O distances: Bridging H: +0.199 O: -0.385 Terminal H: +0.165

*In the columns $H(C_1)$ and $H(C_2)$, the first of each pair of numbers refers to the hydrogen that is cis to the water molecule, the second refers to the hydrogen that is trans.

hydrogen bonding should not be expected to induce a polarity in the epoxide that would make it more attractive to nucleophiles, contrary to a suggestion that has been made [20].

A seemingly inconsistent aspect of our results is the fact that the hydrogen bonding in the $C_2H_4O \cdot H_2O$ complex definitely does not weaken the C-O bonds, whereas such an effect was equally clearly found to be present in the case of the syn diol epoxide of benzene (section III). The existence of a hydrogen bond in the $C_2H_4O \cdot H_2O$ system is not in doubt; indeed it appears to be a stronger one than in the diol epoxide. Furthermore, the variation of the atomic charges as the C-O bond is stretched is extremely similar in the two cases.

There is one difference, however, that we speculate may be the key to explaining the apparent inconsistency. The hydrogen bonding in the syn diol epoxide, VIII, is intra-molecular, and the hydrogen bond can be regarded as part of a six-membered ring. In the $C_2H_4O \cdot H_2O$ system, the hydrogen bond is intermolecular, and no ring formation is possible. It may be that the formation of the six-membered ring confers an extra degree of stability to the distorted structure that results when a C-O bond is stretched in the case of the syn diol epoxide, and therefore makes it easier to stretch and break this bond. This extra stability would not be present, however, when a C-O bond in $C_2H_4O \cdot H_2O$ is stretched, and so no bond-weakening effect due to the hydrogen bonding would be observed. (Our results actually indicate that the presence of the water slightly strengthens the C-O bonds in ethylene oxide; this may simply reflect the increase in the size of the system.)

If the above reasoning is valid, then it may be concluded that hydrogen-bonding does not possess an intrinsic capacity for weakening epoxide C-O bonds and catalyzing ring-opening. It is not a question of differing from protonation in degree but not in kind. The hydrogen bonding may indeed have a bond-weakening catalytic effect, but it is because some other favorable factor is also present, such as the possibility of forming a six-membered ring.

ACKNOWLEDGMENT

We gratefully acknowledge the partial support of this project by BRSG Grant RR07169-02 awarded by the Biomedical Research Support Grant Program, Division of Research Resources, National Institutes of Health. We also express our appreciation for the support of the University of New Orleans Computer Research Center.

REFERENCES

[1] R. T. Morrison and R. N. Boyd (1973) Organic Chemistry, 3rd ed., Allyn & Bacon, Inc., Boston, ch. 17.

[2] A. C. Hopkinson, M. H. Lieu, I. G. Csizmadia and K. Yates (1978) Theoret. Chim. Acta 47 97.

[3] P. Politzer, K. C. Daiker, V. M. Estes and M. Baughman (1978) Internat. J. Quantum Chem., Quantum Biol. Symp. No. 5, 291.

[4] J. E. Ferrell, Jr. and G. H. Loew (1979) J. Amer. Chem. Soc. 101 1385.

[5] S. M. Kupchan, W. A. Court, R. G. Dailey, Jr., C. J. Gilmore and R. F. Bryan (1972) J. Amer. Chem. Soc. 94 7194.

[6] S. M. Kupchan and R. M. Schubert (1974) Science 185 791.

[7] D. H. R. Barton and Y. Houminer (1973) J. C. S. Chem. Comm. 839.

[8] Y. Houminer (1975) J. C. S. Perkin Trans. 1, 1663.

[9] I. B. Weinstein, A. M. Jeffrey, K. W. Jennette, S. H. Blobstein, R. G. Harvey, C. Harris, H. Autrup, H. Kasai and K. Nakanishi (1976) Science 193 592.

[10] P. G. Wislocki, A. W. Wood, R. L. Chang, W. Levin, H. Yagi, O. Hernandez, D. M. Jerina and A. H. Conney (1976) Biochem. Biophys. Res. Commun. 68 1006.

[11] S. K. Yang, P. R. Roller and H. V. Gelboin (1978) in: Carcinogenesis, Vol. 3: Polynuclear Aromatic Hydrocarbons, P. W. Jones and R. I. Frendenthal, eds., Raven Press, New York, p. 285.

[12] M. Koreeda, P. D. Moore, P. G. Wislocki, W. Levin, A. H. Conney, H. Yagi and D. M. Jerina (1978) Science 199 778.

[13] A. M. Jeffrey, K. W. Jennette, S. H. Blobstein, I. B. Weinstein, F. A. Beland, R. G. Harvey, H. Kasai, I. Miura and K. Nakanishi (1976) J. Amer. Chem. Soc. 98 5714.

[14] K. Nakanishi, H. Kasai, H. Cho, R. G. Harvey, A. M. Jeffrey, K. W. Jennette and I. B. Weinstein (1977) J. Amer. Chem. Soc. 99 258.

[15] H. Yagi, D. R. Thakker, O. Hernandez, M. Koreeda and D. M. Jerina (1977) J. Amer. Chem. Soc. 99 1604.

[16] D. M. Jerina, H. Yagi, O. Hernandez, P. M. Dansette, A. W. Wood, W. Levin, R. L. Chang, P. G. Wislocki and A. H. Conney (1976) in: Carcinogenesis, Vol. 1, Polynuclear Aromatic Hydrocarbons: Chemistry, Metabolism and Carcinogenesis, R. I. Freudenthal and P. W. Jones, eds., Raven Press, New York, p. 91.

[17] D. L. Whalen, A. M. Ross, H. Yagi, J. M. Karle and D. M. Jerina (1978) J. Amer. Chem. Soc. 100 5218.

[18] H. Yagi, O. Hernandez and D. M. Jerina (1975) J. Amer. Chem. Soc. 97 6881.

[19] P. B. Hulbert (1975) Nature 256 146.

[20] J. W. Keller, C. Heidelberger, F. A. Beland and R. G. Harvey (1976) J. Amer. Chem. Soc. 98 8276.
[21] H. C. Dalzell, R. K. Razdan and R. Sawdaye (1976) J. Org. Chem. 41 1650.
[22] P. Politzer, K. C. Daiker and V. M. Estes (1979) Internat. J. Quantum Chem., Quantum Biol. Symp. No. 6, in press.
[23] Program developed by W. J. Hehre, W. A. Lathan, R. Ditchfield, M. E. Newton and J. A. Pople; available from Quantum Chemistry Program Exchange, Indiana University, Bloomington, IN, 47401, program #236.
[24] R. Lavery and B. Pullman (1979) Internat. J. Quantum Chem., in press.
[25] C. Y. Yeh, P. P. Fu, F. A. Beland and R. G. Harvey (1978) Bioorg. Chem. 7 497.
[26] G. J. Kasperek, T. C. Bruice, H. Yagi and R. M. Jerina (1972) J. C. S. Chem. Comm. 784.
[27] G. J. Kasperek and T. C. Bruice (1972) J. Amer. Chem. Soc. 94 198.
[28] G. J. Kasperek, T. C. Bruice, H. Yagi, N. Kanbisch and D. M. Jerina (1972) J. Amer. Chem. Soc. 94 7876.
[29] D. L. Whalen, J. A. Montemarano, D. R. Thakker, H. Yagi and D. M. Jerina (1977) J. Amer. Chem. Soc. 99 5522.
[30] R. S. Mulliken (1955) J. Chem. Phys. 23 1833.
[31] J. J. Kaufman, H. E. Popkie, S. Palalikit and P. C. Hariharan (1978) Internat. J. Quantum Chem. 14 793; J. J. Kaufman, private communication.
[32] Tables of Interatomic Distances and Configuration in Molecules and Ions (1958, 1965) L. E. Sutton, ed., The Chemical Society, London, Spec. Pub. Nos. 11 and 18.
[33] P. Politzer and K. C. Daiker in: The Force Concept in Chemistry, B. M. Deb, ed., Van Nostrand Reinhold, New York, ch. 7.
[34] L. Pauling (1960) The Nature of the Chemical Bond, 3rd ed., Cornell University Press, Ithaca, N.Y., ch. 12.

A COMPARISON OF DIOXYGEN ACTIVATION BY BIOCHEMICAL AND SYNTHETIC POLYMERIC CHEMICAL SYSTEMS

E. Bayer
Institut für Organische Chemie, Universität Tübingen
Auf der Morgenstelle 18, 7400 Tübingen/Germany

ABSTRACT

Synthetic soluble polymers can be functionalized by various catalytically active centers, in order to obtain homogeneous catalysts. Especially useful are linear, bi- or monofunctional polymers, like polyoxyethylene. The terminal groups of polyoxyethylene can be functionalized by peptide groups. These polyoxyethylene-peptides can be used as model compounds for imitation of catalytic properties of metal proteins. The synthesis of hemo-polyoxyethylene-peptides is described. These hemo-polymers can be reversibly oxygenated in aqueous solution under physiological conditions. The stereochemistry of the peptide and the shielding effect of the polymer contribute both to the properties of the active polymers.

INTRODUCTION

The selectivity and reactivity of catalysis in biochemistry are mainly associated with the protein part of the enzymes. For this reason it is difficult to imitate biochemical catalysis by low molecular weight model compounds, even if the low molecular weight active center of the enzyme is incorporated into the model compound. Both substrate selectivity and reaction rates are dramatically increased by the protein part. The protein contributes to the reactivity of enzymes by several factors

- different ligands (O, N, S) of different amino acid residues are coordinated to metal atoms in metal proteins, thus changing the type of reaction catalysed by the active center.
- the relatively rigid conformation of the protein creates catalytic sites with high reactivity. In the case of metal proteins deficiencies in coordination number or distortions of coordination geometry may occur.
- the active center of the enzyme is shielded so that only stereochemically fitting substrates are bound and activated by weak interactions such as coordination, dipole interaction, van der Waals interaction or hydrogen bonds.

B. Pullman (Ed.): Catalysis in Chemistry and Biochemistry. Theory and Experiment. 323–333.

- depending upon the amino acid residues surrounding the active center, reaction pockets of the required optimum polarity are created.
- the protein part often solubilises apolar, water insoluble active centers in aqueous solution for homogeneous catalysis.

SOLUBLE SYNTHETIC POLYMERS AS MODELS FOR NATURAL POLYMERS

The imitation of the protein part of an enzyme may be approximated by certain shorter peptide sequences which bear the essential features of the enzyme. Since shorter peptides are often insoluble in both water and organic solvents, they must be solubilized to obtain homogeneous synthetic catalysts. It has been shown that peptides can be bound to the two terminal OH-groups of polyoxyethylene (POE) via the C-terminal carboxylic group of the peptide and that such polyoxyethylene-peptides are soluble in water and organic solvents, even if the peptide alone would be completely insoluble (1,2). The peptides canalso be synthesized using POE as a homogeneous soluble carrier step by step. This method, developed by us as an alternative to the solid-phase peptide synthesis of Merrifield, is known as liquid-phase peptide synthesis (3-5 The advantage of using homogeneous soluble polymers instead of using insoluble cross-linked polymers of the Merrifield-type are obvious:
- The reactive groups of POE are kinetically equivalent and display equa reactivity, in contrast to polyfunctional cross-linked insoluble polymers.
- The reaction does not involve diffusion effects because it is homogeneous.
- The structure of soluble polymer bound catalysts can be more readily investigated and better characterized than insoluble polymers.
- The conformation of a peptide bound to POE is not altered by this polymer fixation (1,6).

In POE-peptides, the POE part normally shows random coil conformation whereas the peptide part possesses the same conformation as the corresponding free peptide. However the ether oxygens of the POE may serve as acceptors of hydrogen bonds or ligands in complex formation and so contribute to the active center. This is not only important for polyoxyethylene-peptides which may serve as model compounds for enzymes.

Also other derivatized polyoxyethylenes may serve as catalysts, and these modified polymers show catalytic activities which differ markedly from their low molecular weight analogues by higher catalytic reacticity. This shows that increase of activity by binding of active sites to polymers is not only a privilege of proteins, but may occur also with many synthetic polymers. In this way many new polymeric synthetic homogeneous catalysts have been synthesized for a variety of reactions (7,8).

Metal halogenides can be bound to polyoxyethylene by simple ligand exchange, as is shown in the reaction scheme for the reaction of dichlor dicyclopentadienyl-titanium(IV):

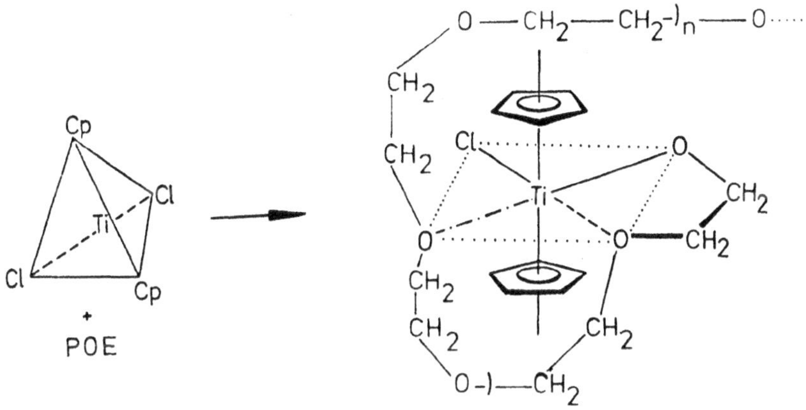

[I]

By the coordination of the ether oxygens of POE to the $TiCP_2Cl_2$, a six coordinate complex [I] is formed. This polymeric titanium complex is an excellent polymerization catalyst, which does not require cocatalysts. In homogeneous solution e.g. poly(acrylo-nitrile), poly-(methylmethacrylate), polystyrene and their copolymers with extremely high molecular weight between $1\text{-}10\cdot10^6$ and relatively narrow molecular weight distribution, and high purity can be synthesized from the monomers with this polymer titanium complex (9). Also for many other reactions like hydroformylation and hydrogenation, new homogeneous polymeric catalysts have been synthesized (7,8,10).

In order to obtain synthetic polymeric model compounds imitating the properties of metal proteins, polyoxyethylene peptides with various peptide sequences have been synthesized (11,12). In general, the specific stereochemical requirements are considered when planning the peptide sequence. The polymer solubilizes the peptide chain, may contribute a shielding effect, and may participate bay way of its end groups in stabilizing the conformation of the peptide.

SYNTETIC HEME POLYMERS

An instructive example of this approach to imitate specificity and reactivity of natural metal proteins is the synthesis of heme polymers capable of reversible oxygenation under physiological conditions.

Myoglobin and hemoglobin both contain the protoporphyrin-iron(II)-complex as active center, which is embedded in a relatively weakly polar protein pocket, and the protein is bound via an imidazole nitrogen to the 5th coordination site of the iron, whereas the 6th coordination site of the iron is free. The low molecular weight active center prophyrin iron complex cannot be reversibly oxygenated. Under exposure of oxygen, the iron is irreversibly oxidized, and simultaneously the dioxygen reduced. Investigations of this irreversible oxidation revealed that an intermediate dimerization occurs (13), which must be prevented if reversible binding of dioxygen is desired. In myoglobin and hemoglobin such a dimerization is prevented by steric reasons because the heme is situated in a protein pocket, such that two hemes cannot come into close contact to form the dimeric intermediates of irreversible oxidation. Only low molecular weight porphyrin iron complexes where one coordination site of the iron is protected by steric hindrance, such as the picket fence porphyrin (14), the capped porphyrins of Baldwin (15) could be reversib oxygenated, however only in organic solvents. It was not possible to achieve reversible oxygenation in aqueous solution under physiological conditions. Another path was followed by immobilization of hemes to cro linked polymers (16-18). In these cases a reversible oxygenation produc as an intermediate of the observed irreversible oxygenation has been postulated. In most cases the kinetics of the oxygenation is extremely slow in comparison to heme proteins, and repetitive cyclic oxygenation is apparently not possible.

We first investigated whether heme, directly bound to soluble polymers without peptide groups, can be reversibly oxygenated. Different bi and polyfunctional polymers have been used, like POE and poly-(vinylpyrrolidinone) (PVP). Fig. 1 shows the derivatization of POE, which is first derivatised to be bi-glycylester [II] , to which the carboxylic groups of the heme are coupled to [III] . In the case of PVP, approximately every 10th pyrrolidinon ring is hydrolyzed to [IV] , and ethylene diamine coupled to the carboxylic groups to give compound [V] . Finally the heme is coupled to form [VI] .

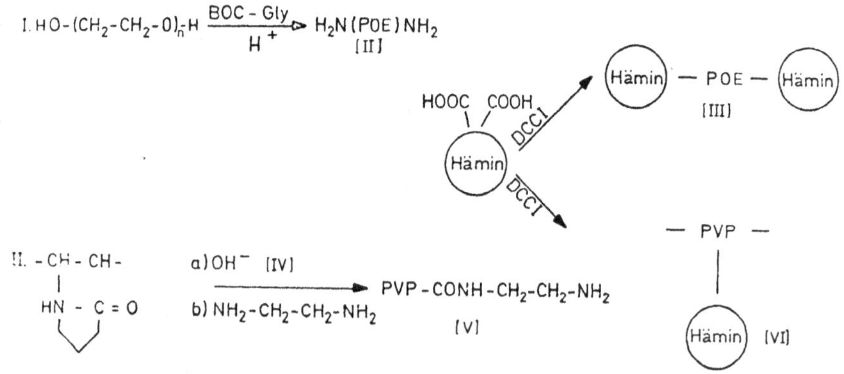

Figure 1. Synthesis scheme for heme polymers.

After the synthesis, the iron is reduced to iron(II) with sodium dithionite. Excess reducing agent should be removed by sephadex chromatograph To these polymers [III] and [VI] , imidazole or methylimidazole is adde and upon exposure to oxygen in aqueous solution we get spectra indicati of oxygen complexes. Very soon, however the spectra of the irreversible oxidized hemipolymers appear. This clearly demonstrates that the shielding effect of the linear polymers is not sufficient to prevent dimerization and subsequent irreversible oxidation. The soluble polymers therefore are not different in this respect from insoluble heme polymers. Remarkable however is the very fast kinetics of the oxidation of the soluble heme polymers in comparison to insoluble heme polymers.

It was necessary to refine our model by peptide groups which, in addition to the shielding effect of the polymer, prevent heme dimerization, We decided to insert between the polyoxyethylene and the heme, short peptide chains which by their special conformational arrangement protect one side of the heme moiety. Different approaches have been followed. The first model synthesized (11) is shown in Fig. 2. The synthesis of this compound is achieved by consecutive coupling of glycine, histidine, heme and 3-(5-imidazolyl)-propylamine (API) by the liquid phase peptide synthesis (3). Both histidine and imidazolyl-propylamine are bound as amides to the propionic acid side chains of heme. In the polymer the 3-(5-imidazolyl)-propylamine can undergo optimal coordination to the iron. However the imidazole of the histidyl residue is sterically in a much less favored situation for coordination to the iron. In the case of low molecular weight histidine-protohemin esters it has been demonstrated that a coordination of the 1-N-imidazole is only possible with great strain (19,20). Molecular models and IR-spectra show that a cis-conformation of the peptide bond between the heme carboxylic group and the histidine is more favorable. This steric restrain leads to a less stable coordinative bond of the histidine to the iron.

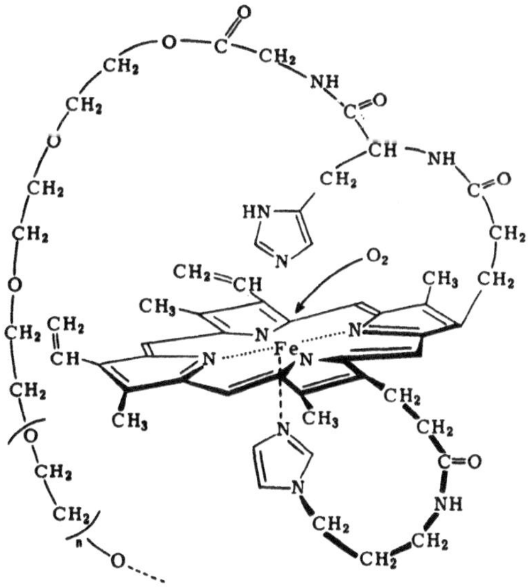

Figure 2. Structure of heme polymer POE-Gly-His-Heme-API.

In this way our synthetic polymers possess a quasi-distal histidine. UV-spectra reveal coordination of this quasi-distal histidine. The coordination is however much less stable than in the case of the imidazolylpropylamine in trans position. The weakly bound histidine prevents the hemo peptide polymer from dimerization and in addition to the polymer shields the heme moiety.

In Fig. 3 a, the orientation of the imidazole rings is illustrated. Now a small ligand like dioxygen or carbon monoxide can replace the weakly coordinated imidazole, and at the same time the cis peptide bond can undergo a change to the preferred trans conformation (Fig. 3 b).

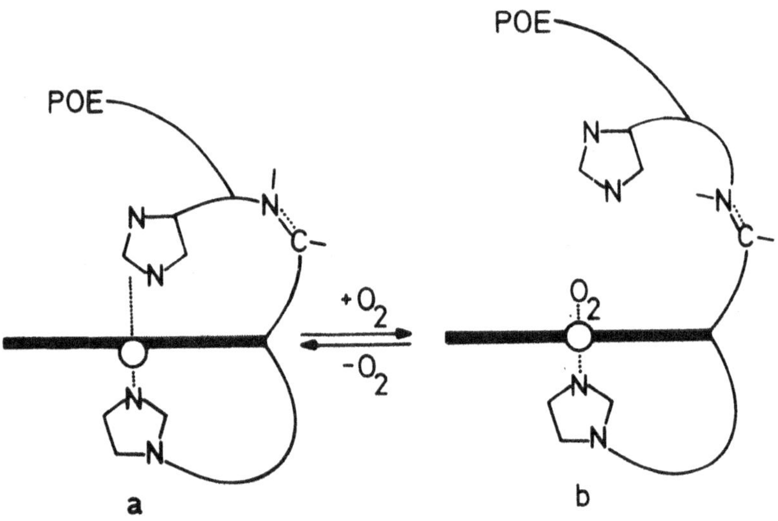

Figure 3. Orientation of imidazole rings in POE-Gly-His-Heme-A a) reduced and b) oxygenated form.

Therefore a conformation change in the peptide polymer is associated with the oxygenation. The electron spectra of the deoxygenated and the oxygenated form are summarized in Table 1. Oxygenation can be achieved in aqueous solution at room temperature under physiological conditions (11). The oxygen can be released again by applying vacuum or flushing with nitrogen or argon, which demonstrates the reversibility of oxygenation under physiological conditions.

Table 1. Absorption maxima (nm) of various complex types of the heme polymer POE-Gly-His-Heme-API.

Reduced	Oxygenated	Irr.Oxidized	CO complex
418	410	405	414
527	524		535
560	554		560

In contrast to other model compounds this synthetic heme polymer is capable of reversible oxygenation in aqueous solution. This goal has been reached by a stereochemical arrangement which is close but not identical to the active center of hemoglobin. In our hemo peptide polymer the shielding effect of the polymer is not sufficient to generate a coordinatively unsaturated five coordinative iron. Instead the distal histidine which is not bound to the iron in hemoglobin is closer to the iron and forms a weak coordinative bond which introduces strain into the peptide. This bond can be replaced by a small ligand like O_2, whereby a conformational change releases the strain in the peptide. Now the deoxygenated form again can be oxygenated. The kinetics of oxygenation and deoxygenation is fast and takes place in fractions of seconds as is the case with hemoglobin, and in contrast with the slow oxygenation of insoluble cross-linked hemo polymers. Our model compounds are very sensitive to prolonged exposure to oxygen. After three to four cyclic reversible oxygenations and deoxygenations considerable irreversible oxygenation occurs. However there is another analogy to heme proteins. Irreversible oxidized myoglobin is reduced in organism by myoglobin reductase to myoglobin. The isolated enzyme myoglobin reductase also reduces irreversibly oxidized hemi peptide polymer to the hemo compound which then can be again utilized for oxygen binding. Also CO can be bound to the synthetic hemo polymer (Table 1). The oxygen sorption curve for the complex is reproduced in Fig. 4. The O_2 partial pressure for half saturation is with 0.10 torr in the region of the values for myoglobins (0.2 - 1.2 torr[15]). The sigmoid shape of the O_2 sorption curve indicates a cooperation between two heme centers. The Hill coefficient (21) n_H eauals 2. The reason for this cooperative effect is not yet known. We assume that the conformation change occuring during oxygenation exposes the active center and makes possible a direct cooperation between two heme moieties (11). Also the decreased stability toward irreversible oxidation in comparison to myoglobin is indicative of this.

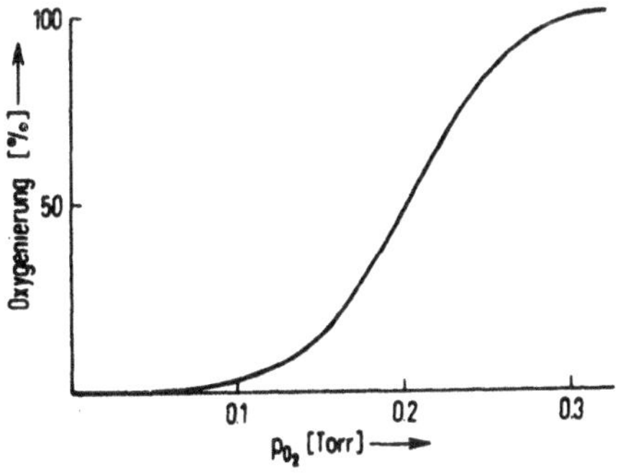

Figure 4. Oxygen sorption of polymer POE-Gly-His-Heme-API at 25° C in 0.05 M phosphate buffer pH = 8.1.

In order to increase the shielding effect we synthesized hemopolymers with peptide chains which better protected the heme group. One way was the synthesis of the following bridged polyoxyethylene hemo peptides with a trifunctional amino acid in the side chain:

H_3CO - POE - Gly - Asp - His - hematoheme(API)(Gly), with Asp and hematoheme bridged through Gly [VII]

This bridged hemo polymer peptide [VII] shows different electron spectra than the hemopolymer in Figure 2. The structure of [VII] is shown in Figure 5.

Obviously the histidine residue can no longer be coordinated to the iron, due to increased steric strain. The spectral data in table 2 demonstrate this.

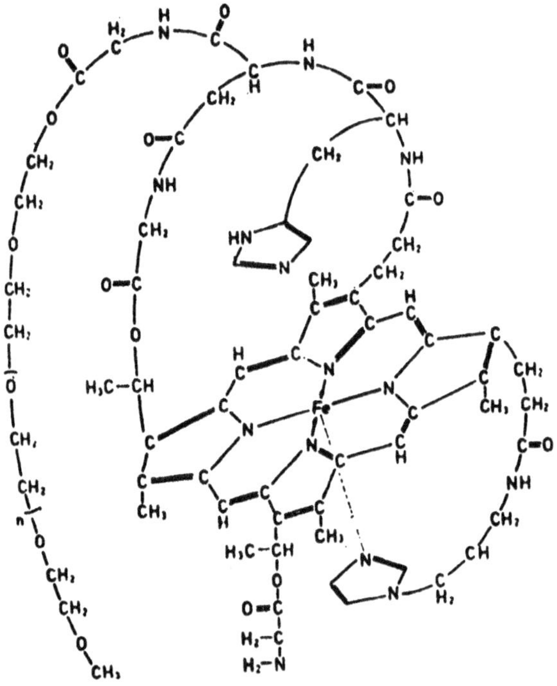

Figure 5. Structure of hemo polymer [VII].

Table 2. Light Absorption maxima (nm) of hemopolymer [VII] (HP) its oxygen and CO adduct in comparison to hemoglobin.

Bridged polymer				hemoglobin (22)			
HP [VII]	545		412	Hb	555		430
HPO_2	564	534	405	HbO_2	577	541	415
HPCO	559	529	409	HbCO	559	540	419

The pattern of the electron spectra are very similar to hemoglobin and its oxygen and CO complexes. Especially the single peak of the deoxygenated form at 545 nm indicates a five coordinated iron species. Upon oxygenation or exposure to carbon monoxide the double peak of the six coordinated species occurs. In this compound the histidine assumes a distal position like in hemoglobin.
Also the corresponding cobalt-(II)-complex of this polymer has been prepared. This complex and the O_2 adduct again show very similar electron spectra as coboglobin, the cobalt analogue of hemoglobin (table 3).

Table 3. Light absorption maxima of the cobalt-hematoporphyrin-analogue of [VII] (CoHP) and its O_2 adduct in comparison to coboglobin.

Bridged Co(II)-hematoprophyrin-polymer				coboglobin (23)			
CoHP	550		397	CoHb	552		402
$CoHPO_2$	561	529	419	$CoHbO_2$	571	538	422

Irreversible oxidation of the hemopolymer [VII] is much slower than of the hemo polymer in figure 2.

An ultimate goal of these investigations would be the use of synthetic heme polymers as artificial blood substitutes. An appealing idea of such blood substitutes could be the lack of immunological properties, which would enable application independent of the blood group of patients. These polyoxyethylene peptide hemes are easily accessible by the liquid phase peptide synthesis (1), and larger quantities of the heme polymer POE-Gly-His-Heme-API have been prepared for immunological and toxicological studies. No immunological properties are observed. If 20 % of the blood of rats are exchanged against an isotonic solution of the hemopolymer no toxicological properties could be detected. However, these studies have to be extented to long term toxicological studies and to higher exchange rates.

Soluble polymer peptides may be of continuing interest for imitating biological catalysis and for a new class of synthetic catalysts. Recently we demonstrated that polyoxyethylene peptides can also be used as model compounds for rhodopsin (12).

Also other polymers besides polyoxyethylene can be utilized. We have synthesized poly-organosiloxanes to which L-amino acids are bound. These chiral polymers interact with chiral adrenergic drugs, the extent of which can be correlated with the pharmacological properties of these drugs (24,25).

At the present time we are at the beginning of the evaluation of the catalytic properties of functionalized, homogeneous soluble polymer catalysts, which may have a promising future.

REFERENCES

(1) Mutter, M., Mutter, H., Uhmann, R., and Bayer, E.: 1976, Biopolyme: 15, pp. 917-927.
(2) Mutter, M., Uhmann, R., and Bayer, E.: 1975, Liebigs Ann.Chem. pp. 901-915.
(3) Bayer, E., and Mutter, M.: 1972, Nature (London) 237, pp. 512.
(4) Mutter, M., Hagenmaier, H., and Bayer, E.: 1971, Angew.Chem. International Ed. 11, pp. 811.
(5) Mutter, M., and Bayer, E.: 1974, Angew.Chem.International Ed. 13, pp. 88-89.
(6) Bayer, E., Mutter, M., Polster, J., and Uhmann, R.: in Y.Wolman, Peptides 1974, John Wiley Sons, New York 1975, pp. 129-136.
(7) Bayer, E., and Schurig, V.: Angew.Chem.International Ed. 1975, 14, pp. 493.
(8) Schurig, V., and Bayer, E.: 1976, Chemical Technology 6, pp. 212-2
(9) Bayer, E., and Kutubuddin, M., unpublished results.
(10) Bayer, E., and Schurig, V.: 1974, German Patent, DOS 2 326 489.
(11) Bayer, E., and Holzbach, G.: 1977, Angew.Chem.Internat.Ed. 1977, 16, pp. 117-118.
(12) Das, P.K., Becker, R.S., Hannak, D., and Bayer, E.: 1979, J.Amer. Chem.Soc. 101, ßß. 239.
(13) Hoffmann, A.B., Collins, D.M., Day, V.W., and Fleischer E.B.: 1973 J.Amer.Chem.Soc. 95, pp. 5810-5814.
(14) Collman, J.P., Gagne, R.R., Reed, C.A., Halbert, T.R., Lang, G., a Robinson, W.T.: 1975, J.Amer.Chem.Soc. 97, pp. 1427-1428.
(15) Almog, J., Baldwin, J.W., and Huff, J.: 1975, J.Amer.Chem.Soc. 97, pp. 227-228.
(16) Wang, J.H.: 1958, J.Amer.Chem.Soc. 80, pp. 1109.
(17) Tsuchida, E., and Honda, K.: 1975, Polymer J., pp. 498-502.
(18) Tsuchida, E., and Nishide, H.: 1977, Advances Polymer Science, 24 pp. 1-88.
(19) Van der Heijden, A.: 1972, Dissertation, University of Leiden, pp. 8-10.
(20) Warme, P.K., and Hager, L.P.: 1970, Biochemistry 9, pp. 1599-1606.
(21) Rossi-Fanelli, A., and Antonini, E.: 1958, Arch.Biochem.Biophys.77 pp. 478-490.
(22) Antonini, E., and Brunori, M.: 1971, Hemoglobin and Myoglobin and their Reactions with Ligands, North-Holland Publ.Comp., Amsterdam.
(23) Yonetani, T., Yamamoto, H., Woodrow, G.V., and Kayne, F.J.: 1974, J.Biol.Chem. 249, pp. 682-691.

(24) Frank, H., Nicholson, G.J., and Bayer, E.: 1978, Angew.Chem.International Ed. 17, pp. 363-365.
(25) Bayer, E., and Frank, H. in Carraher, E.: 1979, Functionalized Polymers, ACS Monograph, Am.Chem.Soc., Washington, in print.

SURFACE AND END EFFECTS IN THE ELECTRONIC STRUCTURE OF CRYSTALS AND POLYMERS

O. Fromm and J. Koutecký
Institut fuer Physikalische Chemie und Quantenchemie
Freie Universitaet Berlin, 1000 Berlin 33
West Germany

ABSTRACT

A method suitable for the determination of the electronic properties of locally perturbed large periodic systems is described and the relation to other approaches based on the Green function technique clarified. Representative examples of applications for crystal surfaces and polymer ends are shown.

I. INTRODUCTION

The basic idea of how to treat complicated problems concerning the electronic structure of solid state surfaces is a very common one in quantum mechanics: The knowledge obtained by solving a simpler, physical system is often used to solve a more complicated physical problem which is of actual interest.

In the theory of crystal surfaces it is often assumed that the problem of the infinite unperturbed crystal is solved in a given approximation. The crystal translational symmetry namely makes - for this instant - the solution easier than in the case where the symmetry is lowered due to the existence of a surface region.

Use of the Green function, or resolvent, technique enables the symmetry properties of the bulk crystal to be employed in surface calculations. The basic idea behind this strategy is that the electronic properties of the bulk crystal play a more decisive role in the determination of the electronic structure than those of the individual atoms in the surface.

Similarily, in the theory of chemisorption and catalysis,

B. Pullman (Ed.): Catalysis in Chemistry and Biochemistry. Theory and Experiment. 335–346.

one can start from the knowledge of the isolated crystal surface and gas phase molecules or atoms and consider their mutual interaction as a perturbation.

Because of its easy interpretation from the chemical point of view, the LCAO approach (or tight binding approximation in the language of physicists) will be discussed here

GREEN'S OPERATOR

The Green function technique may be employed in many different ways, some will be demonstrated later, however initially a few general concepts will be discussed (compare e.g. ref. 1).

Green operator $\hat{G}$ assigned to a Hamiltonian operator $\hat{H}$ describing the investigated system S is defined as

$$\hat{G}(z) = (z\hat{I} - \hat{H})^{-1} \equiv \hat{\Delta}^{-1}(z) \tag{1}$$

where z is in general a complex number. It is evident that the properties of Green operator, or of the convenient matrix representation, can be used for discussion of the properties of the eigenvalue problem of $\hat{H}$:

$$\hat{H}|E_j,a> = E_j|E_j,a> \tag{2}$$

The Green function, or the matrix elements of Green's operator, can be used in the framework of the 1-el. approach to determine such properties as the density of states, or the local density of states, which yield information about the distribution of electrons in respect to energy and space

To use the Green operator enables one to construct a bridge leading from the "known" country of solid state theory to the unknown country of the surface quantum chemistry. Let us consider a known system S_o with the Hamiltonian

$$\hat{H}_o = \hat{H} - \hat{V} \tag{3}$$

Green operator $\hat{G}_o$ associated with $\hat{H}_o$ is related to Green's operator $\hat{G}$ for the investigated system S via the Dyson equation. The formal solution of the Dyson equation has the form

$$\hat{G}(z) = \hat{L}^{-1}(z)\ \hat{G}_o(z) \tag{4}$$

with

$$\hat{L}(z) = \hat{I} - \hat{G}_o(z)\hat{V} \tag{5}$$

The Schrödinger equation for the system S can be rewritten in the form

$$\hat{L}(E_j)\ |E_j\rangle = |0\rangle \tag{6}$$

if the operator $\hat{G}_0(E_j)$ exists, it means, that E_j is not included within the spectrum of $\hat{H}_0$. The condition to find the E_j's which are eigenvalues of $\hat{H}$ but not of $\hat{H}_0$ can then be written in form of a determinantal equation

$$\det \underline{L}(E) = \det (\langle\mu|\hat{L}(E)|\nu\rangle = 0 \tag{7}$$

where the vectors $|\mu\rangle$ and $|\nu\rangle$ are those from the complete basis set which can be used to express the eigenstates of $\hat{H}$ and $\hat{H}_0$.

An application of this concept is possible if a subspace Q in the space spanned by vectors $|\mu\rangle$ can be found in such a way that $\langle\mu|V|\nu\rangle = 0$ if $|\nu\rangle$ does not belong to the subspace Q, then the determinantal equation can be simplified to

$$\det (\langle\mu|\hat{Q}\hat{L}(E)\hat{Q}|\nu\rangle) = 0 \tag{8}$$

the order of the determinant is then equal only to the dimension of the space Q.

This method has been used extensively in the theory of the one-electron localized surface states, in the framework of the one-electron approximation. In this case, the Hamiltonians $\hat{H}$ and $\hat{H}_0$ are the effective one-electron Hamiltonian, the vectors $|\nu\rangle$ of the basis are the atomic orbitals and the region Q can be approximatively identified with the surface region.

PARTITIONING TECHNIQUE

The other application of the Green function technique is the direct determination of the matrix elements of Green's operator in the appropriate representation which allows one to use the advantage of the localized perturbation, caused by differences between the specific surface properties, in comparison with the bulk properties. We will use the one-electron LCAO approach for simplicity, because the partitioning of the vector space of the one-electron wave function can be easily interpreted as the partitioning of the real space in which the atomic orbitals are localized. Because the partitioning technique is very useful in this respect we shall now briefly describe it (compare e.g. ref. 2). Let us consider the system S as originated by the interaction

$\hat{V}(A,B) = \hat{V}_{AB} + \hat{V}_{BA}$ from the system S_O, composed from two noninteracting parts A and B;

$$\hat{H}=\hat{H}_O+\hat{V}(A,B)=\hat{H}_A+\hat{H}_B+\hat{V}_{AB}+\hat{V}_{BA} \tag{9}$$

The formal solution of the Dyson equation (4) can be written

$$\hat{G}(z) = \hat{L}^{-1}(z)(\hat{G}_A^O + \hat{G}_B^O) \tag{10}$$

Using the definition of the operator $\hat{L}^{-1}(z)$ the projection of Green's operator in the subspace A,$G_A(z)$ can be expressed in the form

$$\hat{G}_A(z)=(z\hat{A} - \hat{H}_A^O - \hat{V}_{AB}\hat{G}_B^O\hat{V}_{BA})^{-1} \tag{11}$$

and the "mixed" Green's operator $\hat{G}_{AB}(z)$ is given by

$$\hat{G}_{AB}(z) = (\hat{\Delta}(z)-\hat{V}_{AB}\hat{G}_B^O\hat{V}_{BA})^{-1}\ \hat{V}_{AB}\hat{G}_B^O \tag{12}$$

The relations for $\hat{G}_B(z)$ and $\hat{G}_{BA}(z)$ can be obtained immediately from equations (11) and (12) by exchanging the role of the labels A and B. The operator equations can be, of course, rewritten easily in an appropriate matrix form.

The important feature of the relation (11) is that the global influence of one part of the system, mainly of B, is included in the Hamiltonian of part A of the investigated system.

This partitioning procedure can be used, for example, with advantage if the Hartree-Fock procedure needs to be explicitly carried out in a relatively small part A of the whole system only. The electronic structure of part B being determined mainly by translation symmetry.

A further application the partitioning procedure is attractive for treating large systems S which are essentially one-dimensional and which have a periodic behaviour for regions away from the locally perturbed region A. Cutting the part A of the infinite system, S results in another infinite system B which has the property that B remains unchanged in its physical properties due to its infinity if successive cutting of the equivalent parts C,C',C",... is performed. The C's can be for example one or more crystal layers parallel to the crystal surface. Let R be the subsystem of B obtained by removing the first part C. The periodicity of the infinite system B implies, that the rectangular matrix $\underline{V}_{CR}$ contains as its essential non-zero part only a square matrix $\underline{V} = \underline{V}_{CC'} = \underline{V}_{C'C''} = \ldots$, which can, of course, have some zero rows or zero columns.

Now the Dyson equation is applied for the partitioning the system B in the subsystem C and R. Its projection in the subsystem C gives

$$(\underline{G}_B)_C = \underline{G}_C + \underline{G}_C \underline{V}_{CR} \underline{G}_R \underline{V}_{RC} (\underline{G}_B)_C \qquad (13)$$

where G_C and G_R is the Green matrix of the isolated system C and R, respectively. Referring to the form of the matrix $\underline{V}_{RC}$, the circumstance that the matrix $\underline{G}_R$ projected in the part C' i.e. $(\underline{G}_R)_{C'}$, is identical with $\underline{G}_B$ projected in the part C i.e. $(\underline{G}_B)_C$ simplifies the equation (13) to

$$(\underline{G}_B)_C = \underline{G}_C + \underline{G}_C \underline{V} (\underline{G}_B)_C \underline{V}^+ (\underline{G}_B)_C \qquad (14)$$

Multiplication of this equation with $\underline{\Delta}_C \equiv \underline{G}_C^{-1}$ from the left and with $\underline{V}^+$ from the right leads to [3]

$$\underline{VX}^2 + \underline{\Delta}_C \underline{X} + \underline{V}^+ = \underline{O} \qquad (15)$$

where

$$\underline{X} = -(\underline{G}_B)_C \underline{V}^+ = -(\underline{G}_R)_{C'} \underline{V}^+ \qquad (16)$$

is introduced.

The matrix $\underline{X}$ is describing relations between Greenmatrix elements in the bulk region; e.g.

$$(\underline{G}_B)_{C(\nu)C} = \underline{X}^\nu (\underline{G}_B)_C \qquad (17)$$

If $\underline{X}$ is found by solving the matrix equation (15) then the Green matrix $(\underline{G}_S)_A$ for the surface region, A, of the complete system S can be determined by using the original relation (11) for partitioning the system S in A and the rest B:

$$(\underline{G}_S)_A = (\underline{\Delta}_A + [\underline{VX}])^{-1} \qquad (18)$$

Care must be taken that the dimension of the surface region A is chosen in such a way that the rectangular matrix $\underline{V}_{AB}$ is composed by $\underline{V}$ and zero elements only. $[\underline{VX}]$ in equation (18) refers to a matrix of the same dimension as $\underline{\Delta}_A$ with zero matrix elements where necessary:

$$[\underline{VX}]_{ss} = \begin{pmatrix} \underline{O} & \underline{O} \\ \underline{O} & (\underline{VX})_{nn} \end{pmatrix} \qquad (19)$$

s and n are the dimension of the surface region A and of the region C, respectively.

DETERMINATION OF MATRIX $\underline{X}$ [3]

If the matrix $\underline{X}$ can be diagonalized in the form

$$\underline{X} = \underline{WYW}^{-1} \tag{20}$$

with the diagonal matrix

$$Y_{jk} = \xi_j \, \delta_{jk} \tag{21}$$

then the solution of the matrix equation $\underline{X}$ can be constructed from the roots of the determinantal relation

$$\det\,(\underline{V}^{+} + \underline{\Delta}_{C}\xi + \underline{V}\xi^{2}) = 0 \tag{22}$$

only such solutions of equation (22) can be used in relation (21) which satisfy the inequality

$$|\xi_1| < 1 \tag{23}$$

The reason for this limitation is that the relation (17) for $|\xi_1| > 1$ would lead to the nonrealistic increase of the Green function $(\underline{G}_B)_{C^{(\nu)}C}$ with increasing distance between the regions $C^{(\nu)}$ and C.

The diagonalization procedure of the Hamiltonian matrix for a semiinfinite system S with periodic local properties at greater distances from the surface, carried out by the difference equation method, leads to an equation of exactly t same form as Eq. (22) for the parameter ξ as in the case of a determination of Green's operator matrix elements. The parameter ξ has, in the context, an other meaning. It descibes the propagation of the AO coefficients between different layers with increasing distance from the surface if the same two-dimensional translational symmetry of the surface and of the layers parallel to the surface exists. Because of the close analogy between these two forms of the theory, only Green's operator form is discussed in this contribution even when the other form gives interesting information in a very clear way, for example when the problem of the different hybridyzation in different crystal layers is concerned.

The calculation of the parameter ξ represents a generalized eigenequation problem for given energy E of the one-electron wave-function for the semiinfinite system. The numerical evaluation makes no difficulties and it is similarily efficient as the normal eigenvalue procedure. Once the ξ's are determined, the appropriate behaviour of the one electron functions may be obtained by using either the fitting condition (18) if the Green matrix is applied, or the usual bonbary conditions if a diagonalization procedure for the Hamiltonian is used involving the difference equation method.

APPLICATION

Two examples will be discussed in this report. Figure 1 shows the projected local density of states in points of the two-dimensional Brillouin zone as a function of the energy E for the (111) surface of Si with reconstruction of the surface layer following from the rehybridyzation according the ideas of Appelbaum [4] and other authors [5]. The LCAO Hückel-type approach is used with the parametrization according to the procedure of Koster and Slater [6] with respect to only the nearest neighbour interaction among the sp^3 hybrids of Si atoms.

It is possible to see how the form of the energy dependence of the projected LDS changes with position in the two dimensional Brillouin zone. These energy dependencies of the projected LDS show even more characteristic differences if different sp^3 hybrids are compared. The surface states of the Shockley type indicating the dangling bonds in the Si (111) surface are located in the gap between the conduction and valence bands.

The surface states due to the backbonds are possible to see in the gaps between individual bands which contribute to the global valence and conduction bands. In our example the transition from localized surface backbond bands to resonances within the band is also noticible.

The calculation of the $\underline{X}$ matrix for given E and $k_{\parallel}$ can easily be performed which makes it possible to determine, relatively quickly, the total LDS. It is worth mentioning that the $\underline{X}$ matrix, which characterizes the crystal bulk, can once calculated, be used for other systems which differ in properties in the surface region only.

The second example is a polymeric chain in the Hückel model with an elementary cell, composed from condensed six and five rings, as shown in Fig. 2.(polymer 1,cf. ref. 7). The LDS for the end atoms, Fig. 1,2 and 3 are shown for the polymers ending with a six ring (Fig.2a) and with a five ring (Fig. 2b). The differences in the structure of the LDS are remarkable, the discontinuities on some band edges being very characteristic. Some of these one-electron end states are similar to the Shockley states because approximately there exist nodes, which are distributed in an almost periodic manner in the polymer. The complicated structure of these end state necessitates the use of a quite elaborate technique for their determination, as the method presented here.

For the sake of qualitative understanding of this type of end states, very simple topological considerations are useful. Let us consider the linear conjugated polymer which has one six ring in the elementary cell, the elementary cell being joined by two bonds (polymer 2, cf. Fig. 3). If the MO has zero coefficients at the AO's number 5 and 6 the polymer chain can be considered as the chain of non-interacting cis-butadienes. The following Hückel-type conditions have, of course, to be fulfilled among the coefficients C_1, C_2, C_3 and C_4 of the prespective butadiene Hückel MO:

$$C_1 + \xi C_2 = 0$$

$$C_4 + \xi C_3 = 0 \qquad ; \ |\xi|<1 \tag{22}$$

This is satisfied for MO energies $E_j = \pm(1+5)\frac{\beta}{2}$ of butadiene . For the energies E_j, end states exist which have exponentially decreasing amplitudes with increasing distance from the polymer end and with coefficients different from zero only for the centers inside the indented elementary cell i.e. cis-butadiene.

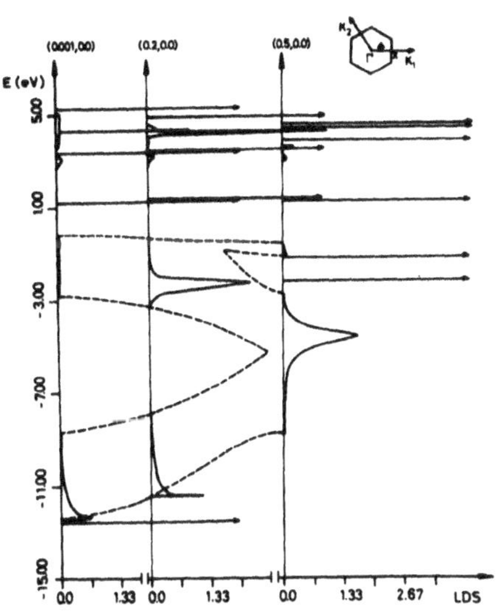

Fig. 1: "Projected" LDS for sp^3 hybrids participating in backbonds of the Si atoms on the relaxed (111) surface in the points $k_1=0.001$, $k_1=0.2$ and $k_1=0.5$ on the line $\Delta(k_2=0)$ of the two-dimensional Brillouin zone (full lines). The valence band structure is indicated by broken lines. The energies of the surface states are shown by arrows.

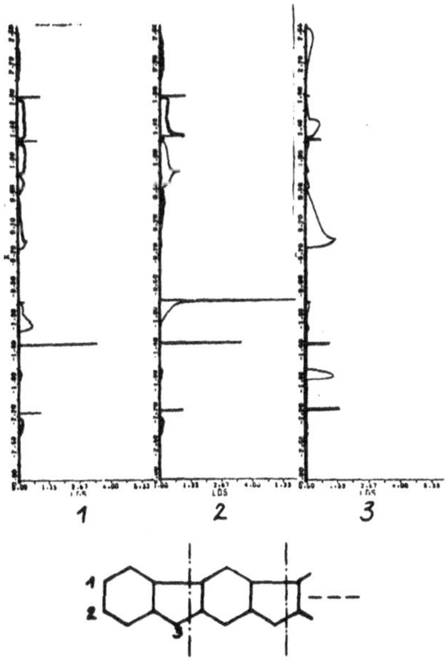

Fig. 2a: LDS for end atoms in the polymer 1 ending with a six ring.

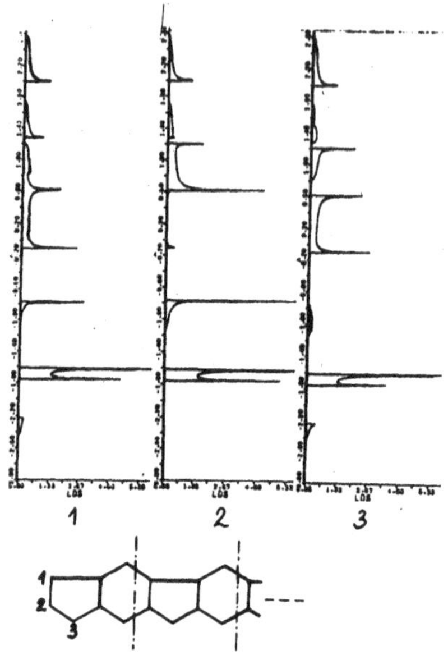

Fig. 2b: LDS for end atoms in the polymer 1 ending with a five ring. Energies of the end states are indicated by vertical lines.

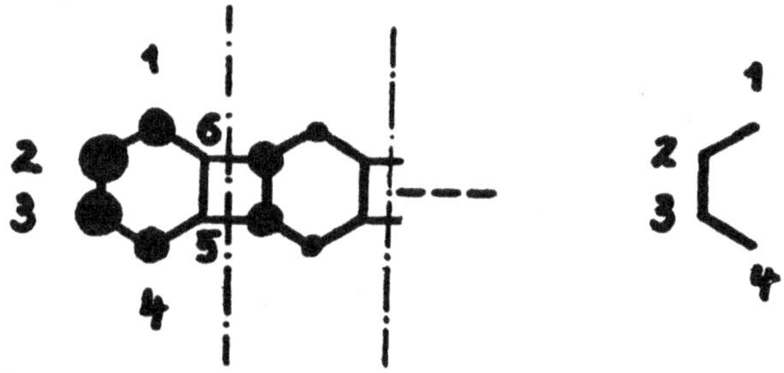

Fig. 3: Schematical picture of the AO coefficients for a Shockley type end state in the first two elementary cells of the polymer 2. The cis-butadiene is the indented elementary cell.

CONCLUSIONS

In this contribution we have described a method based on Green's operator approach and on the solution of a matrix equation (cf. 14) for a matrix which can be interpreted as a quotient of the two appropriate Green matrices (compare Eq. 16). This basically algebraic method is related to the approaches of Bortolani [8] and that of Matsuda [9] and Biczó [10] however the inversion of the matrix, which describes the interaction between two subsystems of the system studied, is avoided. This feature of the method discussed is very important especially if the interaction between two subsystems goes only through a few "channels" so that the inversion of the considered matrix is impossible.

The generality of the method discussed makes it possible for example, to describe the changes in the hybridyzation in the different crystal layers in a relatively simple form. Different types of localized crystal surface or polymer end 1-el. states are obtained and they can be studied also in quite complicated systems even for models where the transparent simple characteristic of these states do not longer exist.

ACKNOWLEDGEMENT

This work has been partly supported by a grant from Deutsche Forschungsgemeinschaft.

LITERATURE

1. Koutecký,J., Prog.Surf.Mem.Sci., 11, 1 (1976).
2. Koutecký, J., in "Excited State in Quantum Chemistry" Eds. C.A.Nicolaides and D.R.Beck, Reidel Publishing Comp.,Dodrecht,Holland, 531, (1978).
3. Fromm, O., to be published
4. Appelbaum, J.A. and Hamann, D.R., Phys.Rev.Letters 31, 106 (1973).
5. Pandey, K. and Phillips, J.C., Solid State Comm., 14, 439 (1974).
   Del Re, G., Preprint No. 27 Cattedra di Chimica Teorica, Universita di Napoli
6. Koster, G.F. and Slater, J.C., Phys. Rev. 95, 1167 (1954).
7. Rolansky, O.E. and Tyutyulkov, N.N., MATCH 3, 149 (1977).
8. Bortolani, V., Celli, V. and Marvin, A., "Phys. of Semiconductors, Proc. Int. Conf. 13th.", Ed. Fumi, F.G., North Holland, 714 (1976).
9. Matsuda, H., Okada, K., Takase, T. and Yamamoto, T., J. Chem. Phys., 41, 1527 (1964).
10. Biczó, G., "Proc. 7th. Int. Vac. Congr. and 3rd Int. Conf. Solid Surfaces (Vienna)", 407 (1977).

SOME ASPECTS OF ASYMMETRIC INDUCTION IN ADDITION REACTIONS TO OLEFINS

Piero Pino, Annibale Stefani and Giambattista Consiglio
Swiss Federal Institute of Technology
Department of Industrial and Engineering Chemistry
Universitätstrasse 6, 8092 Zurich, Switzerland

Abstract

Catalytic and non catalytic asymmetric addition reactions to olefinic double bonds yielding two chiral regioisomers and occurring with "cis" stereochemistry have been investigated. From the enantiomeric excess of the products, enantioface discrimination and regioselectivity on the two faces of the olefins have been calculated.

In most of the cases investigated, the regioselectivity on the two faces of the olefin is different and the contribution to enantiomeric excess connected with the regioselectivity difference is larger than that connected with enantioface discrimination. In two cases the prevailing enantiomer of one regioisomer is opposite to that corresponding to enantioface discrimination.

When a reagent HX is added to a non symmetrically disubstituted (non chiral) olefin, two isomers (A and B) each containing a chiral carbon atom, generally form. The relative amount of the two isomers has been indicated (1) as regioselectivity (RE) (2).

When the two isomers (A and B) are chiral and the reaction occurs with a "cis" stereochemistry, the two antipodes of each isomer (a, a' and b, b' respectively) arise from the addition of the reagent HX on opposite faces of the olefin.

In the case of asymmetric reactions it may be useful, for the investigation of the nature of the diastereomeric transition states and/or intermediates, to split the enantiomeric excess experimentally found for the two isomers ($(e.e.)_A$ and $(e.e.)_B$) in two parts: i.) the extent (6) of discrimination between enantiofaces (8) (D) and ii.) an additional factor which is related to the difference between the regioselectivity of the reaction on one or on the other enantioface of the substrate (Δ(RE)).

B. Pullman (Ed.): Catalysis in Chemistry and Biochemistry. Theory and Experiment. 347–354.

The regioselectivity on each of the two enantiofaces is indicated by the absolute value of the parameters $(RE)_1$ and $(RE)_2$ given by the equations (1) and (2)

$$(RE)_1 = \frac{[a] - [b]}{[a] + [b]} \quad (1) \qquad (RE)_2 = \frac{[a'] - [b']}{[a'] + [b']} \quad (2)$$

in which as in the expression of enantioface discrimination (D) a and b represent the antipodes of the two isomers A and B which arise from the face prevailingly attacked during the reaction (9).

The two contributions to the enantiomeric excess found at the end of the reaction for the two regioisomers A and B ($(e.e.)_A$ and $(e.e.)_B$) can be calculated from equations (3) and (4) where P_A and P_B are two asymmetric induction parameters the absolute values of which correspond to the e.e. of the two regioisomers and x_A and x_B are the molar fractions of A and B.

$$(P)_A = D + \frac{(RE)_1 - (RE)_2}{4x_A} \left[1 - \left(\frac{D}{100}\right)^2\right] \quad (3)$$

$$(P)_B = D - \frac{(RE)_1 - (RE)_2}{4x_B} \left[1 - \left(\frac{D}{100}\right)^2\right] \quad (4)$$

In the present paper some data on the isomeric composition and on the enantiomeric excess observed in catalytic and non catalytic reactions leading to two chiral regioisomers will be presented. Furthermore, for the hydrocarbonylation reactions, the steps in which enantioface discrimination and regioselectivity might be determined will be briefly discussed.

RESULTS

In the table the type and extent of enantiomeric excess of the two isomers obtained in hydroformylation (10,11), hydrocarboxylation (12) and hydroboration (13) of some olefinic substrates are reported. From these data enantioface discrimination (D) and regioselectivity difference $\Delta(RE) = [(RE)_1 - (RE)_2]$ for the reaction on the two prochiral faces of the olefins have been calculated. For cis-2-hexene hydroboration the data on which the calculations are based are shown in Scheme 1.

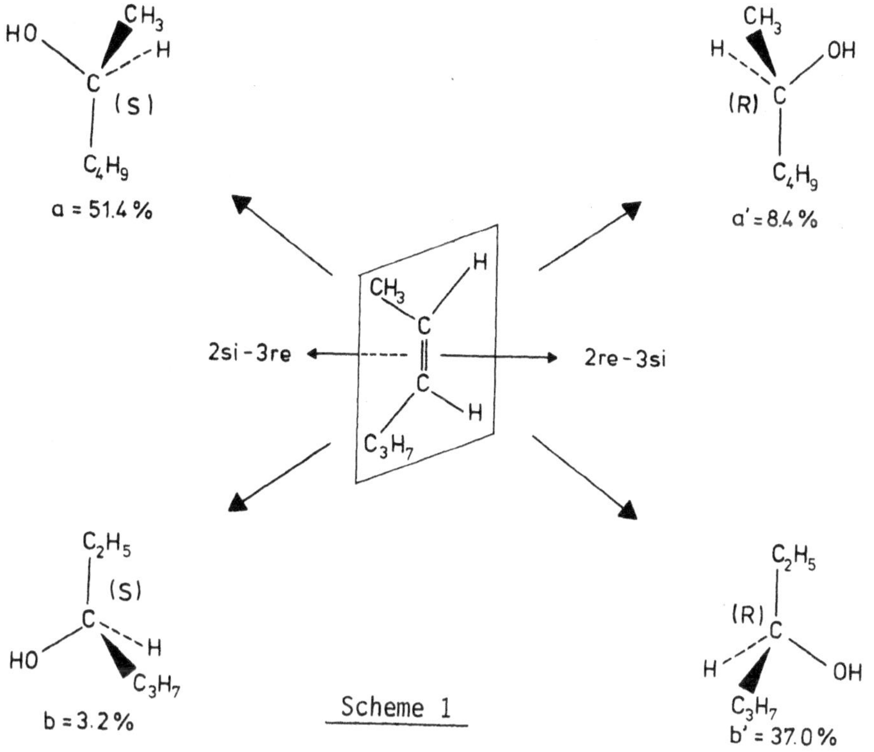

Scheme 1

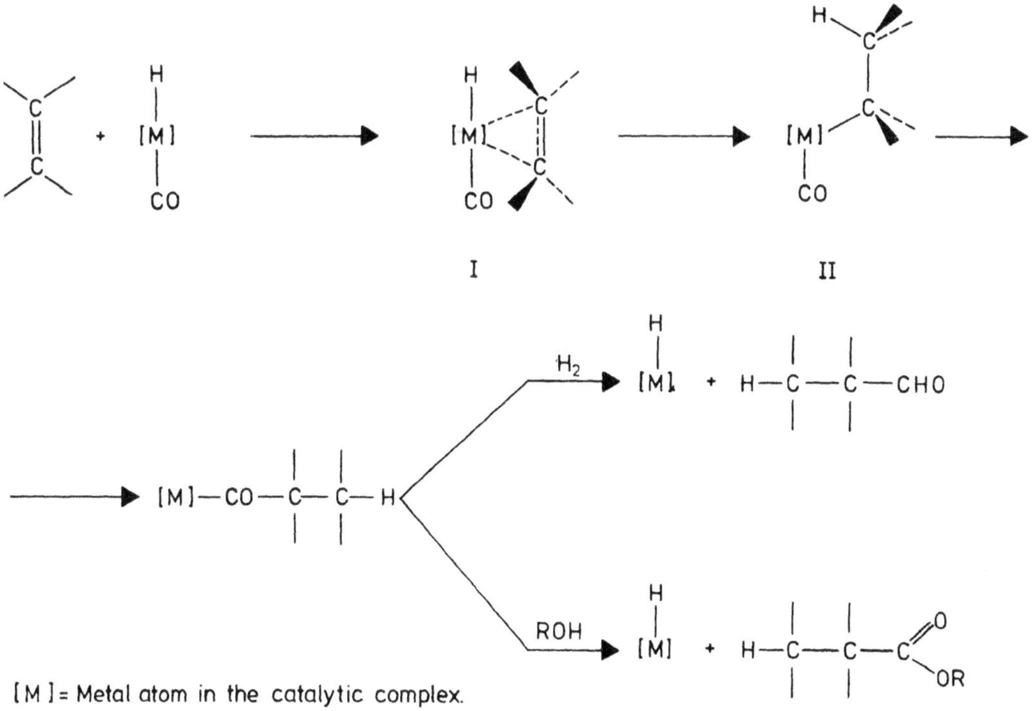

[M] = Metal atom in the catalytic complex.

Scheme 2

Table. Enantiomeric Excess in Some Asymmetric Additions to Olefinic Double Bond in Which Discrimination of the Olefin Enantioface Occurs

Substrate	Reaction product Type	%	Δ(RE)	Enantiomeric excess A Type	A %	B Type	B %	C Type	C %
α-d_1-styrene[1]	2-d_1-2-phenylpropanal	69	≈0	(S)	15	(S)	15	-	0
	3-d_1-3-phenylpropanal	31		(R)	15	(R)	15	-	0
trans-2-hexene[1]	2-methylhexanal	58	1.45	(S)	1.4	(S)	2.0	(R)	0.6
	2-ethylpentanal	42		(R)	2.9	(R)	2.0	(R)	0.9
cis-2-hexene[2]	2-hexanol	59.8	151.3	(S)	71.9	(S)	9.2	(S)	62.7
	3-hexanol	40.2		(R)	84.1	(S)	9.2	(R)	93.3
cis-2-hexene[1]	2-methylhexanal	60	12.9	(S)	7.6	(S)	2.2	(S)	5.4
	2-ethylpentanal	40		(R)	5.8	(S)	2.2	(R)	8.0
2-phenyl-1-butene[3]	ethyl 3-phenylpentanoate	5	2.97	(S)	18.0	(S)	3.2	(S)	14.8
	ethyl 2-methyl-3-phenylbutanoate	95		(S)	2.4	(S)	3.2	(R)	0.8

A = Experimentally found
B = Contribution to e.e. attributed to enantioface discrimination (D in eq. 1 and 2)
C = A-B = Contribution to e.e. related to the regioselectivity difference Δ(RE) according to eq.1&2

1: Hydroformylation 2: Hydroboration 3: Hydrocarbalkoxylation

The above calculations show that

1.) Regioselectivity for the reactions on the two faces of the olefin can be remarkably different; the largest difference has been observed in the hydroboration of cis-2-hexene (Scheme 1), while no appreciable difference has been observed in the case of α-deuteriostyrene hydroformylation.

2.) In general the two isomeric reaction products have different enantiomeric excesses. In our calculations the difference between the enantiomeric excess of the two isomers is related to the different regioselectivity (Δ(RE)) of the reaction on each of the two faces of the substrate. The contribution connected with Δ(RE) is inversely proportional to the molar fraction of the two isomers.

3.) In some cases the contribution to the enantiomeric excess corresponding to enantioface discrimination is overwhelming and determines the prevailing enantiomer for both isomers. In other cases (hydroformylation and hydroboration of cis-2-hexene, hydrocarboxylation of 2-phenyl-1-butene) the contribution mainly dependent on the regioselectivity difference Δ(RE) is much larger.

   In two cases (hydroformylation of cis-2-hexene and hydroboration of cis-2-hexene), in one of the isomers the absolute configuration of the prevailing enantiomer is opposite to that of the enantiomer arising from the preferred enantioface.

DISCUSSION

The results of the table, based on the e.e. and on the i.e. of the final reaction products, correspond to the asymmetric induction and regioselection occurring during the reaction only if the selectivity of the reaction is practically complete. In the case of multisteps reactions (as e.g. the hydrocarbonylation reactions) no racemization or structure isomerization must occur after the step (or steps) in which asymmetric induction and/or regioselection occur.

The first condition is verified in all our examples with the exception of the 2-phenyl-1-butene hydrocarboxylation in which a relatively low isomerization of the substrate (15%) to 2-phenyl-2-butene has been found.

In the case of hydroboration, the reaction is believed to occur in one step (14) and the second condition does not apply. This reaction should occur through 4 independent reaction paths, and shows that indeed asymmetric induction and regioselection can occur simultaneously in a single step.

In the case of hydrocarbonylation a multistep reaction path (Scheme 2) is generally accepted (15). If the isomerization of metal-alkyl (II) can be excluded, no change of (e.e.) or of (i.e.) appears likely in the following steps.

Although the type and extent of (e.e.) and of (i.e.) in the final products correspond qualitatively and quantitatively to regioselection and asymmetric induction, they give no indication on the step (or steps) in which the above selections take place and on their mechanism.

Concerning the step (or steps) in which asymmetric induction takes place, it has been shown in the case of rhodium catalyzed hydroformylation (10) and in palladium catalyzed hydrocarboxylation (16) of butenes that asymmetric induction leading to the formation of optically active aldehydes or esters occurs before or during metal-alkyl (II) formation. In the case of hydrocarboxylation of 3-phenyl-2-butene with palladium catalysts it was shown (17) that asymmetric induction takes place, at least in part, during the coordination of the olefin to the palladium atom of the catalysts.

Concerning the step in which regioselection takes place, as the formation of the metal-alkyl complexes in the above examples is substantially irreversible (10), there is little doubt that it also takes place before or during the insertion of the substrate into the metal-hydrogen bond.

In the first papers on the asymmetric hydroformylation (10) of olefins, enantioface discrimination was indicated as the factor controlling asymmetric induction in C_S olefins while for C_{2V} olefins like cis-2-butene a model based on conformational equilibria in which a discrimination between the two trigonal carbon atoms of the face of the olefin was proposed.

It appears from the table that the contribution to (e.e.) connected with regioselectivity difference on the two olefin faces cannot be in general overlooked and that at least for the examples shown (e.e.) is only seldom a quantitative indication of the extent of asymmetric induction in the step in which D is determined. Sometimes even the type of asymmetric induction occurring in that step is opposite to that shown be (e.e.) in the final products (8).

If enantioface discrimination (D) would be the only source of asymmetric induction, the enantiomeric excess for the two regioisomers formed in the reaction should be same in the two cases.

It appears that in most cases in the two isomeric reaction products the amount of the two antipodes arising from the same face of the substrate is different. In view of the above facts, it seems reasonable to assume that asymmetric induction in these catalytic reactions occurs in two steps, the first involving the postulated π-complexes (I) and leading to enantioface discrimination, and the second involving a regioselection on each of the two faces of the substrate already bound to the catalyst probably during the formation of the metal-alkyl complexes (II).

We thank the "Schweizerischer Nationalfonds zur Förderung der wissenschaftlichen Forschung" for financial support.

REFERENCES AND NOTES

1) Hassner, A.: 1968, J. Org. Chem. 33, 2864.

2) Regioselectivity has been quantitatively expressed (3,4) as [A]/[B] or as [A]/([A] + [B]).

In asymmetric reactions it is convenient to express the regioselectivity as regioisomeric excess (i.e.), that is as the absolute value of a parameter (RE) which is given by the equation

$$(RE) = \frac{[A] - [B]}{[A] + [B]} \cdot 100$$

A and B are the isomers in which the asymmetric carbon atom formed during the reaction corresponds to the trigonal carbon atom of the double bond the position of which is indicated by lower and higher number respectively, in the structural formula of the substrate. (For instance when HCl is added to 2-hexene, 2-chlorohexane will be A and 3-chlorohexane will be B). For 1,1-disubstituted ethylenes, in which the same trigonal carbon atom becomes asymmetric in the two isomers, the isomer A is defined as that one obtained by addition, to the trigonal carbon atom, of the group having the higher priority according to the Cahn, Ingold and Prelog rule (5). (For instance adding HBr to 2-phenyl-1-butene, 2-phenyl-2-bromobutane will be A and 2-phenyl-1-bromobutane will be B).

3) Piacenti, F., Pino, P., Lazzaroni, R., and Bianchi, M.: 1966, J. Chem. Soc. (C), 488.

4) Lai, R., and Ucciani, E.: 1978, J. Mol. Catal. 4, 401.

5) Cahn, R.S., Ingold, C., and Prelog, V.: 1966, Angew. Chem., 78, 413.

6) The extent of discrimination between enantiofaces D is given by

$$D = \frac{([a] + [b]) - ([a'] + [b'])}{[A] + [B]}$$

when a and b are the antipodes of A and B arising from the enantioface prevailingly attacked.

The enantiofaces are indicated according to the Hanson nomenclature (7) specifying the position of the trigonal carbon atoms of the olefin molecule, (i.e., in cis-2-hexene the two enantiofaces are indicated as 2si-3re and 2re-3si, and in 2-phenyl-1-butene as 2si-3s and 2re-3re).

7) Hanson, K.R.: 1966, J. Amer. Chem. Soc. 88, 2731.

8) Izumi, Y., and Tai, A., "Stereodifferentiating Reactions" Kodansha Ltd., Tokio, Academic Press, New York, 1977, pp. 70ff.

9) Of course in the particular case of asymmetric induction in which D=O the indexes 1 and 2 can be given to (RE) arbitrarily.

10) Pino, P., Consiglio, G., Botteghi, C., and Salomon, C.: 1974, Adv. Chem. Ser. 132, 295.

11) Stefani, A., and Tatone, D.: 1977, Helv. Chim. Acta 60, 518.

12) Consiglio, G., unpublished results.

13) Stefani, A., unpublished results.

14) Morrison, J.D., and Mosher, H.S., "Asymmetric Organic Reactions", Prentice-Hall, Inc., Englewood Cliffs, New Jersey, 1971, pp. 220-30.

15) Pino, P., Piacenti, F., and Bianchi, M., in "Organic Synthesis via Metal Carbonyls", Wender, I., and Pino, P., Eds., Wiley, New York, 1977, pp. 136 and 233.

16) Consiglio, G., and Pino, P., 1976, Chimia 30, 193.

17) Consiglio, G., 1976, Helv. Chim. Acta, 59, 124.

18) These facts are true, at least in principle, also when only one of the regioisomers is chiral (e.g. hydroformylation of 1-butene); in this case, as the regioselectivity on the two faces cannot be determined, guessing of the type and extent of D on the base of (e.e.) can give wrong indications.

THE CATALYTIC FUNCTION OF ANTHRAQUINONES IN THE PHOTOOXIDATION OF CHLORIDE TO CHLORINE

Hans-Dieter Scharf[x*] and Robert Weitz
Institut für Organische Chemie der RWTH Aachen
Prof.-Pirlet-Str. 1, D - 5100 Aachen (Germany)

Abstract: The photooxidation of chloride to chlorine in aqueous solution catalysed by anthraquinone derivatives in presence of oxygen proceeds via electron transfer from the chloride to the anthraquinone in its triplet excited state. The obtained anthrasemiquinone radical is oxidized and protonated to ground state anthraquinone ready for another reaction cycle. Quantum yields of chlorine generation with anthraquinonesulfuric and -methanosulfuric acids as well as with new catalysts of the aza-anthraquinone tpye are given. For the reaction, which also passes off well in sunlight, the efficiency of solar energy conversion into chemical energy is calculated.

World's demand for chlorine is exclusively produced by either the diaphragm process or the mercury cell-process[1] under considerable consumption of electrical energy[2] and, in the latter case, enviromental contamination. In search of alternative energy sources for basic chemical processes we came across the photocatalytic oxidation of chloride to chlorine by excited anthraquinonesulfuric acids (AQ) (Fig.1) which has been observed for the first time by A. Eckert[3] qualitatively.

He exposed solutions of α- resp. β-anthraquinonesulfuric acid (α-AQS resp. β-AQS) in aqueous hydrochloric acid in open beakers to sunlight. However the significance of oxygen in this reaction was not recognized by Eckert and thus if one repeats his experiments with oxygen free solutions no chlorine is evolved.

This photocatalytic reaction resembles the Deacon-Process, that is to say the gas-phase oxidation of hydrochloric acid under the catalytic action of Copper chloride (Fig.1) however the reaction enthalpy is negative in the Deacon-Process and this process is remarkably reversible at temperatures above 300 ^{o}C[4].

B. Pullman (Ed.): Catalysis in Chemistry and Biochemistry. Theory and Experiment. 355–365.

Deacon - Process

$$2HCl_{(g)} + \tfrac{1}{2}O_{2(g)} \underset{>350\,^\circ C}{\overset{\text{Cu-Catalyst}}{\rightleftharpoons}} Cl_{2(g)} + H_2O_{(g)}$$

$$\Delta H = -59\ KJ\cdot mol^{-1}$$

- - - - - - - - - - - - - - - -

Photocatalytic-Oxidation

$$2HCl_{(aq)} + \tfrac{1}{2}O_2 \xrightarrow{^3(AQ)^*} Cl_{2(aq)} + H_2O$$

$$\Delta H = +49\ KJ\cdot mol^{-1}$$

Fig. 1: Thermal and photocatalytic oxidation of chloride to chlorine

The photocatalyticly active species is known to be the anthraquinone-sulfuric acid in its triplet excited state ($^3(AQ)^* \equiv cat^*$) which is formed via the excited singlet state by an efficient intersystem crossing (isc) process[5].

In Fig.2 a general term scheme of a photocatalytic redox system is presented where the reactive species of the catalyst "cat" is a triplet state. A/B represent the partners in the redox system:

$2\ HCl_{(aq.)} + \frac{1}{2}\ O_2\ /\ Cl_{2(aq.)} + H_2O$ and ΔE means the reaction enthalpy difference $\Delta H = +49\ KJ\cdot mol^{-1}$.

Fig.3 exhibits on the left hand side the activating steps in detail. The photosubstitution - right hand side - is one of the processes which deactivate the catalyst. It proceeds from the excited singlet-$n\pi^*$-state: $^1(AQ)^*$[6]. The so formed chlorinated anthraquinones are insoluble in aqueous hydrochloric acid and consequently precipitate. From the resulting slurry the chlorine generation decreases. There is however a difference between the isomeric anthraquinone sulfuric acids AQS. The photosubstitution proceeds about 40 times faster from α-AQS but from β-AQS.

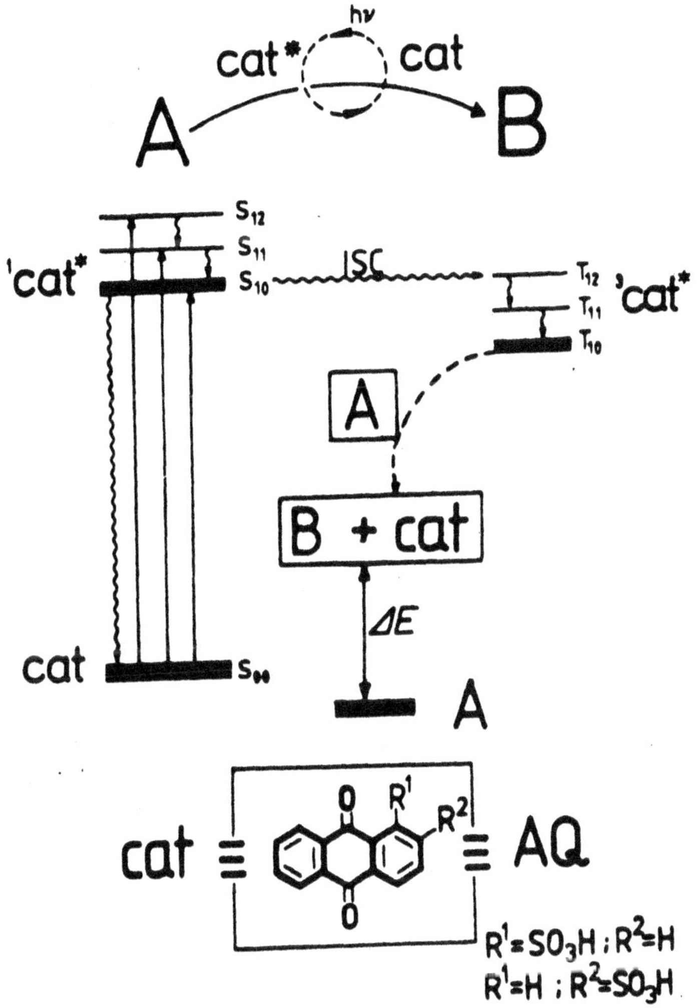

Fig. 2: General term scheme of photocatalytic redox systems

Since the quantum yields of the isc-process are as follows: $\Phi_{isc}^{\alpha\text{-AQS}} \ll \Phi_{isc}^{\beta\text{-AQS}} = 1$ [6,11] the bimolecular process of photosubstitution is discriminated in β-AQS from α-AQS. This is also reflected in the relative quantum yields of phosphorescence emission[7].

No phosphorescence emission[20] and consequently no photocatalytic activity is found in amino- resp. hydroxy-anthraquinones (AQ-NH_2 resp. AQ-OH) because of $\Phi_{isc} \ll 0.1$.

Anthraquinone itself has a $\Phi_{isc} \geq 0.9$. It does act as a photocatalyst in the system under discussion as well but its solubility in aqueous hydrochloric acid is very low.

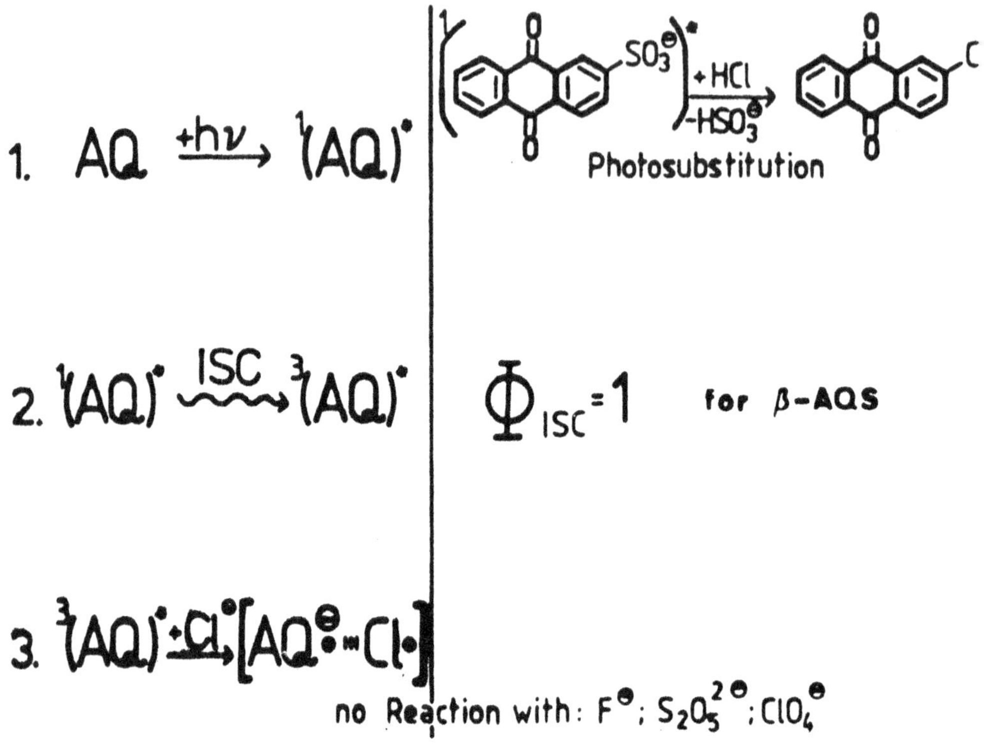

Fig. 3: Electron transfer

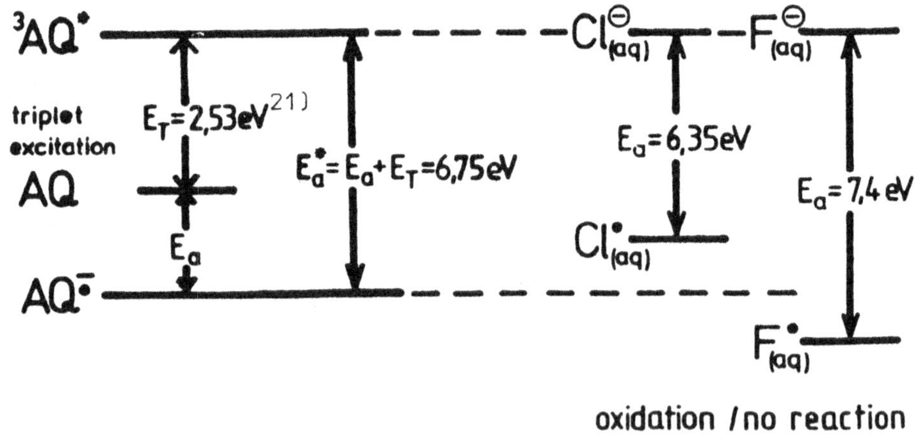

Fig. 4: Electron affinity E_a^* and redox potential of $^3AQ^*/AQ^{\ominus\bullet}$ and $Cl^{\ominus}_{(aq.)}/Cl^{\bullet}_{(aq.)}$

The decisive step (3. Fig.3) is the electron transfer reaction from the $Cl^{\ominus}$ anion to the anthraquinonesulfuric acid $^3(AQS)^*$-state under formation of a radical anion pair. This has been examined by V.A. Kuzmin and A.K. Chibisov[5,8] by means of flash photolytic experiments.

The electron transfer does not proceed from $F^{\ominus}$, $S_2O_5^{2\ominus}$ and $ClO_4^{\ominus}$ because of the limited oxidation potential of the $^3(AQ)^*$-states.

The electron affinity E_a^* and the redox potential of the pairs $^3(AQ)^*/AQ^{\dot{\ominus}}$ resp. $Cl^{\ominus}_{(aq.)}/Cl^{\bullet}_{(aq.)}$ has been estimated by Kuzmin, Chibisov[8] and Berdnikov, Bazhin[9].

According to their calculations the electron affinity E_a^* of tho triplet excited $^3(AQ)^*$ exceeds the oxidation potential of $Cl^{\ominus}_{(aq.)}/Cl^{\bullet}_{(aq.)}$ by 0.4 eV; whereas that of $F^{\ominus}_{(aq.)}/F^{\bullet}_{(aq.)}$ is about 0.65 eV greater than E_a^* of the $^3(AQ)^*$. This would explain the above mentioned empirical facts.

Fig.5 sums up our knowledge about the function of anthraquinones in the photooxidation of chloride to chlorine in accordance to the hitherto known experimental results.

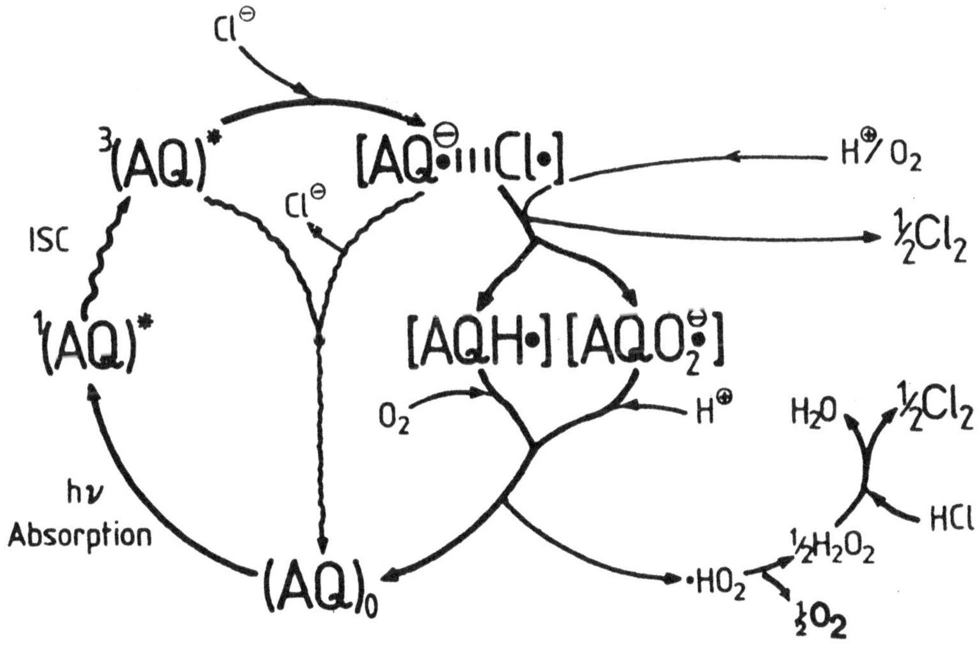

Fig. 5: The reaction cycle in presence of oxygen

The radical anion pair $(AQ^{\ominus\bullet}\ldots Cl^{\bullet})$ is formed as a key intermediate via the excited singlet and triplet state of the anthraquinone derivativ This key intermediate and thus the quantum yield of chlorine formation suffers from two competing deactivation pathways:
i. the collapse to the ground state and i.i. the chloride quenching of the $^3(AQ)^*$-state[5,10,11].

$$\text{i.: } (AQ^{\ominus\bullet}\ldots Cl^{\bullet}) \rightsquigarrow AQ_0 + Cl^{\ominus} \leftsquigarrow \text{ii.: } {}^3(AQ)^* + Cl^{\ominus} \; {}^{5,10)}$$

Subsequently the reaction proceeds via protonation resp. oxygen insertion ejecting a chlorine molecule. Further on the two intermediate species are alternatively oxidized or protonated under formation of a $^{\bullet}HO_2$ radical and AQ in its ground state. After dimerization of the $^{\bullet}HO_2$ radical hydrogenperoxide again is able to oxidize hydrochloric acid under formation of chlorine.

The number of times an AQ-molecule passes through the cycle until it terminates either by photosubstitution or an other photochemical side reaction, is in addition a limiting factor of the photocatalyst and thus the yield of chlorine.

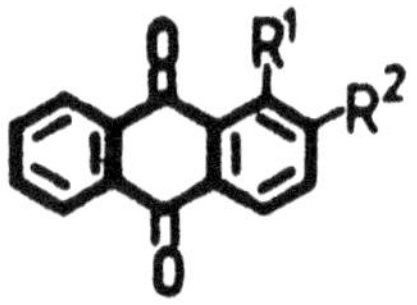

≡

2 λ_g $(n\pi^*)$: 415 nm

$$^{rel}\Phi^{Cl_2}_{(AQ)} = \frac{\Delta[Cl_2(AQ)]}{\Delta[\text{acetone}(AQ)]}$$

for $\Phi^{\text{acetone}}_{(AQ-2)} = 1$

AQ	R^1	R^2	$^{rel}\Phi^{Cl_2}_{(AQ)}$
1	$SO_3^{\ominus}$	H	-
2	H	$SO_3^{\ominus}$	0,1 ± 20
3	H	$-CH_2-SO_3^{\ominus}$	0,13
4	$-CH_2-SO_3^{\ominus}$	H	0,015
5	Cl	$-CH_2-SO_3^{\ominus}$	0,11
6	Br	$-CH_2-SO_3^{\ominus}$	0,02

Fig. 6: Anionic photocatalysts and quantum yield of Cl_2-formation

The relative quantum yields of chlorine formation in dependence on the pattern of substituents which assure the solubility of the catalyst are given in Fig.6.

The quantum yields have been determined relative to the formation of acetone from i-propanol by the same catalyst β-AQS (2, Fig.6). The latter reaction has been studied by Wilkinson[12]. He observed a quantum yield for the acetone formation: $\Phi_2^{acetone} = 1$.

Thus the figures represent the molar amount of chlorine evolved relative to one mol acetone formed from i-propanol under the same conditions.

The syntheses of the anthraquinone-methanosulfuric acid catalysts 3-6 as well as the differences of the quantum yield of chlorine production are described elsewhere[13]. The functional stability of the anthraquinone-methanosulfuric acids 3-6 under the applied conditions is better than that of the anthraquinonesulfuric acids 1, 2 and the relative quantum yields (with 3 and 5) are by comparison even slightly better, but at least they also are decomposed under formation of anthraquinonecarboxylic acids which again are insoluble in aqueous hydrochloric acid.

	UV-Absorption		Triplet energy/ lifetime		Cl_2-Production
	$(\pi\pi^*)\lambda_{max}/nm$ (log ε) 20% aq. HCl at 293K	$(n\pi^*)\lambda_{00}/nm$ C_2H_5OH-HCl 96:4 at 77K	E_T nm / $kJ{\cdot}mol^{-1}$ C_2H_5OH-HCl 96:4 at 77K	τ_{Ph}/ms	rel $\Phi_{AQ}^{Cl_2}$
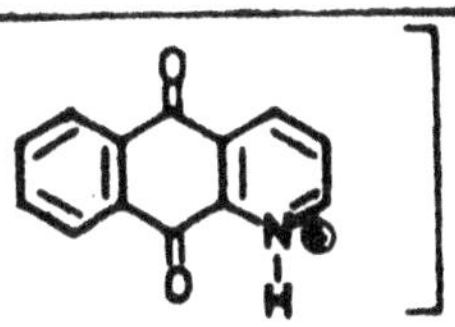 $Cl^\ominus$	343 (3,58)	410 sh	460 / 260	23,6	0,033
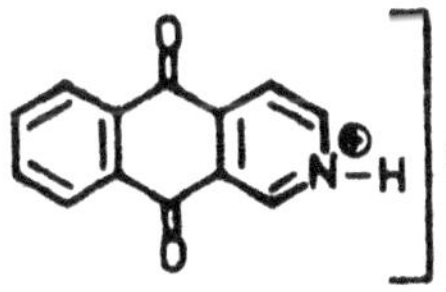 $Cl^\ominus$	343 (3,51) 302 (3,77)	425 sh	474 / 252	5,4	0,023
α-AQS	333 (3,67)	—	456 / 262	15,7	—
β-AQS	333 (3,69)	420 sh	460 / 260	3,6	0,1

Fig. 7: Cationic photocatalysts, spectroscopic data and quantum yield of Cl_2-formation

In order to avoid the destruction of the AQ-photocatalysts (photosubstitution in 1 and 2 or oxidative destruction in 3-6, Fig.6)[13] new cationic photocatalysts have been examined (Fig.7).

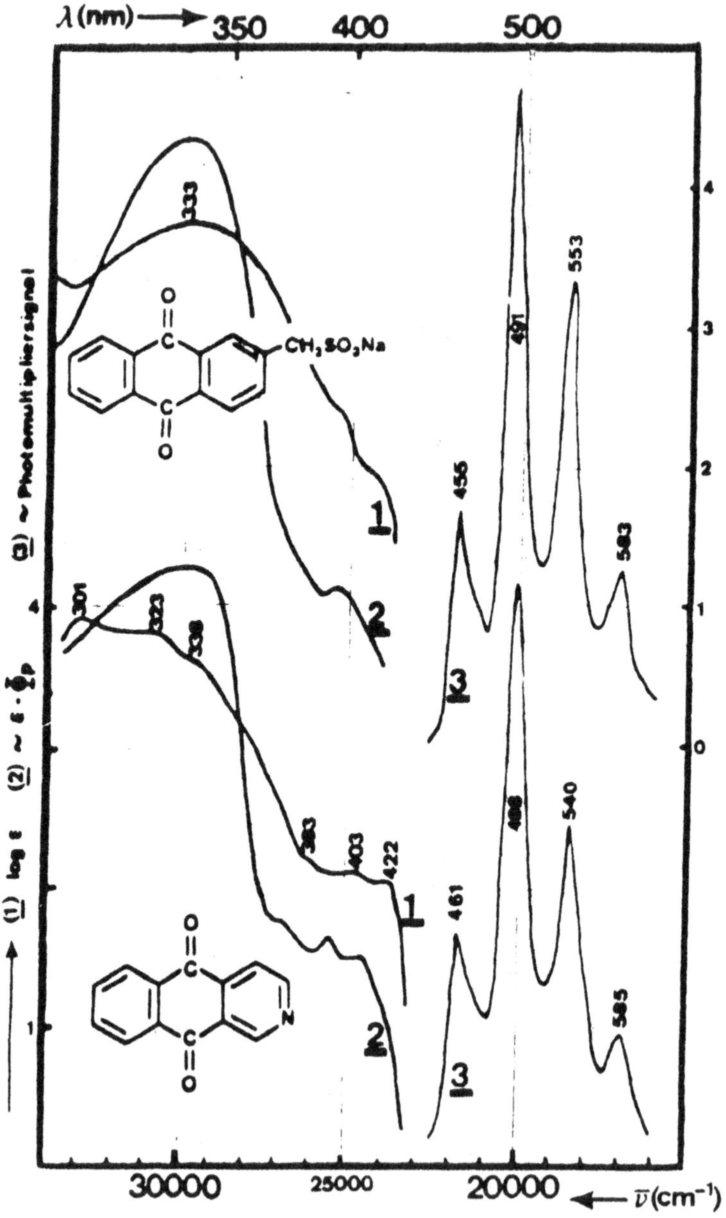

Fig. 8: 1 UV-Absorption in H_2O resp. cyclohexane at 295 K
2 Phosphorescence-excitation at 77 K
3 Phosphorescence-emission at 77 K in methanol/ethanol (5:2 by volume) resp. 2-methyltetrahydrofuran

Benzo(g)quinoline-5,10-dione hydrochloride 7 and Benz(g)-isoquinoline-5,10-dione hydrochloride 8 have been prepared according to loc.cit.[14,15]. The spectroscopic data of absorption and phosphorescence emission as well as the relative quantum yields of chlorine production and lifetimes of the triplet excited states are summarized in Fig.7 in comparison to the anionic photocatalysts 1 resp. 2. Indeed we found the stability of the catalysts 7 and 8 under the reaction conditions to be at least 10^3 times greater than that of the anionic catalysts. However the quantum yields are lower. Triplet energies of both types of catalysts are comparable as demonstrated in Fig.8.

In order to take advantage of the sun radiant energy for the photocatalytic oxidation of chloride the cationic catalysts 7 and 8 in aqueous hydrochloric acid derive benefit from their absorption at longer wavelength.

The efficiency $\eta_W^{A/B}$ of solar energy conversion into chemical energy is given by a product of four distinct factors (Fig.9)[16,17] whereby η_E the "energy conversion factor" represents the fraction of photon energy stored in the chemical energy of the redox system A/B given as the enthalpy difference between A/B.

$$\eta_W^{A/B} = \eta_E \cdot \Phi^{A/B} \cdot \eta_{AM2} \cdot \eta_{ABS}$$

energy conversion factor — quantum yield — fraction of usefull energy — absorption yield

$$\eta_W^{A/B} = \left(\frac{\Delta H^{A/B}}{\frac{N_L \cdot h \cdot c}{\lambda_g}}\right) \cdot \Phi^{A/B} \cdot \left(\frac{1}{\lambda_g} \frac{\int_0^{\lambda_g} I_\lambda^{SUN} \cdot \lambda \cdot d\lambda}{\int_0^{\infty} I_\lambda^{SUN} \cdot d\lambda}\right) \cdot \left(\frac{\int_0^{\lambda_g} I_\lambda^{SUN} (1-10^{-E_\lambda}) \cdot \lambda \cdot \frac{E_{A\lambda}}{E_\lambda} d\lambda}{\int_0^{\lambda_g} I_\lambda^{SUN} \cdot \lambda \cdot d\lambda}\right)$$

$$\eta_W^{Cl^{\ominus}/Cl^{\bullet}} = 0{,}17 \cdot 0{,}15 \cdot 0{,}04 \cdot 1. \cdot 100 = 0{,}1\,[\%]$$

Fig. 9: Effiency of solar energy conversion into chemical energy

$\Phi_{AQ}^{A/B}$ is the quantum yield of the reaction. η_a means the fraction of sun energy[18] absorbed by the photocatalyst and η_{abs} is its absorption yield.

This implies for the present system that 0.1 % of the incident sunlight energy is stored in the enthalpy difference of the system:

$$2\ HCl_{(aq.)} + \frac{1}{2}\ O_2/Cl_{2(aq.)} + H_2O\ .$$

This is only little worse than the efficiency of photosynthesis[19].

Literature

ǂ Dedicated to Professor Dr. E. Havinga in Leiden on the occasion of his 70th birthday

1) Hass, K., and Schmittinger, P.: 1976, Electrochimica Acta 21, pp.111
Argade, S.D., and Leitz, F.B.: 1977, J. Electrochem. Soc. 124, pp.12

2) e.g. The Federal Republic of Germany produced in 1975: $2.3 \cdot 10^6$ t-Chlorine expending $8.2 \cdot 10^9$ KWh electrical energy.

3) Eckert, A.: 1925, Ber. Dtsch. Chem. Ges. 58, pp. 318
1927, Ber. Dtsch. Chem. Ges. 60, pp. 1691

4) Engel, W.F., Waale, M.I., and Müller, S.: 1962, Chem. Ind.(London)pp.

5) Kuzmin, V.A., and Chibisov, A.K.: 1972, J. Chem. Soc. D, pp. 1559

6) Studzinskii, O.P., and El'tsov, A.V.: 1972, Zh. Org. Khim. (UdSSR) 8(4), pp. 774

7) The intensity of the phosphorescence emission at 77 K is for α-AQS more than hundred times weaker than for β-AQS. The quantum yield of phosphorescence emission in general is not necessarily proportional to Φ_{isc} because of competing radiationless deactivations form the singlet- and/or the triplet state. But in the above class of compounds it turns out to be true to a first approximation.

8) Kuzmin, V.A., and Chibisov, A.K.: 1973, Dokl. Akad. Nauk SSSR 212(5), pp. 1146

9) Berdnikov, V.N., and Bazhin, N.M.: 1970, Zh. Fiz. Khim. 44, pp. 7712

10) Schulte-Frohlinde, D., and Sonntag, C.V.: 1965, Z. phys. Chem. (Frankfurt) 44, pp. 314

11) Studzinskii, O.P., and El'tsov, A.V.: 1973, Zh. Org. Khim 9(9)pp.193

12) Wilkinson, F.: 1962, J. phys. Chem. 66, pp. 2569

13) Scharf, H.-D., and Weitz, R.: 1979, Tetrahedron in press.

14) Hopf, H., and Käppeli, V.: 1965, Chimia 19, pp. 228

15) Philips, A.: 1926, Ber. Dtsch. Chem. Ges. 27, pp. 124

16) Archer, M.D.: 1978, Solar Energy 20, pp. 167

17) Scharf, H.-D., Leismann, H., Ressler, I., Weitz, R., Fleischhauer, J., and Schleker, W.: "Criteria for efficiency, stability and capacity of abiotic photochemical solar energy storage" 1979, Angew. Chem. Int. Ed. in press

18) Trivich, D., and Flynn, P.A.: in "Solar Energy Research", Ed. Daniels, F., and Duffie, J.A.: 1955, Thames and Hudson

19) Schumacher, E.: 1978, Chimia 32, pp. 193

20) 2-hydroxyanthraquinone has a Φ_{Ph} of .06 . Allen, N.S., and McKellar, J.F.: 1976, J. Photochem. 5, pp. 317

21) We[13] found for $E_T^{0\to 0}$ of β-AQS: $E_T^{0\to 0}$ = 2.71 eV.

TETRAPHENYLPORPHYRIN COMPLEXES AS CHEMICAL, PHOTO- AND ELECTRO-CATALYSTS

Joost Manassen
Plastics Research Department
Weizmann Institute of Science
Rehovot, Israel

ABSTRACT

Quantitative data are given, which illustrate the catalytic activity of tetraphenylporphyrin complexes in redox reactions, whenever activated by thermal energy, an electrical potential or light quanta in the visible range. In the thermal and photochemical reactions p-substituents on the phenyl groups influence catalytic activity as expected from their electron donating or withdrawing power. The phenyl groups are known to be at an almost rectangular angle with the rest of the molecule in the solid state. Calculations would be of great help to find out, whether resonance effects can be transmitted under these conditions, or that the substituent effects are due to oscillating or rotating of the phenyl groups in solution, as can be shown to occur by NMR measurements.

INTRODUCTION

The porphyrin skeleton has been chosen by nature as the active component of two centrally important biologically active materials. One is haemoglobin, which contains a porphyrin-iron complex and takes care of oxygen transport in the blood stream. The other is chlorophyll, which is a reduced porphyrin-magnesium complex, which takes care of converting solar radiation energy into chemical energy. Furthermore the porphyrin unit can be found in many other biologically important materials. Therefore no doubt exists about the importance of the porphyrin macrocycle in biochemistry. We want to describe here however the use of porphyrin complexes, not as catalysts in natural processes, but as a synthetic catalyst in chemical, electrochemical and photochemical reactions.

To this end we shall first discuss the versatility and ease of preparation of tetraphenylporphyrin complexes, followed by a description of our older and more recent research on their catalytic activity. Figure 1 gives the computer drawn model in perspective of the Sn^{IV}-complex of tetraphenylporphyrin at the hand of which we can discuss the more general features of the molecule.

B. Pullman (Ed.): Catalysis in Chemistry and Biochemistry. Theory and Experiment. 367–376.

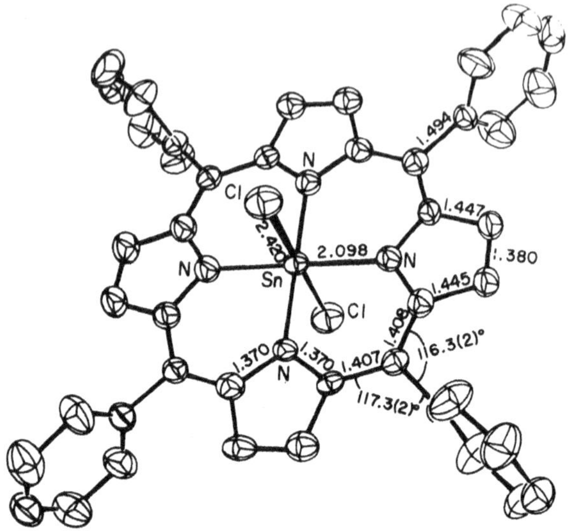

Figure 1. Computer drawn model in perspective of the Cl_2(Sn)tetraphenylporphyrin molecule (hydrogen atoms omitted).

We have to distinguish between three separate components of the molecule.

1. The porphin macrocycle, which is a resonance stabilized flat unit, which is aromatic and can be compared to benzene.

2. The $(SnCl_2)^{2+}$ unit, which can be replaced by almost any other bivalent metal ion. If it is replaced by two protons we get the neutral free base, which is a dihydroaromatic molecule and can be compared with hydroquinone. Two additional protons can be added giving the doubly charged conjugate acid. The two chlorine ions are charged ligands, but with central ions like bivalent iron or cobalt many non charged ligands can occupy the equatorial positions.

3. The four phenyl groups, which are out of plane of the porphin unit In the free base the angle between the phenyl group and the porphin-plane is about 60°, in most metal complexes near to 90° and in the conjugate acid about 20°.

Figure 2 gives a scheme of the calculated energies of the top filled and lowest empty orbitals for complexes of the first row transition metal ions.

Of importance for their catalytic activity is the proximity of the top filled π-orbitals of the ligand and the d-electron levels of the metal ions in the complexes of Mn to Ni. The synthesis of the molecule is simple. By condensation of benzaldehyde and pyrrole in propionic

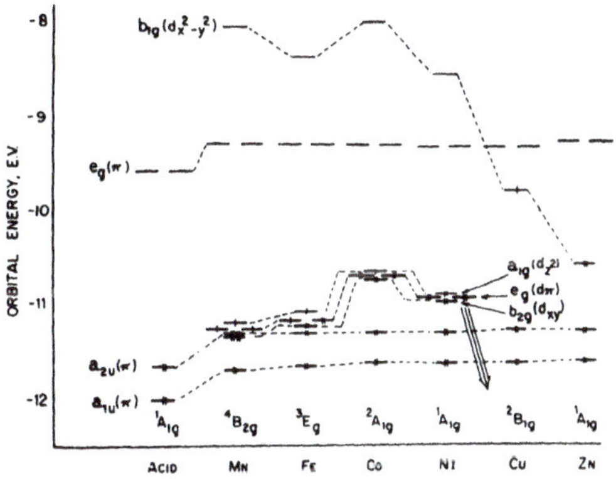

Figure 2. Calculated energies of the top filled and lowest empty orbitals of porphyrin and its first row transition metal complexes.

acid solution it is formed in about 20% yield. If substituted benzaldehydes are taken, we get tetraphenylporphyrins, which carry substituents on the phenyl groups. We understand now, why this molecule is so suited for catalysis. Many metal ions can be coordinated within a flat π-electron system and electron density at this metal ion "site" can be regulated by substituents on the phenyl groups or by equatorial ligands. Moreover the pyrrole units can be partly hydrogenated, conserving the aromaticity of the molecule, but changing the electronic pathway. One of the conditions for catalysis is the availability of an adsorption site for the reacting molecule, for which the equatorial sites are open.

Porphyrin complexes are electron donors as well as acceptors and can be electrochemically oxidized and reduced to their respective cation- and anion-radicals. The redox potentials are a measure for the capacity of the molecule to accept or donate electrons. With metal ions of variable valency, a distinction can be made between metal oxidation/reduction and ligand oxidation/reduction (1).

HETEROGENEOUS CHEMICAL CATALYSIS

In this case the reactants are led in their gaseous state at elevated temperature over the porphyrin complex either in its crystalline state in bulk or precipitated from its solution on an inert carrier.

We shall discuss two reactions: oxidative dehydrogenation and valence isomerisation.

a. Oxidative dehydrogenation

Dehydrogenation, the conversion of a saturated carbon to carbon bond into an unsaturated one and hydrogen, is an endothermic reaction. In order to make it exothermic an oxidative agent is often added, which combines with the hydrogen formed. In our case we took the formation of benzene from cyclohexadiene, while nitrobenzene acted as the oxidative agent and was converted to aniline and water in the process.

In order to check the influence of the central metal ion on the catalytic activity, conversions into benzene and aniline were measured as a function of the ion, for which we took the first row transition metal ions from manganese to zinc. This was compared with the oxidation potential of the complexes as measured by electrochemical oxidation (2). The results are given in Figure 3. (This graph actually applies to a related macrocyclic ring system phthalocyanine instead of tetraphenylporphyrin, which does not alter our conclusions however).

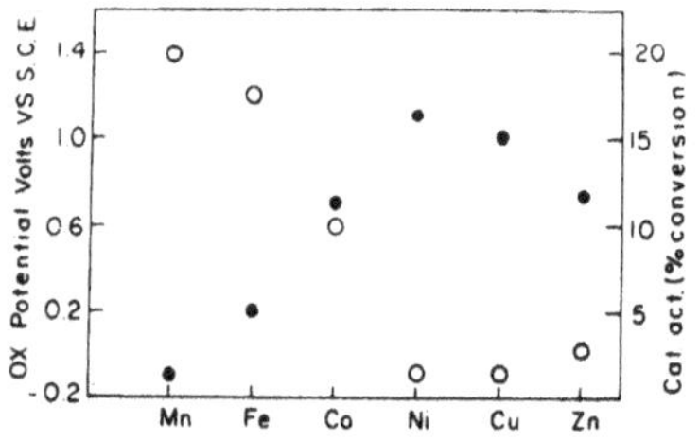

Figure 3. Catalytic activity for oxidative dehydrogenation and oxidation potentials as a function of the central metal ion arranged according to the d-electron configuration of the ion. (o) Catalytic activity; (•) Oxidation potential.

These two curves, which are also found with air as the oxidant, show minimum activity at nickel which corresponds with maximum oxidation potential. The reason for this maximum could be found by ESR measurements on the products of electrochemical oxidation. It appeared that with Mn, Fe and Co the central metal ion is oxidized, while with Cu and Zn the ligand is oxidized. With Ni both processes occur at the same potential. Therefore two details of the mechanism of oxidative dehydrogenation over these complexes can be learned from the graph:

1. Because lower oxidation potential gives higher activity, electron transfer from the complex to the oxidant is the rate determining step.
2. Because activity at the right side of Ni (Cu and Zn) increases less with decreasing potential than is found at its left side (Mn, Fe and Co) oxidation/reduction of the central metal atom is more efficient for the reaction than oxidation/reduction of the π-ligand.

In order to find the influence of phenylsubstituents on catalytic activity, the same reaction was checked with substituted Co-tetraphenylporphyrin as the catalyst (3). Table 1 gives the results:

Table 1. Relative catalytic activities and I.R. frequencies related to the substituent on the phenyl group in Co-tetraphenylporphyrin.

Substituent	Catalytic activity heterogeneous	Catalytic activity homogeneous	IR, cm^{-1}, homogeneous	IR, cm^{-1}, heterogeneous
p-$N(CH_3)_2$		1.44	997	996
p-OH	1.38	1.54	998	991
p-ONa	1.54	1.70	998	991
p-OCH_3	1.12	1.18	996	993
p-CH_3	1.08	1.44	993	1002
H	1.00	1.00	991	1000
3-Pyridine			987	992
4-Pyridine	0.98	0.94	988	992
3-NO_2	0.92	0.88	985	994
4-NO_2	0.88	0.82	984	994
Pentafluoro		0.80	980	985

The first column gives the substituents arranged in such a way, that electron donating power decreases from top to bottom. If electron transfer from the catalyst to substrate is the rate determining step, catalytic activity must also decrease from top to bottom as can be seen to be true in the second column. The third column, which indicates homogeneous catalytic activity relates to an arrangement, where the catalyst is dissolved in a high boiling liquid within the pores of a carrier and under these conditions catalytic activity also decreases from top to bottom.

The last two columns are somewhat speculative, but are of special importance within the framework of this conference. Column 4 gives a frequency in the infrared, which is thought to be indicative of the strength of the Co-N bond. When measured in solution this frequency decreases as expected from top to bottom, but this is not found, when this same frequency is measured on the solid crystals. In the solid state almost all substituents, whether electron donating or withdrawing, cause a decrease in frequency when compared with H. This indicates a difference in the effects caused by the substituent on dissolved and solid porphyrin complex. From Figure 1 we remember that the phenyl group makes an angle of almost 90° with the rest of the molecule, which must be a serious barrier for resonative effects. On the other hand the phenyl groups are known from NMR evidence to oscillate or even rotate in solution (4), which would lower this barrier. This difference may be of great importance as a model for enzymatic reactions, where is found that the structure measured by X-ray analysis cannot always fully explain the kinetic behaviour in solution.

b. Valence isomerization

Figure 4 shows the valence isomerization of quadricyclane to norbornadiene. This is a symmetry restricted reaction.

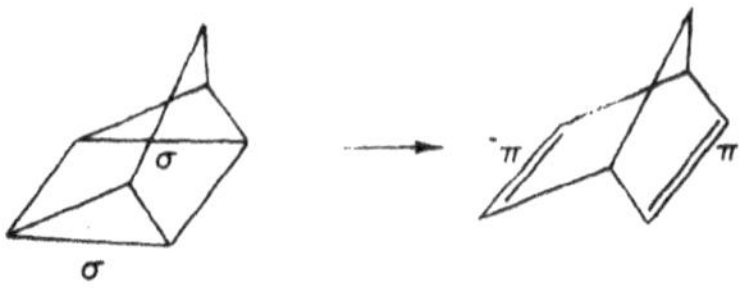

Figure 4. Conversion of quadricyclane to norbornadiene.

Cobalt and nickel tetraphenylporphyrin remove this restriction and conversion is 100%. Copper and zinc tetraphenylporphyrin are entirely inactive under the same conditions (5). Two explanations have been given for this catalytic effect. One is that the catalyst furnishes an empty orbital of the right symmetry, because of which the reaction becomes symmetry allowed. The other, which is the more modern one, assumes the reaction to go in two steps by way of an intermediate complex, which is formed by oxidative addition of quadricyclane. This complex has never been proven to exist however.

This reaction has become actual again, because it opens the possibility for conserving energy by forming quadricyclane from norbornadiene photochemically and decomposing it again catalytically with the formation of heat.

ELECTROCATALYSIS

Since the development of fuel cells there has been a need for catalysts, which can activate the reduction of oxygen at an electrode:

$$O_2 + 4\,e^- + 4\,H^+ \longrightarrow 2H_2O$$

Porphyrin complexes have been shown to be good catalysts and activities follow more or less the same interdependence between activity and metal ion in that $Fe^{2+} > Co^{2+} > Cu^{2+}$. In our own studies we found that activity is almost not influenced by substituents on the phenyl groups (6) although others have found such a dependence (7). Electrocatalysis is measured at room temperature and the catalyst is in its solid state. There exist the possibility therefore that contrary to the heterogeneous catalysis at elevated temperature, the phenyl groups here are relatively at rest and that therefore no influence of substituents on catalytic activity is found.

PHOTOCATALYSIS

In heterogeneous catalysis we were concerned with redox reactions of the porphyrin complex and central atoms of variable valency with uncompleted d-shells worked best. The same applied to electrocatalysis. In photocatalysis we are concerned with the donor and acceptor properties of the excited states. As central metal ions with uncompleted electron shells give complexes with shortlived excited states, we are concerned in photochemistry with closed shell metal ions, like Mg^{2+}, Zn^{2+}, Cd^{2+}, Sn^{4+} and Ge^{4+}.

If ($SnCl_2$) tetraphenylporphyrin or ($GeCl_2$) tetraphenylporphyrin is irradiated with visible light in benzenesolution in the presence of a tertiary amine, one of the pyrrole units in the porphyrin ring is reduced and the amine is oxidized. In this way we can produce the chlorin (one pyrrole unit reduced, the base of the chlorophyll system) and if the reaction is pursued longer we get isobacteriochlorin (two adjacent pyrrole units reduced). Chlorin as well as isobacteriochlorin when irradiated in the presence of nitrobenzene are again oxidized to porphyrin and nitrobenzene is reduced (8). Figure 5 shows fluorescence spectra of the tin-porphyrin complex as a function of amine concentration (9).

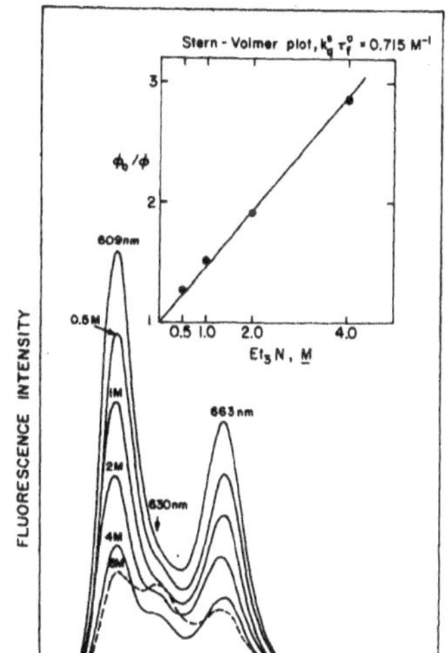

Figure 5. Fluorescence quenching of $5x10^{-5}$ M tintetraphenylporphyrin by various concentrations of triethyl-amine in benzene solution. The dotted line for 6M amine shows clearly a new emission band, λ_{max}=630 nm, attributed to exciplex fluorescence.

Not only that the fluorescence is seriously quenched by amine, but at high concentration a new emission band appears, which is assigned to exciplex fluorescence. Figure 6 shows a similar quenching spectrum of the tinisobacteriochlorin complex by nitrobenzene.

We see that the excited singlet states of porphyrin and isobacteriochlorin are quenched by amine and nitrobenzene respectively, presumably under the formation of an exciplex. Electron transfer in this exciplex is thought to be responsible for the oxidation and reduction reactions which occur.

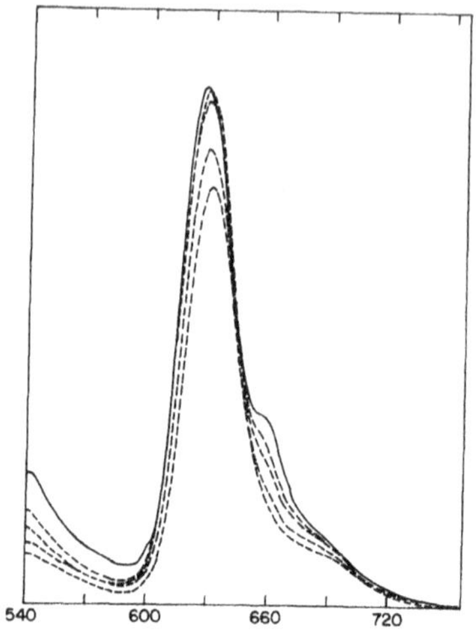

Figure 6. Fluorescence quenching of $5x10^{-5}$M tintetraphenylisobacteriochlorin by nitrobenzene in benzene solution. —— fluorescence spectrum without quencher, ---- fluorescence spectra with increasing nitrobenzene concentration 1,2,4 and 6M.

It is interesting that in these excited state reactions also the influence of the phenyl substituent is felt in a consistent manner as can be seen from Table 2.

Table 2. One electron reduction potentials[1] and photo-reduction rates[2] for tintetraphenylporphyrin and chlorin complexes.

Porphyrin	$E_{1/2}$, V	Rate $M \cdot sec^{-1}$	Chlorin	$E_{1/2}$, V	Rate $M \cdot sec^{-1}$
$pClSnTPPCl_2$	-0.700	$4.4x10^{-6}$	$pClSnTPCCl_2$	-0.712	$2.4x10^{-6}$
$pHSnTPPCl_2$	-0.806	$6.0x10^{-7}$	$pHSnTPCCl_2$	-0.811	$3.5x10^{-7}$
$pOMeSnTPPCl_2$	-0.837	$2.0x10^{-7}$	$pOMeSnTPCCl_2$	-0.870	$1.3x10^{-7}$

[1]Half-wave one electron reduction potentials, in CH_2Cl_2 vs. SCE.
[2]From initial slopes for 1M NMP, 1M Ac_2O in benzene; $\lambda > 500$ nm.

Going from p-substituent Cl via H to OMe the reduction potential gets more negative, which means that in this order the groundstate molecules are less capable of accepting electrons. For porphyrin as well as chlorin reduction rates are seen to decrease in this order. This means therefore that in the excited state substituent effects are felt as they are in the groundstate.

We have seen that porphyrin is photochemically reduced by amine to chlorin and isobacteriochlorin, by which the amine is oxidized. We also know that chlorin and isobacteriochlorin are photochemically oxidized by nitrobenzene, by which the nitrobenzene is reduced. Therefore when a mixture of amine and nitrobenzene is irradiated in the presence of either the porphyrin, chlorin or isobacteriochlorin, amine is oxidized and nitrobenzene is reduced, while a steady state is established of the three reduction steps of the porphyrin. This steady state can be followed spectrophotometrically and is a fine probe of a catalyst intermediate stage during reaction. Table 3 gives the steady state composition as a function of p-substituents.

Table 3. Steady state concentrations and rates of product formation for $2.5x10^{-4}$M porphyrin, 1M nitrobenzene, 1M triethylamine, 1M aceticanhydride in benzene at room temperature.

Porphyrin	Steady-state concentration in %			Rates of product formation in M/sec	
	Sn(IV) TPP	Sn(IV) TPC	Sn(IV) TPisoB	Aniline	Acetanilide
pCl $SnTPPCl_2$	1.7	95.6	2.7	$6.55x10^{-7}$	$1.495x10^{-7}$
pH $SnTPPCl_2$	22	78	0.0	$2.67x10^{-7}$	$7.0\ x10^{-8}$
pOMe $SnTPPCl_2$	85.5	14.5	0.0	$1.54x10^{-7}$	$6.4\ x10^{-8}$
pCl $SnTPPCl_2$[1]	8.0	78.5	13.5	$1.35x10^{-6}$[2]	$9.5\ x10^{-7}$

[1] NMP instead of Et_3N. [2] N-methyl pyrrole formed.

Again from p-substituent Cl via p-H to p-OMe the porphin molecule accepts electrons less readily (Table 1) and donates better. Therefore the photochemical reduction step is slowed down in this order and the oxidation step becomes faster. This results in a steady state composition which becomes poorer in isobacteriochlorin and richer in porphyrin, as is dramatically shown in the table. By taking as the amine N-methylpyrrole instead of triethylamine the steady state composition moves in the direction of isobacteriochlorin, because N-methylpyrrole is a much better electron donor. Similar effects are obtained by changing nitrobenzene for a better electron accepting nitroaromatic molecule.

CONCLUSION

The tetraphenylporphyrin entity is a suitable substrate for studying catalysis. Its complexes catalyze hydrogen transfer from a donor to an acceptor under input of thermal, electrochemical or radiant energy.

In thermal reactions it is mainly the valency change of the central ion, which is responsible for catalytic activity. In photochemical reactions it is the excited singlet state, which by means of exciplex formation facilitates electron transfer to donor and acceptor. In both cases substituents at the paraposition of the phenyl groups influence catalytic activity in a predictable manner.

Calculations can be very important here in order to verify the influence of the parasubstituent as a function of the angle between phenyl group and porphyrin ring. It is not clear yet whether the influence of the p-substituents is felt because of oscillations or rotations of the phenyl groups in solution or at high temperature in the solid, or that also at an angle of 90° resonative effects can be transmitted from the phenyl group to the porphyrin macrocycle.

REFERENCES

1. Wolberg, A., and Manassen, J.: 1970, J. Amer. Chem. Soc. 92, pp. 2982-2991.
2. Manassen, J.: 1974, Cat. Rev.-Sci. Eng. 9, pp. 223-243.
3. Bar Ilan, A., and Manassen, J.: 1974, J. Catal., 33, pp. 68-73.
4. Eaton, S.S., and Eaton, G.R.: 1974, J. Chem. Soc., Chem. Commun., 576-577.
5. Manassen, J.: 1970, J. Catal., 18, pp. 38-45.
6. Manassen, J.: 1974, J. Catal., 33, pp. 133-137.
7. Alt, H., Binder, H., and Sandstede, G.J.: 1973, J. Catal., 28, pp. 8-19.
8. Harel, Y., and Manassen, J.: 1977, J. Amer. Chem. Soc., 99, pp. 5817-5818.
9. Harel, Y., and Manassen, J.: To be published.